How Mathematicians Think

*

How Mathematicians Think

USING AMBIGUITY, CONTRADICTION, AND
PARADOX TO CREATE MATHEMATICS

*

WILLIAM BYERS

PRINCETON UNIVERSITY PRESS

PRINCETON AND OXFORD

Contents *

CHAPTER 9
Conclusion: Is Mathematics Algorithmic
or Creative?

* Acknowledgments *

IT IS A PLEASURE to acknowledge the assistance of many people who helped and supported me in the writing of this book. First of all, thanks to my wife, Miriam, and my children, Joshua and Michele. I would like to thank David Tall and Joe Auslander, who read earlier versions of this manuscript and made many valuable suggestions. Thanks to Hershy Kisilevsky for the many stimulating discussions about mathematics, to Monique Dumont for reading a preliminary version of some chapters, and to John Seldin for his help. I am also pleased to acknowledge the help of Gianna Venettacci for the transferring of the diagrams to the computer and the general organization of the manuscript, and to my editor at Princeton, Vickie Kearn, for her valuable help. I appreciate the support I received from David Graham, Dean of the Faculty of Arts & Science, and to John Capobianco, Vice-Dean Research, at Concordia University.

I would especially like to thank Albert Low. A fascinating man, Low is a veritable fount of strikingly original ideas about, among other things, creativity and its relation to the human condition. In the course of many years now, we have had fascinating discussions over lunch about mathematics and about life. The book owes a great deal to those conversations. It was Low's way of looking at ambiguity that led me to ask the question, "Is there ambiguity in mathematics?" which ultimately flowered into this book. His faith in this project, and his encouragement, have meant a great deal to me.

Turning on the Light

A FEW YEARS AGO the PBS program *Nova* featured an interview with Andrew Wiles. Wiles is the Princeton mathematician who gave the final resolution to what was perhaps the most famous mathematical problem of all time—the Fermat conjecture. The solution to Fermat was Wiles's life ambition. "When he revealed a proof in that summer of 1993, it came at the end of seven years of dedicated work on the problem, a degree of focus and determination that is hard to imagine."[1] He said of this period in his life, "I carried this thought in my head basically the whole time. I would wake up with it first thing in the morning, I would be thinking about it all day, and I would be thinking about it when I went to sleep. Without distraction I would have the same thing going round and round in my mind."[2] In the *Nova* interview, Wiles reflects on the process of doing mathematical research:

> Perhaps I can best describe my experience of doing mathematics in terms of a journey through a dark unexplored mansion. You enter the first room of the mansion and it's completely dark. You stumble around bumping into the furniture, but gradually you learn where each piece of furniture is. Finally after six months or so, you find the light switch, you turn it on, and suddenly it's all illuminated. You can see exactly where you were. Then you move into the next room and spend another six months in the dark. So each of these breakthroughs, while sometimes they're momentary, sometimes over a period of a day or two, they are the culmination of—and couldn't exist without—the many months of stumbling around in the dark that precede them.

This is the way it is! This is what it means to do mathematics at the highest level, yet when people talk about mathematics, the elements that make up Wiles's description are missing. What is missing is the creativity of mathematics—the essential dimension without which there is no mathematics. Ask people about mathematics and they will talk about arithmetic, geometry, or statistics, about mathematical techniques or theorems they have learned. They may talk about the logical structure of mathematics, the nature of mathematical arguments. They may be impressed with the precision of mathematics, the way in which things in mathematics are either right or wrong. They may even feel that mathematics captures "the truth," a truth that goes beyond individual bias or superstition, that is the same for all people at all times. Rarely, however, do most people mention the "doing" of mathematics when they talk about mathematics.

Unfortunately, many people talk about and use mathematics despite the fact that the light switch has never been turned on *for them*. They are in a position of knowing where the furniture is, to use Wiles's metaphor, but they are still in the dark. Most books about mathematics are written with the aim of showing the reader where the furniture is located. They are written from the point of view of someone for whom the light switch has been turned on, but they rarely acknowledge that without turning on the switch the reader will forever remain in the dark. It is indeed possible to know where the furniture is located without the light switch having ever been turned on. "Locating the furniture" is a relatively straightforward, mechanical task, but "turning on the light" is of another order entirely. One can prepare for it, can set the stage, so to speak, but one can neither predict nor program the magical moment when things "click into place." This book is written in the conviction that we need to talk about mathematics in a way that has a place for the darkness as well as the light and, especially, a place for the mysterious process whereby the light switch gets turned on.

Almost everyone uses mathematics of some kind almost every day, and yet, for most people, their experience of mathematics is the experience of driving a car—you know that if you press on the gas the car will go forward, but you don't have any idea why. Thus, most people are in the dark with respect to the mathematics that they use. This group includes untold numbers of students in mathematics classrooms in elementary schools, high schools, and universities around the world. It even includes intelligent people who use fairly sophisticated mathematics in their personal or professional lives. Their position, with respect

2

to the mathematics they use every day, is like that of the person in the dark room. They may know where certain pieces of furniture are located, but the light switch has not been turned on. They may not even know about the existence of light switches. Turning on the light switch, the "aha!" experience, is not something that is restricted to the creative research mathematician. Every act of understanding involves the turning on of a light switch. Conversely, if the light has not gone on, then one can be pretty certain that there is no understanding.

If we wish to talk about mathematics in a way that includes acts of creativity and understanding, then we must be prepared to adopt a different point of view from the one in most books about mathematics and science. This new point of view will examine the processes through which new mathematics is created and old mathematics is understood. When mathematics is identified with its content, it appears to be timeless. The new viewpoint will emphasize the dynamic character of mathematics—how it is created and how it evolves over time. In order to arrive at this viewpoint, it will be necessary to reexamine the role of logic and rigor in mathematics. This is because it is the formal dimension of mathematics that gives it its timeless quality. The other dimension—the developmental—will emerge from an examination of situations that have spawned the great creative advances in mathematics. What are the mechanisms that underlie these advances? Do they arise out of the formal structure or is there some other factor at work?

This new point of view will turn our attention away from the content of the great mathematical theories and toward questions that are unresolved, that are in flux and problematic. The problematic is the context within which mathematical creativity is born. People are so motivated to find answers that they sometimes neglect the boundaries of the known, where matters have not settled down, where questions are more meaningful than answers. This book turns matters around; that is, the problematic is regarded as the essence of what is going on. The consequence is that much of the traditional way of looking at mathematics is radically changed. Actual mathematical content does not change, but the point of view that is developed with respect to that content changes a great deal. And the implications of this change of viewpoint will be enormous for mathematics, for science, and for all the cultural projects that get their worldview, directly or indirectly, from mathematics.

Now, not everyone thinks that such a change in viewpoint is necessary or even desirable. There are eminent spokespeople for an opposing point of view, one that maintains that "The ultimate goal of mathematics is to eliminate all need for intelligent thought."[3] This viewpoint, one that is very influential in the artificial intelligence community, is that progress is achieved by turning creative thought into algorithms that make further creativity unnecessary. What is an algorithm? Webster's New Collegiate Dictionary defines it to be "a procedure for solving a mathematical problem in a finite number of steps that frequently involves the repetition of an operation." So an algorithm breaks down some complex mathematical task into a sequence of more elementary tasks, some of which may be repeated many times, and applies these more elementary operations in a step-by-step manner. We are all familiar with the simple mathematical algorithms for addition or multiplication that we learned in elementary school. But algorithms are basic to all of computer programming, from Google's search procedures to Amazon's customer recommendations.

Today the creation of algorithms to solve problems is extremely popular in fields as diverse as finance, communications, and molecular biology. Thus the people I quoted in the above paragraph believe that the essence of what is going on in mathematics is the creation of algorithms that make it *unnecessary* to turn on the light switch. There is no question that some of the greatest advances in our culture involve the creation of algorithms that make calculations into mechanical acts. Because the notion of an algorithm underlies all of computer programming, algorithms are given a physical presence in computers and other computational devices. The evident successes of the computer revolution have moved most people's attention from the creative breakthroughs of the computer scientists and others who create the algorithms to the results of these breakthroughs. We lose track of the "how" and the "why" of information technology because we are so entranced with what these new technologies can do for us. We lose track of the human dimension of these accomplishments and imagine that they have a life independent of human creativity and understanding.

4

The point of view taken in what follows is that the experience Wiles describes is the essence of mathematics. It is of the utmost importance for mathematics, for science, and beyond that for our understanding of human beings, to develop a way of talking about mathematics that contains the entire mathematical experience, not just some formalized version of the results of that experience. It is not possible to do justice to mathematics, or to explain its importance in human culture, by separating the content of mathematical theory from the process through which that theory is developed and understood.

DIFFERENT WAYS OF USING THE MIND

Mathematics has something to teach us, all of us, whether or not we like mathematics or use it very much. This lesson has to do with thinking, the way we use our minds to draw conclusions about the world around us. When most people think about mathematics they think about the logic of mathematics. They think that mathematics is characterized by a certain mode of using the mind, a mode I shall henceforth refer to as "algorithmic." By this I mean a step-by-step, rule-based procedure for going from old truths to new ones through a process of logical reasoning. But is this really the only way that we think in mathematics? Is this the way that new mathematical truths are brought into being? Most people are not aware that there are, in fact, other ways of using the mind that are at play in mathematics. After all, where do the new ideas come from? Do they come from logic or from algorithmic processes? In mathematical research, logic is used in a most complex way, as a constraint on what is possible, as a goad to creativity, or as a kind of verification device, a way of checking whether some conjecture is valid. Nevertheless, the creativity of mathematics—the turning on of the light switch—cannot be reduced to its logical structure.

Where *does* mathematical creativity come from? This book will point toward a certain kind of situation that produces creative insights. This situation, which I call "ambiguity," also provides a mechanism for acts of creativity. The "ambiguous" could be contrasted to the "deductive," yet the two are not mutually exclusive. Strictly speaking, the "logical" should be contrasted to the "intuitive." The ambiguous situation may contain elements of the logical and the intuitive, but it is not restricted to such elements. An ambiguous situation may even involve the contradictory, but it would be wrong to say that the ambiguous is necessarily illogical.

Of course, it is not my intention to produce some sort of recipe for creativity. On the contrary, my argument is precisely that such a recipe cannot exist. This book directs our attention toward the problematic and the ambiguous because these situations so often form the contexts that produce creative insights.

Normally, the development of mathematics is reconstructed as a rational flow from assumptions to conclusions. In this reconstruction, the problematic is avoided, deleted, or at best minimized. What is radical about the approach in this book is the assertion that creativity and understanding arise out of the problematic, out of situations I am calling "ambiguous." Logic abhors the ambiguous, the paradoxical, and especially the contradictory, but the creative mathematician welcomes such problematic situations because they raise the question, "What is going on here?" Thus the problematic signals a situation that is worth investigating. The problematic is a potential source of new mathematics. How a person responds to the problematic tells you a great deal about them. Does the problematic pose a challenge or is it a threat to be avoided? It is the answer to this question, not raw intelligence, that determines who will become the successful researcher or, for that matter, the successful student.

The Importance of Talking about Mathematics

In preparing to write this introduction, I went back to reread the introductory remarks in that wonderful and influential book, *The Mathematical Experience*. I was struck by the following paragraph:

> I started to talk to other mathematicians about proof, knowledge, and reality in mathematics and I found that my situation of confused uncertainty was typical. But I also found a remarkable thirst for conversation and discussion about our private experiences and inner beliefs.

I've had the same experience. People want to talk about mathematics but they don't. They don't know how. Perhaps they don't have the language, perhaps there are other reasons. Many mathematicians usually don't talk about mathematics because talking is not their thing—their thing is the "doing" of mathematics. Educators talk about teaching mathematics but rarely about mathematics itself. Some educators, like scientists, engineers, and many other professionals who use mathematics, don't talk about mathematics because they feel that they don't possess the expertise that would be required to speak intelligently about mathematics. Thus, there is very little discussion about mathematics going on. Yet, as I shall argue below, there is a great need to think about the nature of mathematics.

What is the audience for a book that unifies the content with the "doing" of mathematics? Is it restricted to a few interested mathematicians and philosophers of science? This book is written in the conviction that what is going on in mathematics is important to a much larger group of people, in fact to everyone who is touched one way or another by the "mathematization" of modern culture. Mathematics is one of the primary ways in which modern technologically based culture understands itself and the world around it. One need only point to the digital revolution and the advent of the computer. Not only are these new technologies reshaping the world, but they are also reshaping the way in which we understand the world. And all these new technologies stand on a mathematical foundation.

Of course the "mathematization" of culture has been going on for thousands of years, at least from the times of the ancient Greeks. Mathematization involves more than just the practical uses of arithmetic, geometry, statistics, and so on. It involves what can only be called a culture, a way of looking at the world. Mathematics has had a major influence on what is meant by "truth," for example, or on the question, "What is thought?" Mathematics provides a good part of the cultural context for the worlds of science and technology. Much of that context lies not only in the explicit mathematics that is used, but also in the assumptions and worldview that mathematics brings along with it.

The importance of finding a way of talking about mathematics that is not obscured by the technical difficulty of the subject is perhaps best explained by an analogy with a similar discussion for physics and biology. Why should nonphysicists know something about quantum mechanics? The obvious reason is that this theory stands behind so much modern technology. However, there is another reason: quantum mechanics contains an implicit view of reality that is so strange, so at variance with the classical notions that have molded our intuition, that it forces us to reexamine our preconceptions. It forces us to look at the world with new eyes, so to speak, and that is always important. As we shall see, the way in which quantum mechanics makes us look at the world—a phenomenon called "complementarity"—has a great deal in common with the view of mathematics that is being proposed in these pages.

Similarly, it behooves the educated person to attempt to understand a little of modern genetics not only because it provides the basis for the biotechnology that is transforming the world, but also because it is based on a certain way of looking at human nature. This could be summarized by the phrase, "You are your DNA" or, more explicitly, "DNA is nothing less than a blueprint—or, more accurately, an algorithm or instruction manual—for building a living, breathing, thinking human being."[4] Molecular biology carries with it huge implications for our understanding of human nature. To what extent are human beings biological machines that carry their own genetic blueprints? It is vital that thoughtful people, scientists and nonscientists alike, find a way to address the metascientific questions that are implicit in these new scientific and technological advances. Otherwise society risks being carried mindlessly along on the accelerating tide of technological innovations. The question about whether a human being is mechanically determined by their blueprint of DNA has much in common with the question raised by our approach to mathematics, namely, "Is mathematical thought algorithmic?" or "Can a computer do mathematics?"

The same argument that can be made for the necessity to closely examine the assumptions of physics and molecular biology can be made for mathematics. Mathematics has given us the notion of "proof" and "algorithm." These abstract ideas have, in our age, been given a concrete technological embodiment in the form of the computer and the wave of information technology that is inundating our society today. These technological devices are having a significant impact on society at all levels. As in the case of quantum mechanics or molecular biology, it is not just the direct impact of information technology that is at issue, but also the impact of this technological revolution on our conception of human nature. How are we to think about consciousness, about creativity, about thought? Are we all biological computers with the brain as hardware and the "mind" defined to be software? Reflecting on the nature of mathematics will have a great deal to contribute to this crucial discussion.

The three areas of modern science that have been referred to above all raise questions that are interrelated. These questions involve, in one way or another, the intellectual models—metaphors if you will—that are implicit in the culture of modern science. These metaphors are at work today molding human beings' conceptions of certain fundamental human attributes. It is important to bring to conscious awareness the metascientific assumptions that are built into these models, so that people can make a reasonable assessment of these assumptions. Is a machine, even a sophisticated machine like a computer, a reasonable model for thinking about human beings? Most intelligent people hesitate even to consider these questions because they feel that the barrier of scientific expertise is too high. Thus, the argument is left to the "experts," but the fact is that the "experts" do not often stop to consider such questions for two reasons: first, they are too busy keeping up with the accelerating rate of scientific development in their field to consider "philosophical" questions; second, they are "insiders" to their fields and so have little inclination to look at their fields from the outside. In order to have a reasonable discussion about the worldview implicit in certain scientific disciplines, it would therefore be necessary to carry a dual perspective; to be inside and outside simultaneously. In the case of mathematics, this would involve assuming a perspective that arises from mathematical practice—from the actual "doing" of mathematics—as well as looking at mathematics as a whole as opposed to discussing some specific mathematical theory.

The "Light of Reason" or the "Light of Ambiguity"?

What is it that makes mathematics mathematics? What are the precise characteristics that make mathematics into a discipline that is so central to every advanced civilization, especially our own? Many explanations have been attempted. One of these sees mathematics as the ultimate in rational expression; in fact, the expression "the light of reason" could be used to refer to mathematics. From this point of view, the distinguishing aspect of mathematics would be the precision of its ideas and its systematic use of the most stringent logical criteria. In this view, mathematics offers a vision of a purely logical world. One way of expressing this view is by saying that the natural world obeys the rules of logic and, since mathematics is the most perfectly logical of disciplines, it is not surprising that mathematics provides such a faithful description of reality. This view, that the deepest truth of mathematics is encoded in its formal, deductive structure, is definitely *not* the point of view that this book assumes. On the contrary, the book takes the position that the logical structure, while important, is insufficient even to begin to account for what is really going on in mathematical practice, much less to account for the enormously successful applications of mathematics to almost all fields of human thought.

This book offers another vision of mathematics, a vision in which the logical is merely one dimension of a larger picture. This larger picture has room for a number of factors that have traditionally been omitted from a description of mathematics and are translogical—that is, beyond logic—though not illogical. Thus, there is a discussion of things like ambiguity, contradiction, and paradox that, surprisingly, also have an essential role to play in mathematical practice. This flies in the face of conventional wisdom that would see the role of mathematics as eliminating such things as ambiguity from a legitimate description of the worlds of thought and nature. As opposed to the formal structure, what is proposed is to focus on the central ideas of mathematics, to take ideas—instead of axioms, definitions, and proofs—as the basic building blocks of the subject and see what mathematics looks like when viewed from that perspective.

The phenomenon of ambiguity is central to the description of mathematics that is developed in this book. In his description of his own personal development, Alan Lightman says, "Mathematics contrasted strongly with the ambiguities and contradictions in people. The world of people had no certainty or logic."[5] For him, mathematics is the domain of certainty and logic. On the other hand, he is also a novelist who "realized that the ambiguities and complexities of the human mind are what give fiction and perhaps all art its power." This is the usual way that people divide up the arts from the sciences: ambiguity in one, certainty in the other. I suggest that mathematics is also a human, creative activity. As such, ambiguity plays a role in mathematics that is analogous to the role it plays in art—it imbues mathematics with depth and power.

Ambiguity is intrinsically connected to creativity. In order to make this point, I propose a definition of ambiguity that is derived from a study of creativity.[6] The description of mathematics that is to be sketched in this book will be a description that is grounded in mathematical practice—what mathematicians actually do—and, therefore, must include an account of the great creativity of mathematics. We shall see that many creative insights of mathematics arise out of ambiguity, that in a sense the deepest and most revolutionary ideas come out of the most profound ambiguities. Mathematical ideas may even arise out of contradiction and paradox. Thus, eliminating the ambiguous from mathematics by focusing exclusively on its logical structure has the unwanted effect of making it impossible to describe the creative side of mathematics. When the creative, open-ended dimension is lost sight of, and, therefore, mathematics becomes identified with its logical structure, there develops a view of mathematics as rigid, inflexible, and unchanging. The truth of mathematics is mistakenly seen to come exclusively from a rigid, deductive structure. This rigidity is then transferred to the domains to which mathematics is applied and to the way mathematics is taught, with unfortunate consequences for all concerned.

Thus, there are two visions of mathematics that seem to be diametrically opposed to one another. These could be characterized by emphasizing the "light of reason," the primacy of the logical structure, on the one hand, and the light that Wiles spoke

11

of, a creative light that I maintain often emerges out of ambiguity, on the other (this is itself an ambiguity!). My job is to demonstrate how mathematics transcends these two opposing views: to develop a picture of mathematics that includes the logical and the ambiguous, that situates itself equally in the development of vast deductive systems of the most intricate order and in the birth of the extraordinary leaps of creativity that have changed the world and our understanding of the world.

This is a book about mathematics, yet it is not your average mathematics book. Even though the book contains a great deal of mathematics, it does not systematically develop any particular mathematical subject. The subject is mathematics as a whole—its methodology and conclusions, but also its culture. The book puts forward a new vision of what mathematics is all about. It concerns itself not only with the culture of mathematics in its own right, but also with the place of mathematics in the larger scientific and general culture.

The perspective that is being developed here depends on finding the right way to think about mathematical rigor, that is, logical, deductive thought. Why is this way of thinking so attractive? In our response to reason, we are the true descendents of the Greek mathematicians and philosophers. For us, as for them, rational thought stands in contrast to a world that is all too often beset with chaos, confusion, and superstition. The "dream of reason" is the dream of order and predictability and, therefore, of the power to control the natural world. The means through which we attempt to implement that dream are mathematics, science, and technology. The desired end is the emergence of clarity and reason as organizational principles of the entire cosmos, a cosmos that of course includes the human mind. People who subscribe to this view of the world might think that it is the role of mathematics to eliminate ambiguity, contradiction, and paradox as impediments to the success of rationality. Such a view might well equate mathematics with its formal, deductive structure. This viewpoint is incomplete and simplistic. When applied to the world in general, it is mistaken and even dangerous. It is dangerous because it ignores one of the most basic aspects of human nature—in mathematics or elsewhere—our aesthetic dimension, our originality and ability to innovate. In this regard let us take note of what the famous musician, Leonard Bernstein, had to say: "ambiguity . . . is one of art's most potent aesthetic functions. The more ambiguous, the more expressive."[7] His words apply not only to music and art, but surprisingly also to science and mathematics. In mathematics, we could amend his remarks by saying, "the more ambiguous, the more potentially original and creative."

If one wishes to understand mathematics and plumb its depths, one must reevaluate one's position toward the ambiguous (as I shall define it in Chapter 1) and even the paradoxical. Understanding ambiguity and its role in mathematics will hint at a new kind of organizational principle for mathematics and science, a principle that includes classical logic but goes beyond it. This new principle will be *generative*—it will allow for the dynamic development of mathematics. As opposed to the static nature of logic with its absolute dichotomies, a generative principle will allow for the existence of mathematical creativity, be it in research or in individual acts of understanding. Thus "ambiguity" will force a reevaluation of the essence of mathematics.

Why is it important to reconsider mathematics? The reasons vary from those that are internal to the discipline itself to those that are external and affect the applications of mathematics to other fields. The internal reasons include developing a description of mathematics, a philosophy of mathematics if you will, that is consistent with mathematical practice and is not merely a set of a priori beliefs. Mathematics is a *human* activity; this is a triumph, not a constraint. As such, it is potentially accessible to just about everyone. Just as most people have the capacity to enjoy music, everyone has some capacity for mathematics appreciation. Yet most people are fearful and intimidated by mathematics. Why is that? Is it the mathematics itself that is so frightening? Or is it rather the way in which mathematics is viewed that is the problem?

Beyond the valid "internal" reasons to reconsider the nature of mathematics, even more compelling are the external reasons—the impact that mathematics has, one way or another, on just about every aspect of the modern world. Since mathematics is such a central discipline for our entire culture, reevaluating what mathematics is all about will have many implications for science and beyond, for example, for our conception of the nature of the human mind itself. Mathematics provided humanity with the ideal of reason and, therefore, a certain model of what thinking is or should be, even what a human being should be. Thus, we shall see that a close investigation of the history and practice of mathematics can tell us a great deal about issues that arise in philosophy, in education, in cognitive science, and in the sciences in general. Though I shall endeavor to remain within the boundaries of mathematics, the larger implications of what is being said will not be ignored.

Mathematics is one of the most profound creations of the human mind. For thousands of years, the content of mathematical theories seemed to tell us something profound about the nature of the natural world—something that could not be expressed in any way other than the mathematical. How many of the greatest minds in history, from Pythagoras to Galileo to Gauss to Einstein, have held that "God is a mathematician." This attitude reveals a reverence for mathematics that is occasioned by the sense that nature has a secret code that reveals her hidden order. The immediate evidence from the natural world may seem to be chaotic and without any inner regularity, but mathematics reveals that under the surface the world of nature has an unexpected simplicity—an extraordinary beauty and order. There is a mystery here that many of the great scientists have appreciated. How does mathematics, a product of the human intellect, manage to correspond so precisely to the intricacies of the natural world? What accounts for the "extraordinary effectiveness of mathematics"?

Beyond the content of mathematics, there is the *fact of mathematics*. What is mathematics? More than anything else, mathematics is a way of approaching the world that is absolutely unique. It cannot be reduced to some other subject that is more elementary in the way that it is claimed that chemistry can be reduced to physics. Mathematics is irreducible. Other subjects may use mathematics, may even be expressed in a totally mathematical form, but mathematics has no other subject that stands in relation to it in the way that it stands in relation to other subjects. Mathematics is a *way of knowing*—a unique way of knowing. When I wrote these words I intended to say "a unique *human* way of knowing." However, it now appears that human beings share a certain propensity for number with various animals.[8] One could make an argument that a tendency to see the world in a mathematical way is built into our developmental structure, hard-wired into our brains, perhaps implicit in elements of the DNA structure of our genes. Thus mathematics is one of the most basic elements of the natural world.

From its roots in our biology, human beings have developed mathematics as a vast cultural project that spans the ages and all civilizations. The nature of mathematics gives us a great deal of information, both direct and indirect, on what it means to be human. Considering mathematics in this way means looking not merely at the content of individual mathematical theories, but at mathematics as a whole. What does the nature of mathematics, viewed globally, tell us about human beings, the way they think, and the nature of the cultures they create? Of course, the latter, global point of view can only be seen clearly by virtue of the former. You can only speak about mathematics with reference to actual mathematical topics. Thus, this book contains a fair amount of actual mathematical content, some very elementary and some less so. The reader who finds some topic obscure is advised to skip it and continue reading. Every effort has been made to make this account self-contained, yet this is not a mathematics textbook—there is no systematic development of any large area of mathematics. The mathematics that is discussed is there for two reasons: first, because it is intrinsically interesting, and second, because it contributes to the discussion of the nature of mathematics in general. Thus, a subject may be introduced in one chapter and returned to in subsequent chapters.

It is not always appreciated that the story of mathematics is also a story about what it means to be human—the story of beings blessed (some might say cursed) with self-consciousness and, therefore, with the need to understand the natural world and themselves. Many people feel that such a human perspective on mathematics would demean it in some way, diminish its claim to be revealing absolute, objective truth. To anticipate the discussion in Chapter 8, I shall claim that mathematical truth exists, but is not to be found in the content of any particular theorem or set of theorems. The intuition that mathematics accesses the truth is correct, but not in the manner that it is usually understood. The truth is to be found more in the fact than in the content of mathematics. Thus it is consistent, in my view, to talk

15

simultaneously about the truth of mathematics and about its contingency.

The truth of mathematics is to be found in its human dimension, not by avoiding this dimension. This human story involves people who find a way to transcend their limitations, about people who dare to do what appears to be impossible and *is* impossible by any reasonable standard. The impossible is rendered possible through acts of genius—this is the very definition of an act of genius, and mathematics boasts genius in abundance. In the aftermath of these acts of genius, what was once considered impossible is now so simple and obvious that we teach it to children in school. In this manner, and in many others, mathematics is a window on the human condition. As such, it is not reserved for the initiated, but is accessible to all those who have a fascination with exploring the common human potential.

We do not have to look very far to see the importance of mathematics in practically every aspect of contemporary life. To begin with, mathematics is the language of much of science. This statement has a double meaning. The normal meaning is that the natural world contains patterns or regularities that we call scientific laws and mathematics is a convenient language in which to express these laws. This would give mathematics a descriptive and predictive role. And yet, to many, there seems to be something deeper going on with respect to what has been called "the unreasonable effectiveness of mathematics in the natural sciences."[9] Certain of the basic constructs of science cannot, in principle, be separated from their mathematical formulation. An electron *is* its mathematical description via the Schrödinger equation. In this sense, we cannot see any deeper than the mathematics. This latter view is close to the one that holds that there exists a mathematical, Platonic substratum to the real world. We cannot get closer to reality than mathematics because the mathematical level *is* the deepest level of the real. It is this deeper level that has been alluded to by the brilliant thinkers that were mentioned above. This deeper level was also what I meant by calling mathematics irreducible.

Our contemporary civilization has been built upon a mathematical foundation. Computers, the Internet, CDs, and DVDs are all aspects of a digital revolution that is reshaping the world. All these technologies involve representing the things we see and hear, our knowledge, and the contents of our communications in digital form, that is, reducing these aspects of our lives to a common numerical basis. Medicine, politics, and social policy are all increasingly expressed in the language of the mathematical and statistical sciences. No area of modern life can escape from this mathematization of the world.

If the modern world stands on a mathematical foundation, it behooves every thoughtful, educated person to attempt to gain some familiarity with the world of mathematics. Not only with some particular subject, but with the culture of mathematics, with the manner in which mathematicians think and the manner in which they see this world of their own creation.

Why Am I Writing This Book?

What is my purpose in writing this book? Where do the ideas come from? Obviously, I think that the ideas are important because the point of view from which they are written is unusual. But putting aside the content of the book for a moment, there is also an important personal reason for me. This book weaves together two of the most important strands in my life. One strand is mathematics: I have spent a good part of the last forty years doing the various things that a university mathematician does—teaching, research, and administration. When I look back at my motivation for going into mathematics, what appealed to me was the clarity and precision of the kind of thinking that doing mathematics called for. However, clarity was not a sufficient condition for doing research. Research required something else—a need to understand. This need to understand often took me into realms of the obscure and the problematic. How, I asked myself, can one find one description of mathematics that unifies the logical clarity of formal mathematics with the sense of obscurity and flux that figures so prominently in the doing and learning of mathematics?

The second strand in my life was and is a strenuous practice of Zen Buddhism. Zen helped me confront aspects of my life that went beyond the logical and the mathematical. Zen has the reputation for being antilogical, but that is not my experience. My experience is that Zen is not confined to logic; it does not see logic as having the final word. Zen demonstrates that there is a way to work with situations of conflict, situations that are problematic from a normal, rational point of view. The rational, for Zen, is just another point of view. Paradox, in Zen, is used constructively as a way to direct the mind to subverbal levels out of which acts of creativity arise.

I don't think that Zen has anything to say about mathematics per se, but Zen contains a viewpoint that is interesting when applied to mathematics. It is a viewpoint that resonates with many interesting things that are happening in our culture. They all involve moving away from an "absolutist" position, whether this means distrust of all ideologies or in rejection of "absolute, objective, and timeless truth." For me, this means that ambiguity, contradiction, and paradox are an essential part of mathematics—they are the things that keep it changing and developing. They are the motor of its endless creativity.

In the end, I found that these two strands in my life—mathematics and Zen—fit together very well indeed. I expect that you, the reader, will find this voyage beyond the boundaries of the rational to be challenging because it requires a change in perspective; but for that very reason I hope that you will find it exciting. Ambiguity opens up a world that is never boring because it is a world of continual change and creativity.

The Structure of the Book

The book is divided into three sections. The first, "The Light of Ambiguity," begins by introducing the central notion of ambiguity. Actually one could look at the entire book as an exploration of the role of ambiguity in mathematics, as an attempt to come to grips with the elusive notion of ambiguity. In order to highlight my contention that the ambiguous always has a component of the problematic about it, I spend a couple of chapters talking about contradiction and paradox in mathematics. These chapters also enable me to build up a certain body of mathematical results so as to enable even readers who are a little out of touch with mathematics to get up to speed, so to say.

The second section is called "The Light as Idea." It discusses the nature of ideas in mathematics—especially those ideas that arise out of situations of ambiguity. Of course the creative process is intimately tied to the birth and the processing of mathematical ideas. Thus thinking about ideas as the fundamental building blocks of mathematics (as opposed to the logical structure, for example) pushes us toward a reevaluation of just what mathematics is all about. This section demonstrates that even something as problematic as a paradox can be the source of a productive idea. Furthermore, I go on to claim that some of the most profound ideas in mathematics arise out of situations that are characterized not by logical harmony but by a form of extreme conflict. I call the ideas that emerge out of these extreme situations "great ideas," and a good deal of the book involves a discussion of such seminal ideas.

The third section, "The Light and the Eye of the Beholder," considers the implications of the point of view that has been built up in the first two sections. One chapter is devoted to a discussion of the nature of mathematical truth. Is mathematics absolutely true in some objective sense? For that matter, what do we mean by "objectivity" in mathematics? Thinkers of every age have attested to the mystery that lies at the heart of the relationship between mathematics and truth. My "ambiguous" approach leads me to look at this mystery from a perspective that is a little unusual. Finally, I spend a concluding chapter discussing the fascinating and essential question of whether the computer is a reasonable model for the kind of mathematical activity that I have discussed in the book. Is mathematical thought algorithmic in nature? Is the mind of the mathematician a kind of software that is implemented on the hardware that we call the brain? Or is mathematical activity built on a fundamental and irreducible human creativity—a creativity that comes from a deep need that we human beings have to understand—to create meaning out of our lives and our environment? This drive for meaning is inevitably accompanied by conflict and struggle, the very ingredients that we shall find in situations of ambiguity.

SECTION I

THE LIGHT OF AMBIGUITY

*

W HAT IS THINKING? If we imagine thinking to be an ordered, linear, and logical progression, then the rigorous thinking that one finds in a mathematical proof or a computer program is the highest form of thinking. Is this the only way to think? More to the point, is this the way mathematicians think? In this section I investigate situations that seem to be at the opposite extreme from logical thought—I look for ambiguities in mathematics. Strangely enough, I find ambiguity everywhere, and not only ambiguity but also its close cousins contradiction and paradox. How strange it is that mathematics, the subject that appears to be the very paradigm of reason, and for this reason the model that other disciplines attempt to emulate, contains as an irreducible factor, the very things that reason ostensibly exists to eliminate from human discourse!

Ambiguity is not only present in mathematics, it is essential. Ambiguity, which implies the existence of multiple, conflicting frames of reference, is the environment that gives rise to new mathematical ideas. The creativity of mathematics does not come out of algorithmic thought; algorithms are born out of acts of creativity, and at the heart of a creative insight there is often a conflict—something problematic that doesn't follow from one's previous understanding. Now one might think that mathematics is characterized by the clarity and precision of its ideas and, therefore, that there is only one correct way to understand a given mathematical situation or concept. On the contrary, I maintain that what characterizes important ideas is precisely that they can be understood in multiple ways; this is the way to measure the richness of the idea.

Ambiguity is the central theme of this book. From beginning to end it is the single thread that unites the disparate subjects that are discussed. We each probably feel that we understand and are familiar with ambiguity. However, in our exploration of ambiguity in mathematics we may find that there is more to ambiguity than meets the eye. Ambiguity is very rich, and so each new aspect of ambiguity we encounter will teach us something not only about mathematics but also about the nature of ambiguity itself—at least about the way in which ambiguity is being used in this book. Since the whole book is, in a sense, a development of the meaning of "ambiguity in mathematics," I ask the reader not to prematurely close accounts with ambiguity.

Of course, start with whatever intuition you have, but hold it in your mind in an open manner, ready to consider new aspects and connotations of ambiguity. An ambiguity is similar to a metaphor, and we shall discover that many facets of mathematics, even the seemingly simplest and most elementary, contain this metaphoric aspect.

Take, for example the number zero. What could be more elementary? Most of us consider zero to be a closed book—we understand it completely. What more is there to say? Yet in recent years there have been a number of books that have been written about the number zero (Barrow 2000; Kaplan; Seife) All these books stress the ambiguous nature of the number zero—"the nothing that is" as one author put it—as well as its importance in mathematical and scientific thought. Normally, ambiguity in science and mathematics is seen as something to overcome, something that is due to an error in understanding and is removed by correcting that error. The ambiguity is rarely seen as having value in its own right, and yet the existence of ambiguity was often the very thing that spurred a particular development of mathematics and science. What, therefore, is the relationship between the ambiguous and the rational, between the ambiguous and mathematics? Is the role of science and mathematics to exorcise the ambiguous or is it an essential part of mathematics? The problematic aspects of "zero" did not stop mathematics from making it into an essential idea. In my view the importance of "zero" is directly proportional to its problematic nature, and not in spite of that nature.

What is true for "zero" is true for many mathematical ideas. The power of ideas resides in their ambiguity. Thus any project that would eliminate ambiguity from mathematics would destroy mathematics. It is true that mathematicians are motivated to understand, that is, to move toward clarity, but if they wish to be creative then they must continually go back to the ambiguous, to the unclear, to the problematic, for that is where new mathematics comes from. Thus ambiguity, contradiction, and paradox and their consequences—conflict, crises, and the problematic—cannot be excised from mathematics. They are its living heart.

Ambiguity in Mathematics

> I think people get it upside down when they say
> the unambiguous is the reality and the ambiguous
> merely uncertainty about what is really unambigu-
> ous. Let's turn it around the other way: the ambig-
> uous is the reality and the unambiguous is merely
> a special case of it, where we finally manage to
> pin down some very special aspect.
> —David Bohm

INTRODUCTION

This chapter begins the process of developing a new way of describing what mathematics is and what mathematicians do. One
might think that this is an easy task—just ask a mathematician
what it is that he or she does. Unfortunately this will not work,
for the business of mathematicians is the doing of mathematics
and not reflecting on the subject of what it is that they do. Davis
and Hersh note that there is a "discrepancy between the actual
work and activity of the mathematician and his own perception
of his work and activity."[1] The only thing I can do is to look
closely at a variety of mathematical concepts and practices, and
base my description of mathematics on what I actually see is
going on.

The most pervasive myth about mathematics is that the logical structure of mathematics is definitive—that logic captures
the essence of the subject. This is the fallback position of many
mathematicians when they are asked to justify what it is that
they do: "I just prove theorems." That is, when pressed, many
mathematicians retreat back to a formalist position. However,
most practicing mathematicians are not formalists: "what they
really want is usually not some collection of "answers"—what
they want is *understanding*."[2] The statistician David Blackwell is
quoted as saying, "Basically, I'm not interested in doing research

and I never have been. I'm interested in *understanding*, which is quite a different thing."[3] Now, understanding is a difficult thing to talk about. For one thing, it contains a subjective element, whereas drawing logical inferences appears to be an objective task that even sophisticated machines might be capable of making. Nevertheless, if one wants to come close to plumbing the depths of mathematical practice, it will be necessary to begin by seeing beyond the formalist approach of equating mathematics with the trinity of definition-theorem-proof.

Logic is indispensable to mathematics. For one thing, logic stabilizes the world of mathematical results so that it presents itself to our minds in the conventional manner—as a body of permanent and absolute truths. However, logic is not the essence of mathematics nor can mathematics be reduced to logic. Mathematics transcends logic. Mathematics is one of the most profound areas of human creativity. Yet the statement that mathematics goes beyond logic needs to be supported. To do this, a number of characteristics of mathematics will be introduced that are clearly not derived from logic. These include a certain form of mathematical ambiguity as well as the related notions of contradiction and paradox.

"Ambiguity" is a central notion, so I shall spend a fair amount of time in explaining what I mean by ambiguity in mathematics. By ranging over a whole host of examples from mathematics and a few from other fields, I hope to show that ambiguity, as I use the term, is a phenomenon which is central to mathematical theory and practice. Ambiguity will give us a way to approach such questions as "What is the relationship between logic and mathematics?" "What is the nature of creativity in mathematics?" "What is meant by understanding in mathematics and what is its relationship to creativity?" Even the old chestnuts, "Is mathematics invented or discovered?" or "What accounts for the 'unreasonable effectiveness' of mathematics in the physical sciences?" Ambiguity will transform the mathematical landscape from the static to one that is dynamic and characterized by the play of ideas.

What I am attempting to develop is nothing less than a paradigm shift in our understanding of the nature of the mathematical enterprise. Once we begin to look at matters in this new "ambiguous" manner, many things suddenly appear in a new light.

These certainly include mathematical practice and the teaching and learning of mathematics. But this manner of looking at things has implications for how we view the scientific enterprise as a whole. These implications extend to the most fundamental of questions, such as "What is (mathematical) truth?" and "What is knowledge?"

With these heady reflections in the back of our minds, I now proceed to take up the basic notion of the meaning of ambiguity (for this book) and proceed to demonstrate its role in mathematics.

What Do I Mean by Ambiguity?

In this book, ambiguity is a key idea whose implications will take some time and effort to flesh out. For me the most elementary mathematical object, like the equation "1 + 1 = 2," for example, is ambiguous. What do I mean by this? I certainly do *not* mean that the statement "1 + 1 = 2" is unclear or incorrect. People often take ambiguity to be synonymous with vagueness or with incomprehensibility. Though this is a possible meaning, it is not the sense in which I shall use the term. What I am trying to accomplish by using the word ambiguous is to point to a certain metaphoric quality that is inherent in even quite simple mathematical situations. When we encounter "1 + 1 = 2," our first reaction is that the statement is clear and precise. We feel that we understand it completely and that there is nothing further to be said. But is that really true? The numbers "one" and "two" are in fact extremely deep and important ideas, as will be discussed in Chapter 5. They are basic to science and religion, to perception and cognition. "One" represents unity; "two" represents duality. What could be more fundamental? The equation also contains an equal sign. Again, in Chapter 5, I discuss various ways in which "equality" can be understood in mathematics. Equality is another very basic idea whose meaning only grows the more you think about it. Then we have the equation itself, which states that the fundamental concepts of unity and duality have a relationship with one another that we represent by "equality"—that there is unity in duality and duality in unity. This deeper structure that is implicit in the equation is typical of a situation of ambiguity. Thus even the most elementary mathematical expressions have a profundity that may not be apparent on the surface level. This profundity is directly related to what I am calling ambiguity.

The word "ambiguity" is actually being used for two main reasons. The first is that the ambiguous is commonly looked at as something to be contrasted with the logical. The second comes from one of the Oxford English Dictionary definitions of "ambiguity"—"admitting more than one interpretation or explanation: having a double meaning or reference." This notion of "double meaning" comes from the prefix "ambi," as can be seen in such words as "ambidextrous" or "ambivalent." However, the definition that I now put forward comes from a definition of creativity that was proposed by the writer Arthur Koestler.[4] He said that creativity arises in a situation where "a single situation or idea is perceived in two self-consistent but mutually incompatible frames of reference." I shall take the above to be the definition of ambiguity. To repeat:

Ambiguity involves a single situation or idea that is perceived in two self-consistent but mutually incompatible frames of reference.

I hasten to add that putting such a precise definition at the beginning of Chapter 1 involves the risk that the reader will assume that ambiguity is now pinned down once and for all. On the contrary, ambiguity is one of those concepts, like "one," "two," and "equality," of which there is always more to say and learn. I am even tempted to say that "ambiguity" is not really a concept at all; it is more like a condition or context that *produces* concepts. If it is not a normal concept, how then do I go about describing it? My strategy is to start with the description above and give it substance by presenting a series of examples each of which will explore some dimension of ambiguity.

This book is an exploration of ambiguity in mathematics. Unfortunately mathematics is usually presented in a linear manner with the simple preceding the complex and assumptions before conclusions. I prefer to think both of mathematics and of this book as explorations. What is the nature of an exploration in mathematics? In the introduction to his textbook, *Transform Linear Algebra*, Frank Uhlig states:

Linear Algebra is a circular subject. Studying Linear Algebra feels like exploring a city or a country for the first time. An overwhelming number of concepts, all intertwined and connected, are present in any first encounter with linear algebra. As with a new city, one has to start discovering slowly and deliberately. Of great help is that linear algebra is akin to geometry, and like geometry, many of its insights have been permanently there within us. We must only explore, look around, and awake our intuition with the reality of this mathematical place.[5]

What a poetic evocation of the spirit of learning and doing mathematics! I'm inviting the reader to enter into an exploration of mathematics in just this spirit. I shall look at mathematics through the lens of ambiguity. In so doing we shall be simultaneously investigating the nature of ambiguity itself. As Uhlig says, many of the basic insights are already there within us, but to discern them we shall have to put aside our habitual point of view and be open to considering a new viewpoint.

SELF-CONSISTENCY, INCOMPATIBILITY, AND CREATIVITY

The definition of ambiguity that I gave above involves a duality—there must be two frames of reference. Now, duality is a familiar idea in mathematics. For example, in projective geometry it is possible to interchange "points" and "lines" so that every statement about lines and points has a dual statement about points and lines. The statement, "Two lines define (meet at) a point" would have the dual statement, "Two points define (determine) a line." This kind of structural duality carries some, but not all of the meaning that I attribute to ambiguity.

Ambiguity, as the term is being used here, is not mere duality. The two frames of reference must be *mutually incompatible*, even though they are individually self-consistent. Yet, in spite of this incompatibility, there exists an over-riding unitary situation or idea. On the one hand, there is the harmony of consistency— things are in peaceful equilibrium. On the other, there is the disorder of incompatibility. Incompatibility is unacceptable in mathematics! It must be resolved! It is this need to resolve incompatibility that makes the situation of ambiguity so dynamic, so potentially creative. There are two perfectly harmonious ways of looking at the situation, yet they are in opposition to one another. So there is a need to resolve this unacceptable situation in order to restore equilibrium. The restoration of equilibrium can only come at a level that is, in a manner of speaking, higher than either of the original frames of reference. The equilibrium condition may not yet exist. It may only come into existence as a result of the need to reconcile the incompatibility of the original situation. Thus, a situation of ambiguity is a situation with creative possibilities.

Ambiguity may seem to be complicated, but its essence can be conveyed very simply. Here is an example of ambiguity. It's a joke—not very funny but with a mathematical connection—and it makes the point about the nature of ambiguity.

> A mathematician is flying non-stop from Edmonton to Frankfurt with Air Transat. The scheduled flying time is nine hours. Some time after taking off, the pilot announces that one engine had to be turned off due to mechanical failure: "Don't worry—we're safe. The only noticeable effect this will have for us is that our total flying time will be ten hours instead of nine." A few hours into the flight, the pilot informs the passengers that another engine had to be turned off due to mechanical failure: "But don't worry— we're still safe. Only our flying time will go up to twelve hours." Some time later, a third engine fails and has to be turned off. But the pilot reassures the passengers: "Don't worry—even with one engine, we're still perfectly safe. It just means that it will take sixteen hours total for this plane to arrive in Frankfurt." The mathematician remarks to his fellow passengers: "If the last engine breaks down, too, then we'll be in the air for twenty-four hours altogether!"

Here you have it—two conflicting frames of reference (one of them implicit) resulting in tension, and then a creative release, laughter. Of course in mathematics the release comes with the birth of a new idea or a new way of looking at the situation but the dynamics of a humorous situation is very similar. A joke is an example of ambiguity and creativity—you have to *get* a joke.

There are two different ways in which an ambiguous situation may manifest itself. On the one hand, there are the two inconsistent points of view that may be reconciled by the creative act of producing the "single situation or idea." On the other, there is the single or unified viewpoint that may be looked at in two different ways. Thus, there is an element of the self-referential when we speak of ambiguity. That is, the definition of ambiguity is itself ambiguous.

THURSTON AND MULTIPLICATION

The mathematician William Thurston tells the following story: "I remember as a child, in the fifth grade, coming to the amazing (to me) realization that the answer to 134 divided by 29 is 134/29 (and so forth). What a tremendous labor-saving device! To me, '134 divided by 29' meant a certain tedious chore, while 134/29 was an object with no implicit work. I went excitedly to my father to explain my major discovery. He told me that of course this is so, a/b and a divided by b are just synonyms. To him it was just a small variation in notation."[6]

Wow! 134/29 is a number! Even if you don't do the division, it is a number that you can use in subsequent computations. In fancier words, you could say that the process of division had been reified, that is, the process has been made into an object. The point is that division has these two perspectives: it is a process and it is a number. The mathematician and mathematics educator David Tall, along with his collaborator Eddie Gray in Warwick, has given this situation a name; he calls it a *procept* (for *process/concept*). In our terms, division is ambiguous. For Thurston, the realization of this ambiguity was a major step in his intellectual development.

$$E = MC^2$$

A few years ago, David Bodanis had the brilliant idea of writing an entire book about an equation. The famous equation was $E = mc^2$, which has literally changed our world. In addition to

31

radically altering the way we think of the world, it had the very practical consequences of establishing the theoretical foundations for the nuclear industry, not to speak of nuclear weapons. Bodanis[7] begins by explaining what an equation is. He does this in a way that is interesting and original. The crux of his approach to the subject of the book, is contained in his unconventional take on the significance of an equation:

> A good equation is not simply a formula for computation. Nor is it a balance scale confirming that two items you suspected were nearly equal really are the same. Instead, scientists started using the = symbol as something of a telescope for new ideas—a device for directing attention to fresh, unsuspected realms. Equations simply happen to be written in symbols instead of words.
>
> This is how Einstein used the "=" in his 1905 equation as well. The Victorians had thought that they'd found all possible sources of energy there were: chemical energy, heat energy, magnetic energy, and the rest. But by 1905 Einstein could say, No, there is another place where you can look where you'll find more. His equation was like a telescope to lead there, . . . He found this vast energy source in the one place where no one had thought of looking. It was hidden away in solid matter itself.

Recall the above definition of ambiguity and its "two self-consistent but mutually incompatible frames of reference." What could be a more self-consistent frame of reference than matter or energy? What could be more incompatible than matter and energy? The terms are practically the opposites of one another. They are conceptually almost antagonistic. According to the *OED*, matter "has mass and occupies space." It refers to "physical substance in general *as opposed to* spirit, mind etc." (you could add energy to this list). Energy, on the other hand, refers to "force or vigor," that is, activity or the potential for activity. The scientific definition of energy is similar; it refers to "the ability of matter to do work." The distinction here is so basic that it is embedded in the basic structure of language itself. Matter refers to things or objects. Language uses nouns to refer to things. To refer to activity or even the potential of activity, that is, energy, language uses verbs. Thus the dichotomy between matter and energy is built into language itself.[8]

How is the gap between these two to be bridged? The first and most obvious way would be to regard matter and energy as complementary. Thus one could regard matter and energy as indispensable aspects of the natural world and maintain that a complete description of nature would involve describing both domains and the laws that govern them. We would go on to describe the relationship between matter and energy. Thus a moving body possesses kinetic energy that is proportional to its mass and the square of its velocity. To look at things in this way would be to miss the radical insight behind Einstein's equation. $E = mc^2$ says that matter *is* energy. It says that these two mutually exclusive ways of describing reality are in fact one—that there is one reality that can be seen as energy when we look at it in one context and as matter when we look at it in another.

Thus, the equation is something that could be called a scientific metaphor. A literary metaphor like Shakespeare's "all the world's a stage, and all the men and women merely players" is a comparison between two different domains—it is really a kind of mapping from one of these domains, here ordinary life, to the other, here the stage. However, a metaphor requires more than a mere correspondence between different domains. "Getting" a metaphor requires an insight: it requires looking at the world in a new way. The power of this particular insight is extraordinary. Its consequence, the atomic bomb, is itself a metaphor for the power of the idea. This equation brings out the full implication of "ambiguity" as the term is being used here. There exist two frames of reference whose incompatibility generates enormous power. This power is then harnessed by the single idea that is represented by the equation $E = mc^2$.

AMBIGUOUS SITUATIONS IN MATHEMATICS

Now let us move on to a more systematic exploration of ambiguity in mathematics. There will be a place for some fairly sophisticated mathematics, but I will begin with a number of elementary examples, very elementary indeed. The reason for including these examples is that they are accessible to everyone. Also, they are here to make the point that no mathematics is completely "trivial." Even elementary arithmetic, algebra, and geometry, when looked at from a fresh perspective, can manage to surprise you.

The Equations of Arithmetic

Let's return to the most basic of equations from arithmetic, something like "2 + 3 = 5." Where is the ambiguity here? I remember the way equations were explained in grade school through the metaphor of the balance. If you put a two- and a three-pound weight on one side of a scale and a five-pound weight on the other, then the two sides will balance. Equality, we were told, means balance. Now "balance" is a good way to think of equality, but is it the only way? From the balance metaphor we derive the idea that "2 + 3" and "5" are just two different ways to describe the same thing—that "2 + 3" and "5" are essentially identical and that the equality sign represents this identity. However "=" does not mean identical, as Bodanis pointed out in the paragraphs I quoted above. Thinking of equations as merely linking two otherwise identical quantities would not explain the power of equations to open up unsuspected relationships between things that were not necessarily connected a priori.

Where is the creative element in "2 + 3 = 5"? Where is the insight, the possibility for an aha! experience? In order to appreciate what is going on, we may have to listen to intelligent people who are less sophisticated than we are—children, for example. Various researchers in mathematics education (e.g., Kieren 1981) have pointed out children's propensity to understand the equality sign in operational terms; that is, "2 + 3 = 5" is understood as an action "2 added to 3 *makes* 5." The sum "2 + 3" is a process, a verb. Children learn what addition is about through the process of counting. Yet the right-hand side is an object, the number 5. What the equation "2 + 3 = 5" is doing is identifying a process with an object. This is similar to the moral of the Thurston story above, where the process of division was seen as a numerical object. To see that a process can be an object or, looking at it the other way around, that the object can be thought of as a process, entails a discontinuous leap—an act of understanding that is in essence a creative act. We all made this creative leap so long ago that we don't remember having done so. But it was an essential step in our development. And what was the essence of this act of understanding? It is that process and object are one ambiguous idea. Thus the ambiguity here is seeing that the two contexts of process and object are unified by this one idea that is captured symbolically by the equation "2 + 3 = 5." All the elements of ambiguity are present here: the two contexts that are in conflict until the conflict is resolved by an act of understanding. Subsequent to the act of understanding, what used to be a conflict becomes a flexible viewpoint where one is free to freely move between the contexts of number as object and number as process. I will return to this same ambiguity in a less elementary situation when I come to discuss infinite decimals.

THE SQUARE ROOT OF TWO

To our contemporary way of understanding things the square root of 2 is no mystery. It is a perfectly well-defined number. In what way, then, can $\sqrt{2}$ be called ambiguous? By our definition, ambiguity required "a single situation or idea"—precisely the fact that $\sqrt{2}$ is well defined. But it also required that $\sqrt{2}$ can be perceived in two self-consistent contexts which are somehow in conflict with one another.

This latter requirement can best be understood historically. In fact $\sqrt{2}$ has an interesting history. It appears, in Euclidean geometry, as a consequence of the Theorem of Pythagoras, as the length of the hypotenuse of a right-angled triangle with sides of unit length.

Thus, $\sqrt{2}$ existed for the Greeks as a concrete geometric object. On the other hand, they were able to prove that this (geometric number) was not rational, that is, it could not be expressed as the ratio of two integers, like 2/3 or 127/369. Such nonfractions came to be called irrational numbers, and the name "irrational" indicates the kind of emotional reaction that the demonstration of the existence of nonrational numbers produced.

35

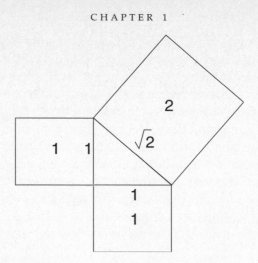

Figure 1.1. $\sqrt{2}$ as a geometric object

There is no question that the demonstration that $\sqrt{2}$ was not rational precipitated a crisis.[9] I shall return to this crisis in a later chapter, but for now let's just say that a whole way of looking at the world, the philosophy of the secret Pythagorean society, was brought into question. Today we would say that $\sqrt{2}$ is an irrational number, but "an irrational number is no number at all ... it is totally man-made," as Leopold Kronecker said, "and thus is of dubious significance philosophically."[10] So is $\sqrt{2}$ a number or not? We can all agree that it is a very different kind of number from the integers and the fractions, the numbers of arithmetic.

William Dunham makes a comment that is relevant here when he says that the irrationality of $\sqrt{2}$ is one instance of "a continuous feature of the history of mathematics ... the prevailing tension between the geometric and the arithmetic."[11] There are two primordial sources of mathematics: counting, which leads to arithmetic and algebra, and measuring, which leads to geometry. Two self-consistent contexts, if you will. Initially these two domains were considered to be identical, but the $\sqrt{2}$ proof brought an inherent conflict between them out into the open. Rational numbers have a consistent meaning in both contexts, but in $\sqrt{2}$ we have a mathematical object that has a clear meaning in a geometric context but is problematic when considered as an arithmetic object, in this case a rational number. A number is a number is a number, to paraphrase Gertrude Stein, but is a geometric number really a number? At the very least there is a tension, an incompatibility, between the geometric and the arithmetic. It is this incompatibility that made $\sqrt{2}$ ambiguous for the Pythagoreans. This does not mean that it was viewed as vague or imprecise. The term ambiguity highlights the problematic aspects of $\sqrt{2}$ for the Greeks.

There are two possible reactions to this sort of ambiguous situation. One can abandon one of the seemingly inconsistent contexts or one can build a new context that is general enough to reconcile the conflict. Both reactions are interesting and can lead to new mathematics.[12] The Greeks chose the former and essentially abandoned algebra for geometry. Even so, the irrationality of $\sqrt{2}$ was a great blow to those, like the Pythagoreans, whose entire worldview was based on the rationality (in the sense of rational numbers or fractions) of the natural world (see Chapter 7). In fact, large portions of Euclidian geometry (the books on ratio and proportion) had been developed on the assumption that any two lengths are commensurable. That means that for any two line segments there is a (smaller) segment that divides into both segments evenly. This amounts to saying that the ratio of the lengths of the two segments is rational. Thus all these proofs (and also, it has been conjectured, the proof of the Pythagorean Theorem itself) that depended on this assumption had to be redone in a different way. This task was, in fact, accomplished successfully. In this activity one can see the need to resolve the incompatibility raised by the ambiguity and therefore the role of the ambiguity as a generator of mathematical activity.

It might be interesting to take a moment and discuss the theory of ratio and proportion. A ratio is the quotient of two numbers. Let's call them x and y. Today we would say that the ratio is the quotient, the number x/y. However the Greeks did not do this—in fact human beings did not do this for the next two millennia. The problem is in a way the same problem Thurston had with $134/29$—is the *process* of division the same as the number that *results* from that process? In the case of commensurable numbers, $x = nz$ and $y = mz$ for some integers m and n, then x/y can be identified with n/m, but what does the quotient mean when x and y are incommensurable, like 1 and $\sqrt{2}$, for example? How can you call this kind of ratio a number? This was the major problem that necessitated a complete reworking of the theory of ratio and proportion. It is fascinating to see what kind of solution the Greeks came up with.

The problem is to give meaning to the ratio x/y. Euclid does this in Book V of the *Elements* by defining what it means for two ratios, x/y and z/w, to be equal; that is, he defines the expression "x is to y as z is to w." Essentially Euclid said that "$x/y = z/w$" means that $mx > ny$ if and only if $mz > nw$ for all integers n and m. That is the ratios are the same if we get the same result when we compare them to every possible rational number—x/y is greater than (or less than) n/m for exactly those fractions for which z/w is also greater than (or less than) n/m. Thus, if we imagine x/y and z/w to be a points on the real line, then Euclid is claiming that even if these points are not rational they are precisely located by their relationship with the rational numbers. This fact that the rational numbers "determine" the real numbers is very close to our modern idea of the relationship between rationals and irrationals. The same idea is behind the "Dedekind cut" approach to rigorously developing the real numbers from the rational numbers. It is fascinating that this idea is present in Greek mathematics and that it was produced by the need to deal with the fallout from the crisis of the irrationality of the square root of two.

In a sense the Greeks never completely resolved the ambiguity of $\sqrt{2}$. It was not until the real number system was rigorously developed in the nineteenth century that we could say that the problems were resolved in a satisfactory way. The real numbers provided a context within which the geometric and arithmetic properties of $\sqrt{2}$ could be reconciled and understood. On the real line $\sqrt{2}$ is just another number/point that is on an equal footing with any rational number such as -5 or $2/3$. In the Cartesian plane, that is, from the point of view of Cartesian geometry, the geometric and the algebraic are two frames of reference that are both valid and consistent. Thus, for example, the straight line joining the points $(0,0)$ and $(3,3)$ *is* the equation $y = x$. This is a classic "resolution" of an ambiguity—the creation of a new context that contains both of the original frames of reference and yet is a "higher" frame of reference in its own right.

In a larger sense the ambiguity between the algebraic and the geometric is a theme that mathematics never tires of. A variation on this theme is the ambiguity between the discrete and the continuous. Is the discrete an approximation to the continuous or, on the contrary, is the continuous a kind of idealization of what is essentially a discrete world? This remains an important question. To give but one example, the mathematician Alexandre Grothendieck claims that his work seeks a unification of two worlds, "the arithmetic world, in which there live the (so-called) spaces having no notion of continuity, and the world of continuous size, where live the 'spaces' in the proper sense of the term, accessible to the methods of the analyst."[13]

The story of $\sqrt{2}$ is relevant to our discussion in many ways. It shows that situations of ambiguity exist in mathematics. Viewed historically, the problematic aspects of a situation of ambiguity are resolved by a new context within which one can move from one context to the other with a new freedom. But it would be wrong to think that $\sqrt{2}$ was ambiguous in the past but not in the present. Ambiguous situations always have these two points of view—before and after. Before you "get it" the situation is problematic because the two frames of reference are seen to be in conflict; afterward there is a flexibility that comes from being able to move freely from one point of view to the other in the realization that you are still in the same situation. The whole situation is what I am talking about when I refer to the ambiguity of $\sqrt{2}$. The problematic aspects of the past reappear in the present as the learning difficulties of students of mathematics. $\sqrt{2}$ was and is ambiguous. This ambiguity was a spur to the development of mathematics just as it is a potential spur to the mathematical development of students.

DECIMAL NUMBERS

Our next example comes from the world of real numbers. Consider decimal notation for real numbers. For example, we are all taught in school that the fraction 1/3 when written as a decimal number is .333. . ., where the dots indicate that the sequence of 3s has no end. Thus

$$\frac{1}{3} = .3333. \ldots$$

Multiplying both sides by 3 we get

$$1 = .999. \ldots$$

Now what is the meaning of these equations? What is the precise meaning of the "=" sign? It surely does not mean that the number 1 is identical to that which is meant by the notation .999. ... There is a problem here, and the evidence is that, in my experience, most undergraduate math majors do not believe this statement. I remember putting this question, "does 1 = .999. . .?" to the students in a class on real analysis. Something about this expression made them nervous. They were not prepared to say that .999. . . is equal to 1, but they all agreed that it was "very close" to 1. How close? Some even said "infinitely close," but they were not absolutely sure what they meant by this. These students may be quite advanced in certain ways, but this statement is still an obstacle[14] for them. What is the obstacle? In my opinion it is the ambiguity contained in equating an infinite decimal to an integer.

The notation .999. . . stands for an infinite sum. Thus

$$.999 \ldots = \frac{9}{10} + \frac{9}{100} + \frac{9}{1000} + \ldots$$

Now an infinite sum is a little more complicated than a finite sum, and this complexity is revealed by the fact that the notation is deliberately ambiguous. Thus this notation stands both for the *process* of adding this particular infinite sequence of fractions and for the *object*, the number that is the result of that process. As was the case of the equations of arithmetic, the two contexts (in the above definition of ambiguity) are again those of process and object. Now the number 1 is clearly a mathematical object, a number. Thus the equation 1 = .999. . . is confusing because it seems to say that a process is equal (identical?) to an object. This appears to be a category error. How can a process, a verb, be equal to an object, a noun. Verbs and nouns are "incompatible contexts" and thus the equation is ambiguous. Similarly, all infinite decimals are ambiguous. Students have a problem because they think of .999. . . only as a process. They imagine themselves actually adding up the series term by term and they "see" that this process never ends. So at any finite stage the sum is "very close" but not equal to 1. They don't see that this infinite process can be understood as a single number.

You can even go through a "proof" with them, something like:

$$\text{Let } x = .999.\ldots$$
$$\text{Then } 10x = 9.999.\ldots \text{ (shifting the decimal point).}$$
$$\text{Thus } 9x = 10x - x = 9.999.\ldots -.999.\ldots = 9.$$
$$\text{So } x = 1.$$

The reaction is interesting. For the most part, the students will now agree that 1 is indeed equal to .999. . . . That is, they now accept it but, in my opinion, something of the old perplexity still remains. They have not resolved the ambiguity. They still do not "understand" the representation for infinite decimals. Understanding requires more than accepting the validity of a certain argument. It requires a creative act, which is what I mean when I refer to the resolution of an ambiguity.

I hasten to add that this ambiguity is a strength, not a weakness, of our way of writing decimals. To understand infinite decimals means to be able to move freely from one of these points of view to the other. That is, understanding involves the realization that there is "one single idea" that can be expressed as 1 or as .999. . ., that can be understood as the process of summing an infinite series or an endless process of successive approximation as well as a concrete object, a number. This kind of creative leap is required before one can say that one understands a real number as an infinite decimal.

VARIABLES

One of the most basic aspects of mathematics involves the reduction of the infinite to the finite. Mathematics has been called the science of the infinite, yet mathematicians are human beings and therefore intrinsically restricted to the finite. Thus one of the great mysteries of mathematics is the manner in which the process of making the infinite finite occurs. This question will be examined in great detail in the discussion of infinity. For the moment, consider the notion of "variable." Most people are introduced to the idea of variable in high school algebra, where they learn to manipulate expressions such as "$3x + 2$." They are told that the "x" is not a number but can represent *any* number. In fact x is usually restricted to some particular set of numbers: natural numbers, integers, rational numbers, real or complex numbers. It may even be a subset of one of these sets of numbers. The domain of the variable may not even be specified explicitly but only inferred by the context. In this sense the notion of a variable is a little ambiguous.

However there is another and more serious way in which the idea of a variable is ambiguous. Let us suppose that we are talking about the positive integers. Then the expression $3x + 2$ actually stands for the whole set of numbers: $3(1) + 2 = 5, 3(2) + 2 = 8, 3(3) + 2 = 11, 14, 17, 20, \ldots$. So $3x + 2$ is a short-hand for the whole set of numbers $\{5, 8, 11, \ldots\}$. However when we work with the expression $3x + 2$ we do not carry around the whole set of potential values in our head. We think of $3x + 2$ as some specific but unspecified element of that set. So we imagine x to have been chosen. It is some (one) specific number that can be written as $3x + 2$, but we know nothing about the value of x except that it is an integer. Thus we simultaneously think of x as general and specific. It is precisely this general/specific ambiguity that gives the notion of variable its importance in mathematics. An infinite set of possible values has been replaced by a finite set of values (here one value). It is true this one value is unspecified, but nevertheless something has been gained.

For example, consider the equation

$$3x + 2 = 8$$

and its solution

$$x = 2.$$

Does the "x" in "$3x + 2 = 8$" refer to *any* number or does it refer to the number 2? The answer is both and neither. At the beginning x could be anything. At the end x can only be 2. Of course at the beginning x (implicitly) can only be 2. Yet at the end we are saying that every number $x \neq 2$ is *not* a solution, so the equation is also about all numbers. Thus at every stage the x stands for *all* numbers but also for the *specific* number 2. We are required to carry along this ambiguity throughout the entire procedure of solving the equation. It begins with something that could be anything and ends with a specific number that could not be anything else. What an exercise in subtle mental gymnastics this is! How could this way of thinking be called *merely* mechanical? No wonder children have difficulty with algebra. The difficulty is the ambiguity. The resolution of the ambiguity, solving the equation, does not involve eliminating the double context but rather being able to keep the two contexts simultaneously in mind and working within that double context, jumping from one point of view to the other as the situation warrants.

A variable is general and specific at the same time. It is all values or it is a unspecified "typical" value. In that ambiguity lies its power. By not resolving the ambiguity until the end of the piece of mathematics one is able to use that ambiguity constructively. Thus when considering the function $f(x) = 3x^2 + 2$, we think of x as a typical real number. But we also think of the whole function as being identified with its parabolic graph. Then we can say that its derivative, for example, is the function $6x$. Again, we think of this in two ways: first, as a formula that is valid for all values of "x" (the derivative at $x = 2$ is 6 times 2 or 12); and second, as a specific (single) point on the graph where the slope of the tangent line is the specific number $6x$.

Without this double or ambiguous point of view, modern mathematics would never have been invented. Remember that Greek mathematics was geometric and not algebraic. Algebraic thought requires the use of the idea of variable. This was not as explicit in Greek thought as it would later become. Again, we can only speculate that it was the Greeks' reverence for clarity and harmony and their distrust and repugnance for ambiguity that prevented them from developing their mathematics in this direction.

The algebraic equation $3x + 2 = 8$ is ambiguous in yet another way. In solving this equation I am really making the following assertion: "Assuming that there exists a number x such that $3x + 2 = 8$, it follows that this number must be 2." Thus in setting out to solve an equation we have taken for granted that the solution exists. That is, the solution is both unknown and (implicitly) known at the same time. This ambiguity between being known and unknown is similar to the ambiguity of a variable that I mentioned earlier and is essential to equations.

I said above (with respect to the equation $3x + 2 = 8$) that we start by assuming that the solution exists and only then determine what it is. What if the solution does not exist? What hap-

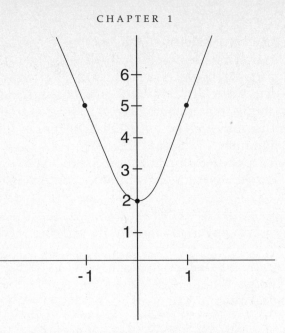

Figure 1.2. Graph of "$f(x) = 3x^2 + 2$"

pens to the ambiguity? Consider the equation $x + 1 = 0$, and assume that there are only the nonnegative integers $(0, 1, 2, 3, \ldots)$ at our disposal. Then the solution $x = -1$ is not available, so there is no solution within the system we are working in. What happens? Remember that I said that ambiguous situations were dynamic; the two incompatible contexts may generate their own creative resolution. Here the incompatibility resides in the fact that the equation is a form that implicitly assumes a solution exists (all the terms in the equation belong to the known system of nonnegative integers), yet no solutions exist (within the system of nonnegative integers). The creative resolution of this dilemma *generates the required solutions.* You could say that the equation $x + 1 = 0$ brings the negative numbers into existence! Thus in order to give meaning to the equation $x + 1 = 0$ (and more generally $x + n = 0$) in a situation where only the nonnegative integers are available, we are forced to invent a new class of numbers. How exciting! Similarly the equation $x^2 + 1 = 0$ produces the complex numbers.[15]

There is power in this ambiguity even if the existence of the solution is not guaranteed. In fact, in this case we can see the generative power of ambiguity to creatively produce new ideas.

Equations can be seen as metaphors, and this way of looking at them explains the generative power of mathematics to produce new structures in response to new situations. Of course there are other ways to introduce larger number systems. Nevertheless, let us not underestimate the power of equations to evoke novel situations. The equation $x^2 + 1 = 0$ makes perfect sense even if we only have integers around, but the latent ambiguity of $x^2 + 1 = 0$ as an equation cries out for a solution even it requires inventing a whole new number system to produce it.

FUNCTIONS

The notion of a function is ambiguous. There are many equivalent definitions, but let me focus on two. There is the ordered pair, graphical definition of a function. This is a static definition: the function is a set (of ordered pairs) or a picture (the graph) or a table (figure 1.3a). However there is also the mapping definition, which is related to the black box, input-output definition (figure 1.3b). This latter is a dynamic definition. Here the x is transformed into the y. This definition is the one that is used in thinking of a function as an iterative process or a dynamical system or a machine.

Mathematicians go back and forth from one of these representations to the other. "Most of the functions introduced in the seventeenth century were first studied as curves, before the function concept was fully recognized."[16] A curve in the plane is a representation of an implicit relationship between variables, but the dependence of one of these variables on the other is not necessarily clear from the geometric picture.[17] Nevertheless, thinking about a function as a curve makes the whole of the function into one geometric object. The unity of the function as object is maintained in later developments, when the function is seen (by James Gregory, for example) "as a quantity obtained from other quantities by a succession of algebraic operations."[18] This kind of definition leads to thinking of a function as a formula. Here we already have a kind of duality with the function as geometric object (graph) and analytic object (formula). Further developments involved allowing the "formula" to extend to an infinite number of operations (power or Fourier series) for which the

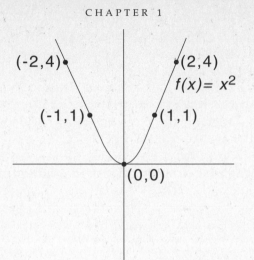

Figure 1.3a. Graphical model of $f(x) = x^2$ (ordered pairs)

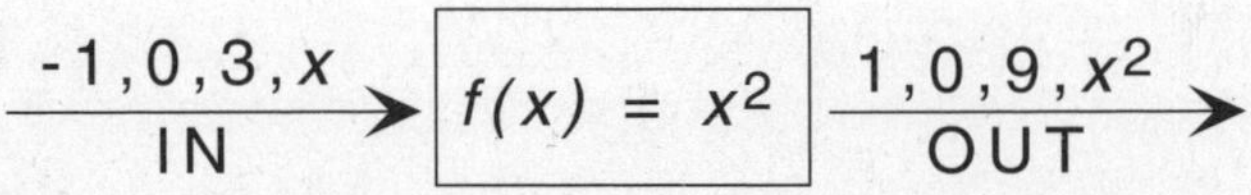

$$\frac{-1, 0, 3, x}{\text{IN}} \longrightarrow \boxed{f(x) = x^2} \frac{1, 0, 9, x^2}{\text{OUT}} \longrightarrow$$

Figure 1.3b. Input-output model

geometric representation was not readily available. Finally with the development of set theory, functions came to be seen as abstract sets of ordered pairs (x, y) where the graph might not even be a geometric object at all.[19] This abstract definition of function is the "single idea" that includes both the geometric and analytic approaches.

In the twentieth century even the set theoretical unification of the idea of function came to be looked at in a new way. New developments in mathematics often entail reworking a concept. The input-output model was crucial to looking at functions as the generators of processes. This new point of view came into its own with the introduction of computers and calculators. A simple mathematical calculator has a series of "function buttons" with names like $\sin x$, $\cos x$, $\exp x$. When we punch a number (the input, x) into the calculator and push one of these buttons, the number is transformed into another (the output, y). The exp button on a calculator is a simple input-output device.

Thinking about a function in this way, we can see that any function can be seen as generating what is called a difference

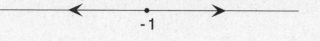

Figure 1.4. "$f(x) = 2x + 1$" generates a dynamical system

equation or dynamical system. Suppose the "generating function" is $f(x) = 2x + 1$. If the process has the value $x_0 = 1$ at time 0, then its value at time 1 would be $x_1 = f(x_0) = 2(1) + 1 = 3$, its value at time 2 would be $x_2 = f(x_1) = 7$, and so on. Thus the function could be considered as a "law" that governs how the process it represents evolves over time. The graph of the function is of little help to us when we think of a function in this way. What we want to know is, if we start with a certain value at time zero, what will happen "in the long run." From this point of view a completely different geometric picture is required. For the generating function $f(x) = 2x + 1$ it would look as shown in figure 1.4.

The picture contains the following information: the "dot" at -1 means that if the system starts at value $x = -1$, then it remains at this value at all future times; the right arrow means that if the initial value is greater than -1 then the future values of the process increase without bound; the left arrow that they decrease without bound if the initial value is less than -1.

So we now have a new way of thinking about the concept of function—a dynamic concept as compared with the versions of functions that were discussed earlier. We now need to reconcile these two definitions by integrating them into a new and more general concept.

At this stage we are thinking about a function as some sort of rule that applies to a whole set of numbers. However, there inevitably comes a time when we want to operate not just on individual numbers but on the rule itself. Thus, if $f(x) = x^2$ and $g(x) = 3x + 1$, we may wish to add $f(x)$ and $g(x)$ and thereby create a new function $h(x) = f(x) + g(x) = x^2 + 3x + 1$. Or we may wish to multiply them and create $k(x) = x^2(3x + 1) = 3x^3 + x^2$. Or again we may wish to consider the result of applying the first rule followed by the second rule $h(x) = g(f(x)) = 3x^2 + 1$. This last operation is called the "composition of functions f and g" and written $h = g \circ f$. It can also be done in the reverse order to obtain $f(g(x)) = (3x + 1)^2$. When we operate on functions in this way we are thinking of the function as one whole, unified object. We have made a function that began as something akin to a pro-

47

cess, a process for operating on numbers, into an object that itself could be operated upon. At one level, one can add or multiply numbers; now, at a higher level, one can add, multiply, or compose functions. This is the process of abstraction at work. Abstraction consists essentially in the creation and utilization of ambiguity. The initial barrier to understanding, that a function can be considered simultaneously as process and object—as a rule that operates on numbers and as an object that is itself operated on by other processes—turns into the insight. That is, it is precisely the ambiguous way in which a function is viewed which is the insight.

At a higher level of abstraction one puts whole families of functions together to form function spaces, for example, all continuous functions defined on the interval of numbers between 0 and 1. Once a function is seen as a point in a larger space, we can talk about the distance between functions, the convergence of functions, functions of functions, and so on. This sort of dual representation is present in a great many mathematical situations.

Fundamental Theorem of Calculus

The Fundamental Theorem of Calculus is one of the great theorems of mathematics. A consideration of this theorem will extend our discussion of ambiguity from the domain of concepts like variables and functions to include the domain of actual mathematical results. How, one might ask, can a mathematical theorem be ambiguous? The *essence* of this theorem is ambiguity; it is asserting that calculus is ambiguous!

Now "differential calculus" and "integral calculus" can be (and historically were) developed independently of one another. They appear, at first glance, to have nothing to do with one another. Integration is a generalization of the idea of area. A typical problem might be to calculate the area between the graph of the curve $y = x^2$ and the x-axis, between 0 and 1 (figure 1.5a). Differential calculus as developed by Newton and Leibniz was concerned with calculating the slope of tangent lines to curves or the related problem of instantaneous change in one variable with respect to another, velocity, for example, as shown in figure

48

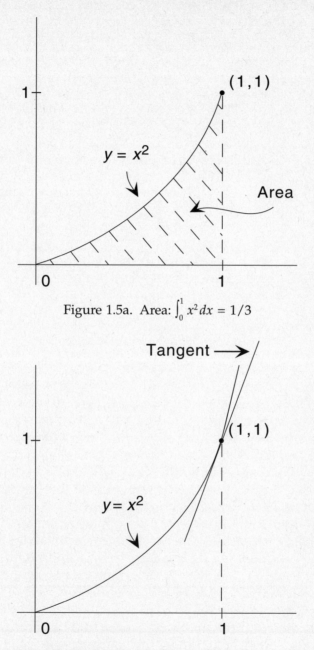

Figure 1.5a. Area: $\int_0^1 x^2\,dx = 1/3$

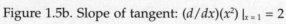

Figure 1.5b. Slope of tangent: $(d/dx)(x^2)\,|_{x=1} = 2$

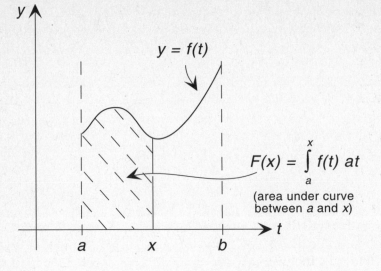

$$\frac{d}{dx} F(x) = F'(x) = f(x)$$

Figure 1.6. Fundamental theorem of calculus

1.5b. The Fundamental Theorem says that these processes are inverses of one another (when the functions involved are "reasonable" as in figure 1.6).

Now it may be possible to start with integration and then develop differentiation or vice versa, but the theorem says that, for functions of one variable, neither process is the more fundamental. Actually, the theorem says that there is in fact one process in calculus that is integration when it is looked at it in one way and differentiation when it is looked at in another. Another way of putting this is that without the Fundamental Theorem there would be two subjects: differential calculus and integral calculus. With it there is just the calculus, albeit with a multiple perspective. This multiple perspective is essential to an understanding of calculus.

How is this multiple perspective used? Well, since differentiating is easier than integrating, we can integrate by taking the inverse of the derivative, that is, by calculating the antiderivative. For example, since the derivative of the function $f(x) = x^2$ is the function $g(x) = 2x$, it follows that the integral of $2x$ is x^2. Whole lists of such antiderivatives may be established and then used to integrate elementary functions.

Fermat's Last Theorem

Perhaps the most famous mathematical problem in the last three hundred years involves the equation,

$$x^n + y^n = z^n$$

for $n \geq 2$. For $n = 2$ this equation represents the relationship between the lengths of the sides of a right-angled triangle according to the theorem of Pythagoras. Thus there exist many sets of solutions, including $x = 3$, $y = 4$, $z = 5$ or $x = 5$, $y = 12$, $z = 13$. The mathematician Pierre de Fermat (1601–1665) claimed that there were no integer solutions to this equation for $n > 2$ and moreover that that he had a "marvelous proof of this." Unfortunately the proof he was thinking of was never found. Building on the work of many talented mathematicians before him, the correct argument was finally obtained by Andrew Wiles in 1993.[20] It was a triumph of human ingenuity and creativity, and the entire story of the work on this conjecture makes fascinating reading for anyone who is interested in mathematics.

The proof hinges on the validity of a conjecture called the Taniyama-Shimura conjecture. This conjecture unifies the seemingly disparate worlds of elliptic curves and modular forms. To understand the power of ambiguity to revolutionize mathematics, one has but to read the comments on this conjecture by the Harvard mathematician Barry Mazur. He compared the conjecture to the Rosetta stone that contained Egyptian demotic, ancient Greek, and hieroglyphics. Because demotic and Greek were already understood, archaeologists could decipher hieroglyphics for the first time.

Mazur said,

It is as if you know one language and this Rosetta stone is going to give you an intense understanding of the other language. But the Taniyama-Shimura conjecture is a Rosetta stone with a certain magical power. The conjecture has the very pleasant property that simple intuitions in the modular world translate into very deep truth in the elliptic world, and conversely. What's more, very profound problems in the elliptic world can be solved sometimes by translating them into the modular world, and discovering that we have insights and tools in the modular world to treat the translated problem. Back in the elliptical world we would have been at a loss.[21]

This is precisely a restatement of what we have come to expect in a situation of ambiguity.

HILBERT'S TENTH PROBLEM

Fermat's last theorem is not the only famous problem concerning Diophantine equations that has been solved in recent years. A Diophantine equation is an equation in a finite number of unknowns, like $x^3 + 2x^2y + xy^2 - 2xy - 4y^2 = 0$, where we seek to find integer values of x and y that solve the equation. Of course Fermat's equation for $n > 2$ is Diophantine but has no integral solutions. David Hilbert's Tenth Problem is another example. Here is the problem in Hilbert's words:

> Given a Diophantine equation with any number of unknown quantities and with rational integral numerical coefficients: *To devise a process according to which it can be determined by a finite number of operations whether the equation is solvable by rational integers.*[22]

Thus he asked whether an algorithm can be devised that would be able to determine in a finite number of steps whether a given Diophantine equation had an integral root. This problem is interesting because of what it tells us about the nature of algorithmic intelligence, a question that will be returned to later on. In 1970 Yuri Matiyasevich showed that Hilbert's Tenth Problem is unsolvable. From our point of view, what is interesting is not only the result but also the manner in which the result was obtained.

In his forward to Matiyasevich's book, Martin Davis, one of the senior researchers in the field, had this to say,

This showed not only that Hilbert's Tenth Problem is unsolvable, but also that two fundamental concepts arising in different areas of mathematics are equivalent. The notion of *recursively enumerable* or *semidecidable* set of natural numbers from computability theory turns out to be equivalent to the purely number-theoretic notion of *Diophantine* set. Dr. Matiyasevich has taken full advantage of the rich interplay between the methods of elementary number theory and computability theory and this equivalence makes it possible to produce an amazing and appealing book.[23]

Here again we have the appearance of the same phenomenon that was observed in the proof of Fermat—a key conceptual step is the recognition that two hitherto unrelated conceptual ideas are in fact equivalent. This reinforces the notion that we are looking here at something that is basic to creative insights in mathematics.

Unsolved Problems and the Mathematical Quest

Our discussion of ambiguity in mathematics can be consolidated and put into its correct perspective through a consideration of the activity of mathematical research, in particular, research into unsolved problems. It seems appropriate to follow up a discussion of Fermat's Last Theorem and Hilbert's Tenth Problem, two of mathematics' most noticeable success stories in recent years, with a discussion of problems that have not been solved. Nothing reveals the huge gap between the general public's perception of mathematics and the view from within the mathematical community more than their respective reactions to open or unsolved questions. Many people are surprised to hear that there "still" remain problems in mathematics that are unresolved—much less that such problems can sometimes be stated quite simply and succinctly. They have the feeling that "surely everything in mathematics is known or, at least, will be known in the near future." For the researcher, mathematics is a vast unknown terrain, a dense forest, in which there are occasional clearings corresponding to developed and well-understood areas of mathematics.

Number theory is an area of mathematics containing many unsolved problems which can sometimes be stated in language that is accessible even to nonspecialists. The only technical word we shall need is the notion of a prime number. *Primes* are positive integers greater than one that have no factors other than themselves and 1. For example the list of primes would begin with 2, 3, 5, 7, 11, 13, 17, 19, 23, 29, 31, and so on. The Greeks developed an argument (which is given in Chapter 5) to show that this list has no largest element, that is, that the number of primes is infinite. All primes with the exception of 2 are odd and, of course, the sum of two odd primes is an even number. But can all even numbers be generated in this way? Goldbach's Conjecture says that they can. It is an unsolved problem in number theory and one of the oldest unsolved problems in all of mathematics. No one has yet come up with a proof. In fact, in the year 2000 the British publisher Tony Faber put up a million-dollar prize for anyone who could come up with a solution before the year 2002. The prize was a way to generate publicity for the novel *Uncle Petros and Goldbach's Conjecture* by Apostolos Doxiadis. The prize was never awarded.

Goldbach's conjecture can be stated simply:

Every even number greater than two can be written as the sum of two prime numbers. For example,

$4 = 2 + 2, 6 = 3 + 3, 8 = 3 + 5, 10 = 3 + 7 = 5 + 5, 12 = 5 + 7, 14 = 3 + 11 = 7 + 7, \ldots$

In his fascinating book on number theory, Daniel Shanks divides unsolved problems into two categories: conjectures and open questions. A conjecture is a "proposition that has not been proven, but is favored by some serious evidence." For an "open question," on the other hand, the "evidence is not very convincing one way or another."[24] There is a great deal of evidence in favor of the validity of Goldbach and most mathematicians believe it to be true. For small values of n (small for a number theorist), $n \leq 6 \times 10^{17}$, the conjecture has been verified by computer.[25] In addition there is a heuristic argument (but not a proof) for the validity of the conjecture based on the formula for the statistical distribution of primes. Even at the level of rigorous proof there have been a number of results that go in the direction of the main conjecture. For example, in 1939 L. G. Schnirelmann proved that every number $n \geq 4$ can be written as the sum of at most 300,000 primes. This showed that the conjecture was true for a large but finite number of primes (instead of two). This result has been improved over time so that it is now known that every even number $n \geq 4$ can be written as the sum of at most six primes.[26] Another result, due to Chen Jingrun (1966), is that every sufficiently large even number n ($n \geq N$, for some N) can be written as the sum of two numbers, the first of which is a prime and the second is the product of two primes. There are many more results that go in the direction of Goldbach's Conjecture.

What is the relevance of unsolved problems to our discussion of ambiguity in mathematics? A well-formulated but unsolved problem has an intrinsic ambiguity both in the problem itself and also in the way one thinks about it or works on it. It may be true or false. It is even conceivable that it cannot be resolved one way or another.[27] While we are working on it we don't know the answer, so we must allow both of these possibilities to live in our minds at the same time. Of course we couldn't work on such a problem without having some intuition, based on some substantial evidence, about the validity of the statement. If, following Shanks, we call it a conjecture, then we are guessing that it is true. If we call the problem "open," then we allow for both possibilities. But whether we call it conjecture or open problem, there are always two possibilities—true or false. If we guess false, we must ask ourselves where we might look for a counter-example. If we guess true, we must ask why is it true, and where we would we look for a proof. Whatever we guess, there is always the possibility that we have guessed wrong. If we feel that the statement of the problem is true, then we are faced with another ambiguity. Is the proof accessible? Many conjectures are felt to be true, and yet one senses that a proof would require new ideas, major new developments in the subject that may not happen for many years. Who would risk wasting their careers and a good portion of their lives working on a problem that is not ripe for solution given the current state of development of the subject? So mathematical research is characterized by an instinct for the right problems: those that are significant yet accessible. These are another pair of conflicting or ambiguous characteristics. Too accessible implies the problem is most likely unimportant; too significant and it may be inaccessible.

The ambiguity of an unsolved problem is mitigated somewhat by the Platonic attitude of the working mathematician. That is, she feels that it is objectively either true or false and that the job of the mathematician is "merely" to discover which of these a priori conditions applies. Psychologically, this Platonic point of view brings the ambiguity of the situation into enough control so that researchers have confidence the correct solution exists independent of their efforts. It moves the problem from the domain of "ambiguity as vagueness" in which anything could happen to the sort of incompatibility that has been discussed in this chapter where there are two conflicting frameworks, true or false.

While the unsolved problem is unresolved, the ambiguity of the situation is there for all to see. Let's spend a few words comparing this situation to one in which the ambiguity of the mathematical situation is hidden from view. In the classroom, for example, the teacher and the student often stand on opposite sides of the ambiguity. In the teacher's perception of the situation, there is no ambiguity—the concept being discussed is clear and precise. For the student the concept is ambiguous in both senses of the word: it both is unclear and may contain various "meanings" that actually conflict with one another. However, and this is what is usually not appreciated, even for the teacher the concept retains its ambiguity. For in addition to its clarity, there is also (if the teacher actually has a deeper understanding of the concept) an openness and flexibility that allows the concept to be applied in a variety of circumstances.

Every situation of ambiguity admits a dual viewpoint that we could characterize as known versus unknown or as teacher versus student. In the case of the unsolved problem the "known" side is missing, so there is no disguising the ambiguity. In the teaching situation the teacher may well deny that the concept is ambiguous, but no one can do this for a problem that is unresolved. We really don't know the true state of affairs.

In fact, famous unsolved problems are often of great importance to the development of mathematics even if they remain unresolved. This is because the effort that is spent unraveling them often results in important developments in the subject. The main conjectures may remain unsolved, but other significant questions that arise in the course of the investigations often are solved. As in the case of the Goldbach Conjecture, different, weaker aspects of the main conjecture may be proved, and this leads to increased evidence for or against the main conjecture. The whole situation requires a state of mind that remains at once rigorous and flexible. It requires the ability on the part of the researcher to develop and sustain a state of ambiguity.

I cannot leave the topic of unsolved problems without commenting on what it tells us about the nature of mathematical research and about the art form that is called mathematics. What kind of person attacks such problems? Working on one of the great unsolved problems of mathematics is like embarking on a quest. The anthropologist Joseph Campbell[28] has written about the mythological Hero's Quest. In it the hero braves great perils in order to make some discovery that he brings back for the benefit of humankind. Working on a great mathematical conjecture is a kind of hero quest. What motivates people to spend their lives on such a quest? Why did Wiles spend seven years in his attic working on Fermat? The true motivation for such activity goes beyond fame and fortune—it must be found in the nature of the activity itself. This is another way by which examining mathematics has something to tell us about the nature of the human condition. It seems to me that the notion of the spiritual quest is the closest one will find to such an explanation. A spiritual quest is something that one is driven to do, driven from the deepest level of one's being. A spiritual quest has no rational explanation, or rather, the rational explanation, the adding up of the pluses and the minuses, always misses the mark. One is just so taken with the question, with the beauty and the excitement of the activity, that the effort and the sacrifice seem a small price to pay. A spiritual quest also has something about it that is self-validating and holds the promise of personal transformation. Its goals are both external and internal—a voyage of both discovery and self-discovery.

Any great quest demands courage. It is a voyage into the unknown with no guaranteed results. What is the nature of this courage? It is the courage to open oneself up to the ambiguity of the specific situation. The whole thing may end up as a vast waste of time; that is, the possibility of failure is inevitably present. To work so hard, in the face of possible failure, is what I mean by working with ambiguity. If we stop to think about it, this quality of ambiguity that one finds in the research environment is no different in kind from the ambiguity that is found in our personal lives. Our lives also have this quality of a quest, the attempt to resolve some fundamental but ill-posed question. In working on a mathematical conjecture, life's ambiguities solidify into a concrete problem. That is, the situation of doing research is isomorphic to some extent with the situation we face in our personal lives. This is one reason that working on mathematics is so satisfying. In resolving the mathematical problem we, for a while at least, resolve that larger, existential problem that is consciously or unconsciously always with us.

$$\bullet \quad \bullet \quad \bullet \quad \bullet \quad \bullet$$

The above discussion should be borne in mind when we think about the learning of mathematics as students, teachers, or just people who are interested in mathematics. Learning something new entails entering into a situation of ambiguity. Situations of ambiguity are difficult by their very nature. Learners need support when they are encouraged to enter into new unexplored ambiguities. A new learning experience requires the learner to face the unknown, to face failure. Sticking with a true learning situation requires courage and teachers must respect the courage that students exhibit in facing these situations. Teachers should understand and sympathize with students' reluctance to enter into these murky waters. After all, the teacher's role as authority figure is often pleasing insofar as it enables the teacher to escape temporarily from their own ambiguities and vulnerability. Thus the value of learning potentially goes beyond the specific content or technique but in the largest sense is a lesson in life itself.

Ambiguity in Physics

Since physics is the science that is closest to mathematics, one might expect to find that the phenomenon of ambiguity is present in this domain as well. Physics involves an explicit duality, namely, the two dimensions of experiment and theory. One of these dimensions is associated with the natural world. It is objective, that is, we feel that it resides "out there," independent of our minds. Theoretical physics, on the other hand, clearly is a subjective human creation, consisting as it does of ideas, concepts, and, of course, mathematics. In mathematics these dimensions of objectivity and subjectivity are not so well delineated. Thus the appearance of ambiguity in physics raises additional questions to those that it raises in mathematics. Does the ambiguity reside in the mathematical formalism or does it reside in the natural world itself?

Earlier the ambiguous nature of the equation $E = mc^2$ was discussed. This brings out a point about the difference between mathematics and physics. The concepts of mass and energy are linked through this equation, and therefore this equation, like other equations, is ambiguous. Nevertheless here we are not only talking about a piece of mathematics—we are talking about the atomic bomb! The stakes are much higher here. Ambiguity seems to tell us something about the nature of reality itself, or at least the manner in which the human mind interacts with the natural world.

ELECTRICITY AND MAGNETISM

Originally electricity and magnetism were studied as separate and distinct phenomena. Then, with the experimental work of Michael Faraday and the theoretical work of James Clerk Maxwell, scientists realized that the two phenomena were intimately related. So intimate is this relationship that one could say, "electricity and magnetism turned out to be not two separate subjects but were different aspects of a single electromagnetic field."[29] This realization had all sorts of practical and theoretical consequences from the invention of electric motors to the realization that light itself is an electromagnetic phenomenon. This breakthrough in understanding has been called the highlight of nineteenth-century scientific thought. This breakthrough is captured in Maxwell's equations but the essential nature of the insight is that electricity and magnetism form a single ambiguous field. Thus we could say that the deeper nature of light itself is that it is one single phenomenon of an intrinsically ambiguous nature.[30]

COMPLEMENTARITY IN QUANTUM MECHANICS

Quantum mechanics, the physics of the subatomic realm, is one of the most novel and successful theories of the last century. It has succeeded in all the ways that a scientific theory can succeed. It predicts the outcome of experiments that can be verified quantitatively to a very high degree of accuracy. It also anticipates the existence of novel phenomena that are then shown to exist through experiment.

Quantum mechanics is a paradigm-breaking scientific theory. It posits a radically new way of viewing the phenomena of the natural world. The aspect of the subatomic world that is most surprising (and not yet completely understood) is called "complementarity." When philosophers and physicists discuss the "meaning" of quantum mechanics, what they are discussing, in the first instance, is complementarity. It is complementarity that makes quantum mechanics not only a "new" scientific theory but also a new *kind* of theory.

We are introducing complementarity into our discussion at this point because the parallels between complementarity and what I have been calling ambiguity are striking. I shall attempt to work out the relationship between the two ideas. In fact I claim that the essence of complementarity—what makes it original and important is precisely what I have been calling ambiguity.

Complementarity is a concept that originates with Niels Bohr in an attempt to explain the seemingly paradoxical world of subatomic particles. Is an electron, for example, a particle or a wave? In certain experimental situations the electron behaves like a localized individual object—it makes sense to say that there is one electron or two or three. In other situations the electron is not localized. It behaves like a wave, with wave-like properties such as diffraction and interference. Both of these, the particle and the wave, are "self-consistent frames of reference," yet, to the normal way of thinking, there appears to be a certain incompatibility between the two descriptions. Many people have regarded the existence of complementary properties like wave/particle duality as paradoxical. Of course this is precisely what one would expect from a situation of ambiguity, but it is not what one usually expects to find in physics. After all, one might ask, what is it *really*, a wave or a particle? We feel that it cannot be both. Yet there is one thing of which we can be sure— there is one electron. When we look at it in one way (in one frame of reference) we observe particle-like properties. When we look in another way we get wave-like properties. The electron is precisely the singular entity that emerges out of this fundamental ambiguity.

If complementarity and ambiguity refer to the same phenomenon, then why not call ambiguity complementarity? I maintain that even though these two ideas are referring to a similar phenomenon there are important differences. Complementarity refers to a situation where there is a duality—two contexts the "sum" of whose complementary aspects "adds up" to the entire actual situation. Ambiguity also involves dual concepts, but each context stands on its own, each context describes the entire situation, so to speak. Thus there would be the "particle" description of nature in which subatomic particles are classical objects with definite attributes. On the other hand there would be the "wave" description in which everything is a cloud of probabilities—what Werner Heisenberg called "tendencies for being," "potentia."

Moreover, an ambiguous situation not only boasts dual contexts but also emphasizes the incompatibility between these contexts. It is this "incompatibility" that most sharply differentiates "ambiguity" from "complementarity." It is this incompatibility that is at work when we read the anguished words of physicists who are trying to make sense of subatomic phenomena. They seem not to make sense! One senses that there exists an obstacle that must be overcome if one is to makes sense of this realm of reality. Using the idea of "ambiguity" brings to the fore the need for this "epistemological obstacle" to be overcome—the need for a new vision, the need for a creative leap.

This "incompatibility" gives the entire situation a dynamic aspect. It is like a force that pushes the situation toward a creative reconciliation of the incompatibility. Thus I prefer to think about the electron as an ambiguous object—not in any vague or mystical sense—but in the sense that the electron is both a particle and a wave and yet it cannot be both at the same time. When it is a particle, it is not a wave, and when it is a wave, it is not a particle. In fact the particle/wave ambiguity is so profound that its implications remain a subject of study and speculation.

Ambiguity may then refer to a phenomenon that is present in the external world of physical phenomena as well as in the interior, cognitive world. Which is primary? Does ambiguity (or complementarity) refer to a property of the natural world, and so it finds its way into the biology of our brains and from there into the world of mathematics? Or is ambiguity a feature of our thinking process, and so the conceptual structures that we create inevitably carry this feature? This is in itself a version of the mind/body problem. What is primary, mind or body?

The dominant view in modern cognitive science is that "mind" is a consequence of "brain." There have also been thinkers and traditions that say that "brain" is a consequence of "mind." The dominant Western tradition going back to Descartes is that there is a mind/body duality. I suggest that there is another possibility—an "ambiguous" possibility. I suggest that the mind/brain and subjective/objective situations are not merely dualities or complementarities but ambiguities. Calling them ambiguities makes all the difference because, whereas a duality may be seen to be a fixed and unchangeable aspect of reality, an ambiguity always allows for a higher-level unification. Thus one could say that there is one unified reality that looks subjective when we approach it in one way and objective when we approach it another (see Chapter 8). If reality itself has this ambiguous nature, then it is not so surprising to see the same ambiguous characteristics arising in both the "subjective" domains of mathematical and physical theory as well as in the "objective" domain of subatomic physics.

It is interesting for our discussion of mathematics that quantum mechanics is a completely mathematical theory. Actually it has two different mathematical formalisms, one discrete and the other continuous. That is, the theory of quantum mechanics is itself ambiguous. Now the two descriptions are mathematically isomorphic or equivalent. This does not mean, however, that there is nothing to be gained by having two different ways to look at the situation. On the contrary, given our previous discussion, one would expect that the subtlety of the phenomenon that we are trying to comprehend would require an ambiguous description.

Finally mathematics has something to learn from the world of quantum mechanics. This involves the normal, "formalist" view that mathematics starts off with "self-evident" ideas and builds up to very complex ones, that there is a movement from simplicity to complexity. In the world of quantum mechanics the elementary objects such as the electron and other subatomic particles are extremely subtle and complex entities. That is, it is conceivable that reality is complex all the way down. There may be a lesson here about mathematical objects. Are they not also complex all the way down? Is there any mathematical object that is "trivial" or "obvious" when viewed from *every* possible mathematical point of view? But more of this later on.

STRING THEORY

String theory (and its generalization M-theory) is an exciting, relatively recent attempt to unify the two most fundamental physical theories of our time, general relativity and quantum mechanics. These "two theories underlying the tremendous progress of physics in the last hundred years . . . are mutually incompatible."[31] Thus the need for string theory arises out of the kind of ambiguous situation that I have been describing in this chapter. Both general relativity and quantum mechanics have been spectacularly successful in their respective domains. Their predictions have been experimentally verified to a very high degree of accuracy. Yet they are incompatible in situations in which both theories apply, for example, black holes and the "big bang." It is this context that created the need for a new theory that would unify the gravitational force with the other physical forces. String theory is the prime candidate for such a unifying theory. It is interesting at this stage of the discussion not only because of the ambiguous context in which it arises but because the theory itself incorporates ambiguity in a profound manner.

String theorists have a word for what I have been calling ambiguity—they call it *duality*. "Physicists use the term duality to describe theoretical models that appear to be different but nevertheless can be shown to describe exactly the same physics." There are "trivial" dualities and "nontrivial" dualities. The former give you nothing new; they are exact correspondences from one language to another. "Nontrivial examples of duality are those in which distinct descriptions of the same physical situation do yield different and complementary physical insights and mathematical methods of analysis." Nontrivial dualities are at the heart of recent developments in string theory. Just as in the discussion of the proof of the Fermat theorem, nontrivial dualities may enable an analysis or calculation that is extremely intractable in one context to be translated to another context in which the calculation is much easier to accomplish.

In the decades before 1995 five distinct versions of string theory were developed. This was a bit of an embarrassment, since one would hope that one characteristic of a unifying physical theory would be its uniqueness. Then Edward Witten "discovered a hidden unity that tied all five string theories together. Witten showed that rather than being distinct the five theories are just five ways of mathematically analyzing a single theory . . . a single master theory [which] links all five string formulations." This proposed master theory is called M-theory and is the subject of much current work. Thus, although I have emphasized that ambiguous situations usually come with two frames of reference, here we have *five* frames of reference that are reconciled by a master theory that encompasses them all.

Thus the ambiguous appears throughout string theory, and it appears in many different guises. One way that interests me is in the interaction between mathematics and physics. Mathematics and physics have distinct points of view. Thus it is conceivable that a certain problem can be looked at from both the purely mathematical and the physical point of view. Such an "ambiguous" point of view is an advantage for it has opened up new ways to attack and solve certain purely mathematical problems. This is another duality or ambiguity, which

> highlights the role that physics has begun to play in modern mathematics. For quite some time, physicists have "mined" mathematical archives in search of tools for constructing and analyzing models of the physical world. Now, through the discovery of string theory, physics is beginning to repay the debt and to provide mathematicians with powerful new approaches to their unsolved problems.

The mechanism at play here is ambiguity harnessed in both directions for the insight and the power that it provides.

SOME PERSPECTIVES FROM MATHEMATICS EDUCATION

Mathematics educators investigate mathematics as it is learned and taught. Therefore they are forced to consider not only the formal, objective aspects of mathematics but also the human dimension of the subject. They are forced to confront such questions as "What is meaning?" "What is understanding?" The result has been that various mathematics educators have developed a rather sophisticated approach to the nature of mathematics. These approaches have in common with my own a desire to free mathematics from an entirely "objectivist" point of view, "objectivist" in the sense that the meaning of mathematics is "out there" in a mind-independent reality. In this section, I shall mention a number of ideas put forward by researchers in mathematics education that have something in common with the notion of ambiguity.

GRAY AND TALL

Eddie Gray and David Tall wrote an excellent paper in 1994 with the suggestive title, "Duality, Ambiguity, and Flexibility: A 'Proceptual' View of Simple Arithmetic." In this paper they discuss "process-product" ambiguities in mathematical notation. In answer to a question about how anything can be a process and an object at the same time, which had been posed by the mathematics educator Anna Sfard,[32] they point to the way that professional mathematicians "employ the simple device of using the same notation to represent both a process and the object of that process." Thus they introduce the word "procept" to represent "the amalgam of three components: a *process* that produces a mathematical *object*, and a *symbol* that represents either the process or the object."[33]

They give a list of examples of ambiguous symbolism which is significant because it includes a great deal of elementary mathematics. These include the notation for addition, multiplication, and division, where, for example, 2/3 stands for division and for the concept of a fraction and −7 can stand for the process of subtracting 7 or the negative number. They also include examples that I have also enumerated: functions, infinite decimals, limits of functions, sequences, and series.

Gray and Tall are well aware that what they are talking about involves the taboo subject of ambiguity when they say, "mathematicians abhor ambiguity and so they rarely speak of it, yet ambiguity is widely used throughout mathematics. We believe that the ambiguity in interpreting symbolism in this flexible way is at the root of successful mathematical thinking." It is true that in a certain way mathematicians abhor ambiguity. Mathematicians in general are masters of the most subtle and elaborate logical arguments. Logic is a form of thinking that conveys power and control that mathematicians use to considerable effect in the classroom, in conversations with one another, and even in their personal lives. Thus there is the feeling that the desirable state of affairs is the rigorous and therefore that ambiguity is a temporary condition that should and ultimately will be replaced by a state of logical certainty. Nevertheless you cannot remain in a state of logical certainty if you wish to do research. Thus the attitude of the mathematician is ambiguous; it consists of two attitudes that are inconsistent with one another. This is connected to the comment I made at the beginning of the chapter, and which has been noted by many authors, to the effect that what mathematicians do is not the same as what they say they do. What they say they do lies in the formal, logical domain. What they actually do lies in what I am calling the domain of ambiguity, and, for the most part, when this is pointed out to them they readily acknowledge it. The problem lies not in the mathematical activity but in the way that activity is described and therefore misinterpreted by people who teach mathematics, who use mathematics in their work, or who find it comforting to hold to a vision of the world that is logically coherent and devoid of ambiguity. Mathematics is mistakenly used as a justification for such a simplistic point of view. The resulting damage to students, to teachers and to society at large, is considerable.

Gray and Tall go on, "We conjecture that the dual use of notation as process and concept enables the more able to 'tame the processes of mathematics into a state of subjection'; instead of having to cope consciously with the duality for product and process, the good mathematician thinks ambiguously about the symbolism for product and process." In a process that reflects what I said above about the use of variables in mathematics, they continue, "We contend that the mathematician simplifies matters by replacing the cognitive complexity of process-concept duality by the notational convenience of process-product ambiguity."[34]

Of course, from my point of view, the key statement is "ambiguity . . . is at the root of successful mathematical thinking." The ambiguity of productive mathematical symbolism is one important instance of the use of ambiguity in mathematics, but certainly not the only way that ambiguity appears in mathematics. Nevertheless Gray and Tall touch upon something of very deep significance for all of mathematics. Their title includes the key ideas "duality, ambiguity, and flexibility." I take ambiguity to be the key idea and, in fact, in the way I have used the term, ambiguity *must* contain both duality and flexibility. Thus Gray and Tall are quite correct in pointing to ambiguity as a (perhaps *the*) key element of mathematical notation. How can it be claimed that the object of mathematics is to eliminate ambiguity when ambiguity is the key ingredient in much mathematical notation?

Sfard and Reification

Sfard in a series of significant articles has highlighted the role of reification, that is, creating objects out of processes, plays in mathematics. She claims that "from the developmental point of view, operational conceptions precede structural; that is familiarity with a process is a basis for reification."[35] She goes on to say that this order of things is not universal. "Mathematicians," she claims, "do not necessarily follow this process-object path." From our point of view the order is not the crucial thing. What is crucial is that in mathematics we are dealing with notions that are ambiguous and that these reifications result in entities that have both a process and an object dimension. It is not as though once we have achieved the notion of 3/4 as a fraction that we can forget 3/4 as division. Rather, the accomplished attitude toward 3/4 is ambiguous in nature.

It is true that when one considers the questions of temporal order, whether process precedes object, one is highlighting the dynamical nature of ambiguity. Sfard makes the important point that "reification, whether it precedes or follows the construction of an operational schema, is often achieved only after strenuous effort, if at all." Whether one is talking about understanding or creativity, one is talking about a dynamic process that requires work. Thus when one uses the term "ambiguity" to refer to a property of some mathematical situation one is not describing a static, objectively fixed situation but rather a situation that has the capacity for change. It is important to stress that ambiguity is not merely a fixed characteristic of a mathematical object or situation; it also refers to the capacity of that situation to give birth to acts of learning or creativity. I have mentioned that situations of ambiguity come with two points of view that can be called "before" and "after" that insight which is the definitive element of the creative act. Thus ambiguous situations are, potentially at least, dynamic—they contain the possibility of changing in time. In a self-referential way, understanding ambiguity, like all acts of creativity, is itself an event not an object. Thus the concept of ambiguity contains within itself this same process/object ambiguity as do the various ambiguous concepts that have already been mentioned.

Farther on Sfard points out how strange the idea of reification is to our normal way of thinking about things. "In fact, the very idea of reification contradicts our bodily experience: we are talking here about creation of something out of nothing. Or about treating a process as its own product. There is nothing like that in the world of tangible entities, where an object is an 'added value' of an action, where processes and objects are separate, ontologically different entities which cannot be substituted for one another."[36] Our mathematical universe is populated by these mysterious, ambiguous entities that cannot be called objects and that appear "out of nothing," like the number zero (which will be discussed in detail in the next chapter). Mathematics unites the dynamic (processes) and the static (objects); it unites the objective (processes and objects) and the subjective (reification). Mathematics is indeed a complex and mysterious domain of human activity.

CONSTRUCTIVISM

There seems to be a general consensus in mathematics education that meaning and understanding are constructed. The papers that we have quoted above on "procepts" and "reification" are attempts to elucidate the mechanisms by which the construction of meaning occurs. This process has also been called "encapsulation" by the mathematics educator Ed Dubinsky.[37] What is this "elusive something that makes us feel that we have grasped the essence of a concept, a relation, or a proof"?[38] Whatever this something is, it belongs not only to the teaching and learning of mathematics but also to a description of mathematics itself. It is constructed both by the individual in acts of understanding or mathematical creation and also by the mathematical culture within which mathematicians are embedded. But to say that mathematics is "merely" constructed is to ignore the sense one has that mathematical truth is universal and "unreasonably effective"—that mathematics opens a window on something that is stable and permanent, in a word, Platonic. Ambiguity seems at first glance to have more in common with constructivism than it does with Platonism, yet this is perhaps just another ambiguity. I shall return to this discussion in Chapter 8.

MATHEMATICS AS METAPHOR

In recent years there has been an ambitious attempt to look at mathematics from the point of view of cognitive science. The central notion in this work is the notion of metaphor.[39] A metaphor is a mapping from one cognitive domain to another. In this view metaphors come in two varieties. *"Grounding metaphors* yield basic, directly grounded ideas. Examples include addition as in the addition of objects to a collection, for example. Linking metaphors yield sophisticated ideas, sometimes called abstract ideas."[40] For example, *the real line,* a line on which every real number has a specific location, is a metaphor. But metaphor is to be understood in a more general way than it is usually viewed, in literature, as a way of comparing two explicit domains. Metaphor, in this view, is literally what "brings abstract concepts into being."[41] To put this in another way, abstract concepts *are* metaphors. This is crucial.

Metaphors are ambiguous. Like the example I gave earlier, "all the world's a stage," they are examples of ambiguity. In fact, they have the following characteristics: (a) duality—there is always a comparison involved; (b) incompatibility—a metaphor is of the form *A* equals *B* (or *A is B*) when it is obvious that *A* does not equal *B*; and (c) creative dynamism—a metaphor must be *grasped*, it requires an insight. When that insight is attained the metaphor, which previously had been flat and uninteresting acquires a certain profundity.

When we become sensitive to the metaphoric dimension of mathematics, the entire subject undergoes a metamorphosis before our eyes. The "real line" is a metaphor that is especially rich in its implications. Numbers become points on the line. Thus we "see" the system of real numbers as being one-dimensional; we "see" that the relationship of "less" between numbers is equivalent to the relation of "to the left of" on the line; we "see" that every infinite decimal has its own unique place on the line. We even say that the absence of a certain number, like the square root of two, would leave a "hole" in the line that is itself a geometric metaphor. Through the metaphor of the real line the reals are conceived of as a new object. This new metaphoric object is now invested with its own reality, for example, that of being a continuum. Even though we could equally imagine a "rational line," the real line carries with it the essential property that a line in the plane that goes from the positive to the negative side of the x-axis must actually cross the real line at a definite point. The geometry is now wedded to the analytic in a productive manner.

When mathematics is seen as a metaphor, it brings to the fore the central role of understanding on the part of both the learner and the expert. Metaphors are not purely "logical" entities. Speaking or reading a metaphor doesn't make that metaphor come alive for you. Grasping a metaphor requires a discontinuous leap. In a later chapter there will be a more complete discussion about the nature of such creative discontinuities. At this point in the discussion it is enough to insist that these creative discontinuous insights be considered as part of the subject of mathematics. A metaphoric description of mathematics will inevitably include a discussion of "doing" mathematics. Sfard[42] includes the following fascinating quote from a mathematician: "To understand a new concept I must create an appropriate metaphor. A personification. Or a spatial metaphor. A metaphor of structure. Only then can I answer questions, solve problems. I may even be able to perform some manipulations on the concept. Only when I have the metaphor. Without the metaphor I just can't do it."

RELATED IDEAS IN OTHER FIELDS

The phenomenon of ambiguity is, of course, not restricted to mathematics. It is not my intention to discuss in any detail the role of ambiguity in other fields. However, it is worthwhile to spend a little time indicating the manner in which various authors have applied the notion of ambiguity in fields unrelated to mathematics. It may be that what is going on here is a deeper characteristic of human thought, so that what is being discussed is how the general phenomenon of ambiguity plays itself out in mathematics. Even if it turns out that ambiguity is something that is present in many different situations, nevertheless ambiguity will emerge in a unique way in a mathematical situation and thus what is being discussed above is "ambiguity in mathematics."

ARTHUR KOESTLER AND CREATIVITY

The definition of ambiguity I gave earlier in the chapter is really Koestler's definition of creativity. What is crucial to the creative act for Koestler is the existence of two self-consistent but habitually incompatible frames of reference. He says, "I have coined the term 'bisociation' in order to make a distinction between the routine skills of thinking on a single 'plane' as it were and the creative act, which, as I shall try to show, always operates on more than one plane."[43] He goes on to identify instances of bisociation in diverse situations. He sees this phenomenon, for example, as the essence of humor.

71

Koestler makes a number of important points that are relevant here. The first is, of course, that ambiguity is linked to creativity. This is crucial to the argument that is being made here, which involves going beyond the formal structure of mathematics and considering how the subject comes into being, how it changes, and how it is learned and understood. All of these are connected to creative acts.

The next point that is derived from Koestler is that there is an "incompatibility" at the base of any creative situation. There is a conflict, something that does not work, that is contradictory or paradoxical. Creativity arises from working with this incompatibility not by denying its existence. This is an especially important point in trying to understand mathematics since the formal aspect of mathematics is set up for the express purpose of establishing consistency, that is, of eliminating incompatibility.

ALBERT LOW AND ZEN BUDDHISM

I was introduced to the idea of ambiguity through the writings of Albert Low. Low is a writer and a teacher of Zen based in Montreal. He describes ambiguity (in the sense that I have used it above) as fundamental to the view of the world that emerges from Zen Buddhism. The Zen koan, those obscure and seemingly paradoxical stories that lie at the heart of Zen training, are seemingly founded on ambiguity. Low goes as far as to propose that the normal Aristotelian logic of science and mathematics is not sufficient to describe the world as envisioned both by Zen and by quantum mechanics, and that this logic should be replaced by what he calls a "logic of ambiguity."

Low's central metaphor for the notion of ambiguity is the famous Gestalt picture of the young woman/old lady (figure 1.7). This picture has two perfectly consistent interpretations, the young woman and the old lady. These are the two self-consistent frames of reference in the definition of Koestler. A certain number of observations should be made about this ambiguous picture. In the first place, the entire picture is really a field of black and white dots; it is this neutral field that is then interpreted as young woman/old lady. Second, each interpretation is an adequate description of the entire field of black and white dots. It

Figure 1.7. Old woman or young lady?

is not as though some of the black and white dots are interpreted as the young woman and some as the old lady. The *entire* field is subject to the given interpretation. Finally, the two interpretations are incompatible with one another. When you see the picture as the young woman you do not see it as the old lady. Thus both interpretations are "true" and explain everything but they are incompatible with one another. Merely saying that the picture is *both* the old and the young woman misses the incompatibility of the two interpretations but of course saying that it is *neither* is also missing something.

73

To Low I also owe a number of fundamental ideas. He says, "Conflict is inherent in any situation, and cannot be resolved in the way a resolution is normally sought; by the elimination of one side over the other, the triumph of one side, or the merging of two into one harmonious . . . whole."[44] Thus conflict is normal; it is not exceptional. Now what has conflict to do with mathematics? Well the conflicts of mathematics include the contradictions, paradoxes, and unresolved ambiguities that have been described above and will be described in the chapters to follow. These also include the multiple ways that we have of looking at related mathematical situations. There may be no "right way" of seeing a given mathematical situation. At any rate the so-called incompatibilities have value! Far from being ignored or suppressed, they need to be seen as opportunities to deepen our understanding. Thus the first lesson here is that we must learn to value the conflicts that arise in mathematics. The other conclusion to be drawn from "conflict is inherent" is that the final victory of reason will never come, nor is it desirable. There is no end to mathematics nor is there even anything we could call a definitive mathematics, that is, a finished and complete formalized mathematics. A perfect mathematical theory that is consistent and complete, with all the bugs worked out and all the problems solved, would be a dead mathematics, an uninteresting object of study. What I am interested in describing is a subject that is vital and alive.

Leonard Bernstein and Music

In 1973, the late conductor and composer Leonard Bernstein was invited to give the Charles Eliot Norton lectures at Harvard University.[45] These lectures, which are now available in written, audio, and video formats, were a brilliant and fascinating discussion of the inner structure of music. In particular, Bernstein suggests that one of the basic mechanisms in music is precisely ambiguity in the same sense that the term has been used in this chapter. A discussion of ambiguity permeates all of his lectures—the fourth of which is even entitled "The Delights and Dangers of Ambiguity."

For Bernstein ambiguity is defined as "capable of being understood in two possible senses."[46] For example, he discusses the ambiguity of diatonicism versus chromaticism. Diatonicism refers to the normal seven-note musical scale, the white notes on a piano, and their natural relationships such as "thirds" and "fifths" that are based on mathematically determined overtones or harmonics of the base tone that determines the key. In other words, the diatonic relationships are the foundations of a piece of music. Chromaticism refers to the process of going beyond the basic seven notes of the diatonic scale to the full twelve notes of the chromatic scale (white and black keys on the piano) and even going farther to notes that are not available on the piano. These extra notes were considered to be dissonant to some because they were not "supposed" to be in a particular key; nevertheless they were often added to a piece of music to give the music more emotional range.

Bernstein says that the growth of chromaticism in musical history was "based on the accretion of more and more remote overtones of the harmonic series.... With that growing chromaticism we found a corresponding growth of ambiguity, and a resulting need to contain that chromaticism, to control it through the basic powers of diatonicism, the tonic-dominant structure of tonal music." He maintains that the "containment of chromaticism-within-diatonicism reached a perfect equilibrium in the music of Bach." "But," he goes on, "we also realized that this perfectly controlled containment is in itself an ambiguity, in that it presents two simultaneous ways in which to hear music, via the contained chromaticism and via the containing diatonicism." In mathematics, the element that supplies a control on the seemingly inexhaustible capacity of the human mind to perceive patterns is, of course, rigorous, logical argument. Also, if ambiguity "presents two simultaneous ways in which to hear music," in mathematics "hearing" should be replaced by "understanding" or "conceptualizing."

Bernstein makes many observations about music that seemingly beg to be translated into corresponding observations about mathematics. For example he claims that the genius of Johann Sebastian Bach "was to balance so delicately, and so justly, the two forces of chromaticism and diatonicism, forces that were equally powerful and presumably *contradictory in nature*" (italics added). Bernstein claims that the ensuing equilibrium remained stable for almost a century, "a century which become a Golden Age." He goes on to say,

> It's a curious thing, and a crucial one, that even through this perfect combination of opposites, chromaticism and diatonicism, there is distilled the essence of ambiguity. Now this word "ambiguity" may seem the most unlikely word to use in speaking of a Golden age composer like Mozart, a master of clarity and precision. But ambiguity has always inhabited musical art (indeed all of the arts), because it is one of art's most potent aesthetic functions. *The more ambiguous, the more expressive*. (italics added)[47]

Of course Bernstein cautions that it is possible to reach a sate of "such increased ambiguity that problems of musical clarity are bound to arise." He goes on to discuss many other musical ambiguities of which the chromaticism/diatonicism cited above is merely the first.

Now parallels between music and mathematics have been made since the time of the Pythagoreans. Here we find a parallel of a different kind, a parallel in the inner workings of music and mathematics. The key point that I take from Bernstein is that ambiguity has a positive function. He says, "these ambiguities are beautiful. They are germane to all artistic creation. They enrich our aesthetic response, whether in music, poetry, painting or whatever by providing more than one way of perceiving the aesthetic surface."[48] Mathematics can and should be added to this list. Ambiguity in mathematics contributes to the profundity and scope of the idea. In music insisting only on clarity at the expense of ambiguity would remove precisely that quality that makes a piece of music great. So in mathematics insisting on logical clarity at the expense of ambiguity leaves us with superficiality and triviality.

Conclusion

In this chapter, I have attempted to establish the ubiquity in mathematics of the phenomenon of ambiguity. The ambiguity that I am talking about is not mere vagueness; it is, as Bernstein emphasizes, a "controlled ambiguity." The control in mathematics is provided by the logical structure, and the power and profundity of mathematics is a consequence of having deep ambiguity under the strictest logical control. We have discovered that not only are fundamental mathematical concepts ambiguous but so are key elements in a number of deep mathematical results. Thus the occurrence of ambiguity is a crucial mechanism in mathematics. Moreover, bringing ambiguity to the fore will force a complete reevaluation of the nature of mathematics.

Many familiar mathematical concepts have an ambiguous, multidimensional nature. For example the Thurston paper lists eight different ways of "thinking about or conceiving of the derivative." He insists that these are not different logical definitions. They are, however, different *insights* into the concept of derivative. Importantly, Thurston warns us that "unless great efforts are made to maintain the tone and flavor of the original human insights, the differences start to evaporate as soon as the mental concepts are translated into precise, formal and explicit definitions." I have stressed repeatedly the dangers of identifying mathematics with its formal dimension. The notion of "derivative" is vast and complex and encompasses many different insights, each of which says something new about what a derivative is. These different insights may reduce to the same formal definition in a specific case, like the definition of the derivative of a real-valued function of one variable. Nevertheless the different points of view retain their value. For example, when applying the concept in a more abstract or general setting one often finds that one way of looking at the concept can be generalized and another cannot. That is, while the precision of formal, logically precise mathematics is valuable; it is at the center of a more loosely defined set of associated ideas that are also mathematically valuable.

This ambiguity is neither accidental nor deliberate but an essential characteristic of the conceptual development of the subject as well as of the person attempting to master the subject. The ambiguity is not resolved by designating one meaning or one point of view as correct and then suppressing the others. The ambiguity is "resolved" by the creation of a larger meaning that contains the original meanings and reduces to them in special cases. This process requires a creative act of understanding or insight. Thus ambiguity can be the doorway to understanding, the doorway to creativity.

It is interesting to point out that the whole of the above discussion is self-referential: not only is ambiguity part of mathematics but mathematics *itself* is ambiguous. Its nature is also multidimensional. There are the logical surface structure and the deeper dimensions of understanding, insight and creativity. It is not possible to imagine mathematics without its computational and formal aspects, but to focus exclusively on them destroys the subject. Ambiguity, even paradox, pushes us out of our airtight logical mental compartments and opens the door to new ideas, new insights, deeper understanding.

The different aspects of mathematics that have been described are in continual interaction, continual evolution. An idea like derivative is formalized. Thus in a sense the multiple possibilities contained in the informal idea are reduced to one. Then the formal idea can be understood in various ways, some of these retrieving some of the viewpoints that were inherent in the original pre-formal situation, others arising out of interpretations of the formal definition of derivative. These new ideas can themselves be formalized and so the whole chain is set in motion again.

Logic moves in one direction, the direction of clarity, coherence, and structure. Ambiguity moves in the other direction, that of fluidity, openness, and release. Mathematics moves back and forth between these two poles. Mathematics is not a fixed, static entity that can be structured definitively. It is dynamic, alive: its dynamism a function of the relationship between the two poles that have been described above. It is the interactions between these different aspects that give mathematics its power.

Mathematics might be viewed as a cultural project, in short, as a culture in itself. A powerful raison d'être of that culture is precisely to stand as a bulwark against the obscure, the contingent, the ambiguous, and especially against paradox and contradiction. The story that mathematics tells about itself is that it has no room for the ambiguous, for example. One can see that from the way that one would use the expression, "it's ambiguous." This would be tantamount to saying that it is wrong or stupid. Yet, as we have seen, ambiguity plays a central yet unrecognized role in mathematics—a role that has in a sense been repressed or, at least, remains unacknowledged. By bringing out this role it will be possible to study its implications for our view of mathematics, of science, of modern culture and its view of the nature of reality.

The Contradictory in Mathematics

... we would know more about life's complexities
if we applied ourselves to the close study of its
contradictions instead of wasting so much time
on similarities and connections, which should,
anyway, be self-explanatory.
—José Saramago, *The Cave*

... how completely inadequate it is to limit the
history of mathematics to the history of what has
been formalized and made rigorous. The un-
rigorous and the contradictory play important
parts in this history.
—Davis and Hersh, *The Mathematical Experience*

THIS CHAPTER is about the role of the contradictory in mathematics. One of the themes of this book is that mathematics does
not inhabit a world that is disjoint from the world of human experience but is in continual interaction with that larger world.
The contradictory is an irreducible element of human life as we
all experience it. It is not only that we often disagree with others;
it is also that we human beings seem capable of simultaneously
sustaining two contradictory points of view. The obscure Bud-
dhist dictum "Life is suffering" is best understood as the claim
that there exists an inner contradiction that resides at the deepest
level of human life. This inner contradiction has been expressed
in many ways, but essentially it involves our dual nature as a
mind that sees itself as powerful and central versus a body that
is subject to death and decay and is powerless in the face of the
vicissitudes of life. When I stop to think about it, I must admit
that, while I have infinite value to myself, I may have marginal
interest to others. Thus while I see myself as having infinite
worth, the person that I see reflected in the eyes of others often
appears to be without value. That is, I am simultaneously central
(to myself) and peripheral (to others). In this way and others

we are all walking contradictions. We cannot escape from the contradictory and the conflicts that arise in its wake. The contradictory is central to the conscious life, and therefore dealing with the contradictory is, consciously or unconsciously, present in all human activity.

How does mathematics deal with this irreducible factor in human experience? The above quotes highlight the value and importance of the contradictory. No description of mathematics would be complete without a discussion of its subtle relationship to the contradictory. Not only does mathematics deal with the contradictory in a unique way, but also the need to escape from the conflicts that the contradictory brings in its wake may be a powerful reason why we find mathematics so attractive. Discussing the role of the contradictory in mathematics will tell us something about the nature of mathematics, but it will also reveal something about the nature of all systematic thought.

In the previous chapter I talked at length about ambiguity and its role in mathematics. You will recall that the definition that was used involved "a single situation or idea that is perceived in two self-consistent but mutually incompatible frames of reference." What if the "two self-consistent frames of reference" were not mediated by the "single situation or idea"? We would then be in a situation of contradiction. Viewed in this way contradiction is related to ambiguity. In fact in a way a contradiction highlights one aspect of ambiguity. A true ambiguity always has the quality of incompatibility. It is this incompatibility that characterizes contradiction. Thus the discussion of the contradictory in this chapter constitutes a deeper exploration of the theme of ambiguity.

The difference between ambiguity and contradiction is, in a sense, a matter of emphasis. It is a question of whether the incompatibility is absolute or not. In the discussion of ambiguity, the incompatibility had (potentially, at least) a resolution in the "single idea or situation." However, when we speak of contradiction we usually imply that there is an absolute quality to the incompatibility—a gap that cannot be bridged. Now absolute incompatibility is situated in the domain of logic, and from this point of view a contradiction is a dead end. Mathematical practice, on the other hand, contains a varied and complex response to contradiction. The emergence of a contradiction may indeed

81

be a negative indication of a mistake or a cul-de-sac. However, it may also have a positive aspect; it may represent a challenge, for example. Far from indicating that the game is over, it may indicate that the game has just begun—that there is something here to be understood. This chapter will go into the many ways in which contradiction enters mathematical practice. To begin with, let me stress again the interrelatedness of contradiction and what I earlier called ambiguity. If a contradiction is a form of ambiguity, then it is equally true that contradiction is present in every situation of ambiguity.

At first glance contradiction is something to be avoided like the plague, especially in mathematics. Formal mathematics, including mathematical proofs, mathematical theories, and logical reasoning, are all within the domain of rationality. Logic and rationality imply consistency, and consistency means the avoidance of contradiction. Thus we would expect mathematics to steer clear of contradiction. A mathematical description of the world means a description that is free of contradiction, does it not? In the normal scientific view it is assumed that the natural world, reality itself, is free of contradiction and so is amenable to a mathematical description that is also contradiction-free. In this view such contradictions as may occur in the description of the world arise because of faulty thinking and not because of the nature of the natural world. Thus in the normal view of things contradiction is a negative characteristic of things, whether these things are in the mind or in external reality.

Any deep investigation into the nature of mathematics will eventually have to deal with the question of consistency. One of the fundamental pillars on which mathematics—not to say all of science—is built is the logical notion of noncontradiction. In such a system of thought it should not be possible to assert that a proposition P is true while simultaneously asserting the truth of its negation, not P. Thus mathematics from the time of Euclid has involved creating domains of noncontradiction.

One of the most basic questions one can ask is whether or not consistency is a fundamental aspect of reality. For many, including almost all scientists, there is a deep belief in the principle of noncontradiction. We choose to believe in consistency because the alternative appears to be incomprehensibility, a world of chaos. Thus the belief in consistency is a foundation block without which any meaningful discourse or thinking is held to be impossible.

Nevertheless, it is possible to take another point of view, one in which noncontradiction is not absolute. In this view consistency would be a local not a global phenomenon. Thus a particular theory would describe a certain range of natural phenomena. It would be a good fit at the centre of its range but might lose coherence as one went toward the periphery of its natural range. At the overlapping periphery of two local theories there might well be contradictions between the two theories. This point of view would fit in well with the way many people regard science. Bohm, for example, says, "all theories are insights which are neither true nor false but, rather, clear in certain domains and unclear when extended beyond these domains."[1] This is not the usual way of looking at mathematics, but it is consistent with the view that sees mathematics as the "science of pattern." How one reacts to the idea of contradictions in mathematics depends very much on what one takes mathematics to be. If one is a strict formalist then, of course, contradictions do not exist in mathematics since the existence of a contradiction, by definition makes something into non-mathematics.

However it is the thesis of this book that it is time to reevaluate the nature of mathematical activity in general and, in particular, the role that is assigned to logic and consistency within mathematical activity. It is not that I am suggesting that logic be removed from mathematics—that would be unimaginable and ridiculous. However, I am suggesting that, before closing the book on the question of logical consistency, it might be worthwhile to take a closer look at what actually happens in mathematics, at mathematical practice itself.

Mathematics *is* mathematical practice. Mathematics is not identical to what people, even mathematicians, will tell you that mathematics is. Many mathematicians will claim to be formalists when pressed by non-mathematicians to define what it is they do. "Oh," they will say, "I just start with assumptions and prove theorems. I don't care what they mean or what they are used for." However what is going on when people actually *do* mathematics is vastly different from what mathematicians or others may *say* is going on. There is a poverty of description that speaks to a need for a philosophy of mathematics that is commensurate with the subject.[2] Perhaps this lack of an adequate description of mathematics and "mathematizing" is due to the reduction of the philosophy of mathematics to a mere discussion of its logical foundations. It is conceivable that there could be a view of mathematics that would hold that there are no (absolute) foundations of mathematics. There is mathematical practice and there are mathematical theories, but there are no foundations of the entire subject in the usual sense of the term. The unifying characteristics of mathematics will have to be found elsewhere.

Such a revision in the way mathematics is described will entail no change at all in the content of mathematics. What will change will be the manner in which we think about mathematics. In particular, it will be necessary to have another look at ambiguity and contradiction.

It would appear that the attitude of mathematics to the notion of contradiction is simple—avoid it at all costs! Yet this is not the case at all. On the contrary, this chapter will discuss the many uses of contradiction in mathematical theory and practice. Like ambiguity, the use of contradiction is so systematic in mathematics that it will force us to reexamine the nature the subject.

Not only do we find contradiction on the boundaries of mathematical theory, where we might expect it, and in informal, preconceptualized mathematics, but we also find contradiction in use as a positive generating principle within formal mathematics. Thus mathematics goes beyond logic. And yet logic is the very language of mathematics. How can this be? It seems to be a contradiction in its own right. Our study of contradiction will take us deeper into this seeming paradox.

EUCLID AND THE DREAM OF REASON

So much of the spirit of modern mathematics has its origins in Euclid's *Elements*. So Euclid is a good place to start thinking about the use of contradiction in mathematics. Of course Euclid was the person who collected all the results of Greek geometry into one unified and consistent deductive development. Starting from axioms and postulates that are considered to be self-evident, and proceeding through logical reasoning, Euclid's *Ele-*

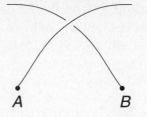

Figure 2.1. Intersecting arcs?

ments were for millennia a model of what systematic thought should be like.

Studying the *Elements* had a profound effect on impressionable young minds. Bertrand Russell, in his autobiography, said, "At the age of eleven, I began Euclid with my brother as tutor. This was one of the great events of my life, as dazzling as first love."

Generations of mathematicians, scientists and philosophers could attest to the intellectual awakening occasioned by their encounter with the *Elements*. What was it that Russell and countless others were awakened to by the geometry of Euclid?

Well they were certainly *not* awakened to logical perfection, for the *Elements* are not without error. In many places the reasoning is incomplete. For a treatise on geometry, it is ironic that so many of the geometric constructions are problematic. It is claimed that the logical arguments are independent of these geometric pictures, yet this is clearly not the case (as has been pointed out by many mathematicians, most notably David Hilbert). There are many places in Euclid where inferences are drawn from an illustration that are not justified by the explicit assumptions. For example, in one of the earliest arguments two arcs of circles are drawn and it is assumed that they intersect in a point (as opposed to just passing through each other without touching) (figure 2.1). There is nothing in the axioms that ensures this intersection. This is just one of many such lacunae.

So if it is not in the perfection of the reasoning, where did Russell's intellectual awakening come from? Perhaps it is the combination of geometrical intuition and logical rigor that one finds in the *Elements* which is irresistible to those with a sensitivity for such things. Euclid awakened in generations of people what we might call the "dream of reason." In his geometry subtle aspects of reality are demonstrated definitively. The geometric world is organized before our eyes and our minds. Why should the interior angles of *every* triangle add up to two right angles? A priori, there is no reason in the world except that miraculously they do, and Euclid shows us why with a simple argument that is utterly convincing. Now we *know that it is so!* What an extraordinary feeling it is—to know and know that you know.

This feeling is as close as the rationalist will ever come to religion. In fact it is similar to what Einstein called "cosmic religious feeling."[3] Einstein felt that "such feeling was the strongest and noblest motive for scientific research." Throughout his writing you sense the fundamental mystery of his life—that the rational mind (his rational mind) was able to see so deeply into the workings of the natural world. This correspondence between the mind and external reality, that is, the power of the mind, is initially sensed by many people in their contact with Euclidean geometry.

It was as though Euclid and his fellow Greek mathematicians had discovered a new way of using the human mind. With this deeply creative insight, a new idea had emerged of what it meant to be human. A window was opened that appeared to look out at Truth and Certainty. Here was a way to emancipate human beings from the chaos and contingency of everyday life by demonstrating the existence of another, deeper realm where everything appeared to be ordered, clear, and definite. Thus Euclid provided a method that went beyond the particular mathematical results he proved. Those results made up a body of knowledge that was impressive enough in its scope to give birth to the "dream of reason." The dream was that all of our knowledge about the natural world and thus the natural world itself might be organized into a vast deductive system of thought based on a small number of intuitively obvious axioms. It is a dream that one gets echoes of to this day in, for example, the so-called "theories of everything" that one hears about in physics.

Euclid's *Elements* comprise a vast body of results including all of the geometric results that were known at the time. The axiomatic development was important for a number of reasons. In the first place, it represented an amazing simplification of thought because in the place of a very large number of independent results, each with their own justifications, we had a unified system all of whose results depended on a small number of initial assumptions. Moreover, the entire deductive system had the following properties:

1. All conclusions were *true*; it was not possible to deduce false statements (assuming that the axioms were true and the reasoning was correct).

2. The system was *consistent*; it was not possible for a statement and its negation to be simultaneously true.

Although this was not guaranteed, it appeared possible that the system was *complete*, that is, every true geometrical proposition could be developed within the system, based on the axioms and propositions that had already been proved to be true, using as a tool simple logical reasoning. Thus the "dream of reason" amounted to the hope that there was a deductive system that satisfied all of these properties, in particular, it was both consistent and complete. Though, as we shall see, this dream has been shown to be unrealistic, nevertheless it continues to provide a great deal of the motivation for even contemporary science.

This "dream of reason" is the first example of what will be one of the themes of this book—the attempts of human beings to construct grand, all-encompassing theories. If the nature of reality is the ultimate mystery, then the history of humankind is the attempt to penetrate and clarify this mystery, to grasp and understand it. It is the attempt, one might say, of mind to control matter. Dostoevsky railed against this attempt in his parable of the crystal palace.[4] He felt that the success of the scientific endeavor would remove mystery from the world and he held mystery to be the highest of truths. Dostoevsky and the myriad thinkers who have shared his concerns may not have appreciated the audacity of the scientific enterprise sufficiently. Perhaps it was naïve to believe that deductive thought could ever definitively capture substantial aspects of reality. Nevertheless, one must approach the attempt to do so with respect and maybe even a little awe. This is *the* quintessentially human act—the attempt to understand. Grasping after what may be incomprehensible. And in that valiant attempt to do the impossible we shall uncover a truth that transcends any particular conceptual development.

The Euclidean System

This section will briefly review Euclid's starting point—the foundation of his system of thought. These included definitions, common notions, and postulates.

The terms that are defined begin with *point*, *line*, and *straight line*. A *point*, for example, is "that which has no part." Such a definition, while it may have conveyed some intuitive feeling to Euclid's Greek audience, conveys little to us today. Modern mathematics understands that any deductive system must include a certain (minimal) number of undefined terms. In Euclidean geometry these would include point and straight line.

Euclid then goes on to define *right angles*, *perpendiculars*, *planes*, *circles*, *triangles*, *equilateral* and *isosceles triangles*, *quadrilaterals*, and so on. A triangle, for example, would be a plane figure that is bounded by three straight lines. He concludes with the definition of *parallel lines* that is discussed on page 92.

After the definitions, Euclid enunciated a series of *common notions*. These were meant to be self-evident truths that were of a very general nature. They included

1. Two things that are equal to a third thing are equal to each other.
2. If equals are added to equals, the sums are equal.
3. If equals are subtracted from equals the differences are equal.
4. Things which coincide with one another are equal to one another.
5. The whole is greater than the part.

All of these except possibly for the fourth are pretty self-evident to most people. What was meant is that if a geometric figure could be moved rigidly and placed on top of a second figure in such a way that the two coincided perfectly, two triangles, for example, then the sides, angles, and area of the first were equal to the sides, angles, and area of the second. The fifth common notion seems perfectly obvious but was put into question in Georg Cantor's theory of "infinite numbers," as we shall see in Chapter 4.

Finally the *Elements* contained a series of purely geometric postulates which we shall state in modern terms:

1. Given any two points, it is possible to draw a line segment with these two points as end points.

2. Any line segment can be extended indefinitely in either direction.

3. Given any point and any length, one can draw a circle with that point as centre and the length as radius.

4. All right angles are equal.

5. The parallel postulate (see page 94).

The first three postulates are self-explanatory, although the second is quite subtle, as we shall see later on. The fourth essentially established a standard measuring stick for angles. The fifth was, of course, extremely controversial, and we shall discuss it separately.

With these axioms in place, Euclid was in a position to deduce conclusions. Some of the early theorems involved the construction of an equilateral triangle (all sides of equal length) on a given base (Proposition 1.1) and the establishment of certain congruence schemes such as:

Proposition: *Given two triangles ABC and DEF for which AB = DE, AC = DF, and ∠ CAB = ∠ FDE, then the triangles are congruent (coincide in the sense of common notion 4). See figure 2.2.*

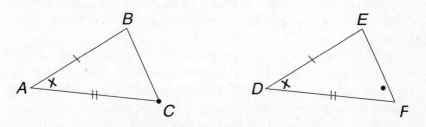

Figure 2.2. Side-angle-side

These initial propositions are proved through a process of straightforward logical reasoning—one starts with the hypotheses and reasons one's way to the conclusions. As a theorem is proved, it is added to the list of known results and may now be used in the proof of subsequent results. Thus a body of proved results is built up systematically. However, at a certain stage this procedure is not sufficient to accomplish what Euclid desires to accomplish. He must add a new technique of proof, namely, proof by contradiction.

PROOF BY CONTRADICTION

Is logic merely an organizing principle, or is it *the* means of arriving at the truth? In the course of this book we shall find ourselves repeatedly coming back to the role of the logic that has been traditionally used in science and philosophy. To begin with, one must understand that mathematics is concerned with the truth, not of individual propositions **P**, but of implications between mathematical propositions, implications of the form "if **P** then **Q**" or "**P** implies **Q**" (also written as **P** $\Rightarrow$ **Q**). Normally one assumes that **P** is the case and works one's way to **Q**. However, the reasoning does not always work in the forward direction. This is because in Aristotelian logic the implication **P** $\Rightarrow$ **Q** is equivalent (that is, has the same truth value as) $\neg$**Q** $\Rightarrow$ $\neg$**P**, where $\neg$**P** stands for the negation of the proposition **P**. For example, suppose we had to establish the truth of the following proposition "If a^2 is an odd integer, then so is a." Then we could just as well establish the related proposition "If a is not odd, then a^2 is not odd," that is, "If a is even, then a^2 is even." This latter formulation (which is much easier to prove) is called the contrapositive. Every mathematical proposition has a contrapositive formulation.

In just the same way, the proposition "all prime numbers are odd" is as true or as false as the contrapositive proposition "all even integers are composite." (Prime numbers were defined in Chapter 1. Composite numbers are the nonprimes, for example, 4, 6, 8, 9, . . . , and therefore can be written as the product of two factors neither of which is 1.)

An argument that is made in the contrapositive mode is an example of a "proof by contradiction." What is a proof by contradiction? We wish to establish the truth of the proposition **P** $\Rightarrow$ **Q**. So we assume that **P** is true and try to reason our way to the truth of **Q**. Instead of working directly toward **Q** we ask the natural question, "What happens if **Q** is false?" That is, we assume **P** and $\neg$**Q**. If the argument is to be successful, we must arrive at a contradiction. What is this contradiction? It may either be local or global, that is, it may either contradict an assumption within the body of the argument or contradict a previous result that is known to be true. We might look at a contrapositive argument in the following way: assume **P** and ¬**Q** and conclude ¬**P**. Thus we have **P** and ¬**P**, a contradiction. We conclude that the assumption ¬**Q** is not so, that is, that **Q** is true, and so **P** ⇒ **Q**. Thus the contrapositive can be considered as a kind of localized proof by contradiction. Examples of local and global contradictions are given below.

Is a proof by contradiction a minor variation on the conventional direct proof that goes directly from assumption to conclusion, or is it an entirely new form of reasoning? For every result that is proved by contradiction, does there exist a direct proof that will also do the trick? As we shall see repeatedly in what follows, there are many significant results in mathematics, like the result that the square root of two is not rational, that are only accessible to reasoning by contradiction. In fact, allowing "proof by contradiction" into our logical arsenal is extremely significant and powerful and determines, to a certain extent, the kind of mathematics we end up with.

A proof by contradiction is a subtle affair. This is seen by considering the difficulties students encounter when attempting to reason in this way. When you understand a proof by contradiction, it may seem like merely another form of argumentation. But from the student's point of view it is not at all obvious that this is a legitimate way to argue. Think about it for a moment. We are trying to argue that **P** ⇒ **Q**. We assume **P**, and ordinarily we would try to derive **Q**. But here we *assume* ¬**Q**. This seems strange since the whole point of the exercise seems to be to prove **Q**. How can you prove something is true by assuming it is false? The whole affair seems a trifle bizarre. It is my experience that this kind of indirect reasoning is difficult to understand and, I believe, legitimately so.

Why is this kind of reasoning so difficult to learn? Why was it (as we shall see) so controversial in the development of mathematics? Remember that in Chapter 1, during the discussion of the use of "variables" in simple algebraic equations, I said that the variable simultaneously represented all possible numbers and a single specific number (the solution of the equation). I maintained that working with variables, that is, algebra, necessitates a way of thinking that I called "ambiguous." Here we are demanding another kind of ambiguous thinking. Is **Q** true or not? In order to show that it is true we assume that it is false. So for a while it is both true and false—at least we must hold both possibilities simultaneously in our mind. Furthermore, in such a proof, the contradiction is deliberately evoked—in fact the creative idea here is to find a way to produce the contradiction— only to be denied. It is said that contradictions are disallowed within mathematics, but, in fact, it is only the results of mathematics that must be consistent. Contradiction is, as we have seen, allowed within the process of mathematical thought. Deductive mathematical systems are founded on the principle of noncontradiction—a seemingly negative prohibition. Ironically, this "negative" principle is used to prove many positive results such as that "most" real numbers are irrational. Thus the use of "proof by contradiction" is fraught with ambiguity in a manner that a direct proof is not. Situations that contain ambiguity are always difficult to master, yet the rewards, as in this case, are often great. The full power and implications of this type of argumentation are first revealed in the work of Euclid.

Parallel Lines

We now return to Euclid's development of the notion of parallelism. For various reasons the notion of parallel lines was the most controversial aspect of Euclidean geometry. Here is Euclid's definition of parallel lines:

> **Definition:**[5] *Parallel* straight lines are straight lines which, being in the same plane and being produced indefinitely in both directions, do not meet one another in either direction.

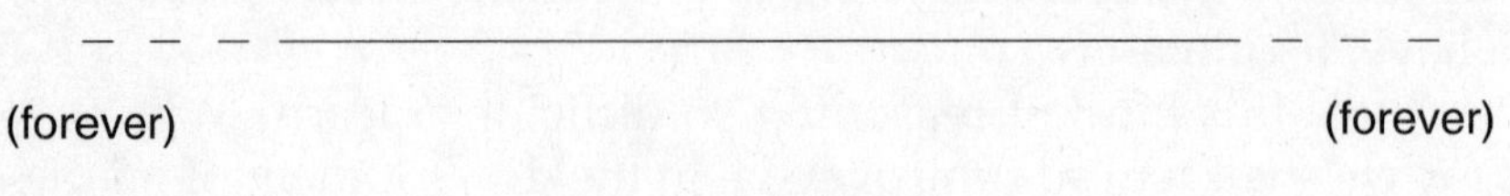

(forever)   (forever)

Figure 2.3. Parallel lines

This definition uses the second geometric postulate, which allows you to extend lines indefinitely. It is this indefinite extension that makes the definition of parallel lines difficult to apply.

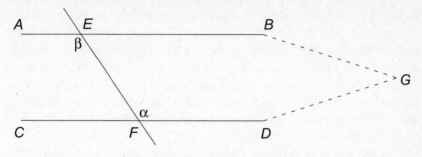

Figure 2.4. "Equal alternate angles" implies lines parallel

It is easy to check two finite line segments to see that they never meet, but how could one ever prove that two *complete* lines never meet. Thus the first thing that Euclid does is to provide a more practical criterion that will ensure parallelism.

Proposition: *If a straight line falling on two straight lines make the alternate angles equal to one another, then the straight lines will be parallel.*

Notice that, since this proposition deals with infinite objects, namely, the infinitely extendible lines, it cannot be proved directly and therefore Euclid must argue by contradiction.

Proof: We proceed in accordance with Figure 2.4, where angle (α) = angle (β). We have to establish that the lines AB and CD are parallel, that is, that they never meet. Assume that the lines are not parallel and do meet at the point G. Then we have a triangle EFG in which the exterior angle β is equal to the interior and opposite angle α. This contradicts a previous result (that the exterior angle is always strictly greater than either of the interior and opposite angles) and allows Euclid to conclude that AB and CD are indeed parallel. ∎

Note that what is contradicted in this argument is a previously obtained result. Thus this is an example of what we called a global contradiction. The contradiction works, not inside this particular proof looked at in isolation, but only when the argument is considered to be a part of a larger deductive system. In fact, the reasoning that is used in the proof of this theorem (and in any proof by contradiction) goes something like this: "Since this assumption leads to a contradiction and contradictions are not allowed within a consistent system, therefore the assumption is false. Since the assumption was the negation of what we assert to be the case, our assertion is true." All proofs by contradiction depend on a logical axiom of consistency: "No contradictions allowed here!" Euclidean geometry and all deductive systems are based on this principle.

It is only when proving the converse of the above result that Euclid is forced to use the famous Parallel Postulate:

> **Postulate 5 (Parallel Postulate)**[6]: If a straight line falling on two straight lines make the interior angles on the same side less than two right angles, the two straight lines, if produced indefinitely, meet on the side on which the angles are less than two right angles. (In other words, if a line crosses *AB* and *CD*, making angles α and β, so that α + β is less than two right angles, then *AB* and *CD* will cross on the same side as α and β.)

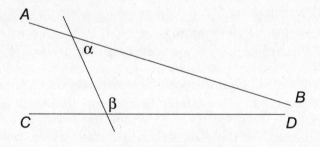

α + β < 2 right angles

Figure 2.5. Parallel postulate

It is interesting that the parallel postulate is framed in the negative. That is, it does not provide a condition for two lines to be parallel; it provides one for two lines to intersect. The reason for choosing to frame the postulate in this peculiar way is clearly to set the stage for the kind of argumentation Euclid had in mind—namely, proof by contradiction. The first use of the postulate is to show that the criterion of having equal alternate angles is both necessary and sufficient for parallelism.

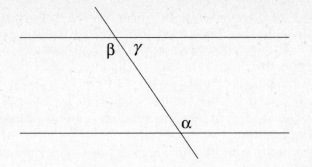

Figure 2.6. Lines parallel imply equal alternate angles

Proposition: *A straight line falling on parallel straight lines makes the alternate angles equal to one another.*

Proof: Here Euclid must convince us that in Figure 2.6 angle α is equal to angle β. Suppose, on the contrary, that angle α is not equal to angle β, for example, angle α is smaller than angle β. Then by adding angle γ we end up with

angle α + angle γ < angle β + angle γ = two right angles.

Thus the "interior angles on the same side" are less than two right angles. The parallel postulate tells us that in this situation the lines must meet, contradicting the assumption that they are parallel. ∎

This is typical of the argumentation one finds in Euclid and which subsequently found its way into all of mathematics. Parallel lines by definition never meet. But lines in Euclidean geometry are infinitely extendible. How then can one hope to *prove* that two lines will never meet *no matter how far they are extended*? Lines are infinite geometric objects. There is a certain intrinsic incompleteness about such objects, and yet mathematics wishes to make definitive statements about them. To prove that lines are parallel requires, at first glance, showing that no matter how far one extends the lines they will never meet. This is a kind of infinite argument in the same sense that showing that the sum of an odd number and an even number is always an odd number is also an infinite argument (it applies to an infinite number of cases). Mathematics characteristically deals with such "infinite" situations. In doing so it must replace an indefinite or infinite property with one that is essentially finite. In the proof, the finite property is the equality of the alternate angles. In the parallel postulate the "infinite" condition of parallelism is replaced by the "finite" condition for nonparallelism. The use of an argument by contradiction is a way of making this essential reduction.

We go on to one more famous result of Greek mathematics.

Theorem: *The square root of two is not a rational number.*

Proof: Suppose that the square root of two is equal to some rational number. Now recall that a rational number is nothing but a fraction, and every fraction is the quotient of two integers. That is,

$$\sqrt{2} = \frac{m}{n}.$$

Now the crucial point in this argument is that the integers m and n can always be chosen so that they have no factors in common. Thus the fraction 12/42 can be rewritten as 2/7, where the 2 and the 7 have no common factors. We assume that we have done so in the case of the rational representation of the square root of 2. By squaring the result we get

$$2 = \frac{m^2}{n^2} \text{ or } m^2 = 2n^2 \quad (*)$$

Thus m^2 is even. It follows that m is also even. (Interestingly this is also an argument by contradiction. For suppose m is odd; then m^2 must also be odd. This is not true, so m is even.) Since m is even it can be written as $m = 2k$. Thus (*) becomes

$$4k^2 = 2n^2 \text{ or } n^2 = 2k^2.$$

Applying the argument above to the latter equation shows that n also must be even. However, this would mean that both m and n are even and so they have the common factor 2, which is contrary to our initial choice of a fraction where the numerator and denominator had no common factors. This contradiction establishes that if $\sqrt{2}$ is a number then it is certainly not a rational one. ∎

This argument uses only certain elementary properties of numbers in order to arrive at its conclusion. Thus it is local, contained within the body of the proof. The result is negative; it states that the geometrical number $\sqrt{2}$ cannot be written in a certain form. It does not say that $\sqrt{2}$ exists, yet it points in that direction, so perhaps it is not entirely negative. $\sqrt{2}$ has been discussed at some length in Chapter 1. The fact that the result is proved using an argument by contradiction highlights the fact that there is indeed something problematic about this result. If $\sqrt{2}$ is an ordinary geometric number, then this contradictory argument tells us that it is a more complicated mathematical object than are the positive integers. There is a depth here, a challenge. There is something that remains to be understood. Then again if we look at $\sqrt{2}$ as a decimal, it is an "infinite," indefinite object. If we look at it as a geometric object, it is "finite" and seemingly definite. This is the nub of the ambiguity of $\sqrt{2}$. The proof by contradiction highlights the complexity of $\sqrt{2}$. It highlights the fact that $\sqrt{2}$ is an interesting mathematical object.

Proof and Counterexample

The vision of mathematics as formal deductive theory tends to overlook the role played by counterexamples. A counterexample is a kind of contradiction in the following sense. A mathematical theorem expresses a pattern or regularity, which is discerned in a certain body of mathematical data. The counterexample sets limits on the range of examples for which the pattern holds. It contradicts the conclusion of the theorem in a particular instance. Now the example might tell you that the conclusion of the theorem is never true, but that would be uninteresting. More interesting is the case where the conclusion is true for a certain range of phenomena but is not always true. Then the question is to determine just exactly when it is true. Mathematics is a balance between theory and counterexamples. Both are equally essential in building up a mathematical theory.

Imre Lakatos[7] developed a major thesis around the manner in which real mathematics—as opposed to formal or computerized mathematics—developed historically. For one thing, he points out that mathematics does *not* develop from axioms to theorems in the manner that one might believe if one only got one's mathematical experience out of Euclid. Lakatos's work will be discussed in more detail in Chapter 5; for the moment let us just say that he captures something very important about mathematics, namely, that more often than not hypotheses, definitions, and conclusions are simultaneously in a state of flux. If one is fortunate, stability finally appears in the guise of a "best" result that simultaneously determines what the appropriate hypotheses (including definitions of terms) and conclusions should be. The center of the "best" result is usually a key mathematical idea. In practice the idea comes first and the hypotheses and conclusions are adjusted so as to produce the "best" result that can be squeezed out of this idea.

I can recall the situation in the Mathematics Department of the University of California at Berkeley in the late 1960s. Stephen Smale had just been awarded the Fields Medal and there was a group of researchers collaborating with him in the investigation of dynamical systems. Every so often, Smale would propose some conjecture concerning the generic behavior of such systems.[8] Some months later, he or someone else in the group would produce a counterexample. This example would then be incorporated into a new, more general conjecture. The investigation would then proceed. This situation was iterated many times in the course of those years. The logical end—the counterexample—was not an end at all but an intrinsic and valuable part of the process that we know as mathematical research.

The point here is that the "contradiction" is not a termination point but a means of refining the ideas and the theory. The logical development is an essential part of the subject, but it does not define the subject. In the above examples the mathematical activity involves attempting to understand some body of mathematical data. Understanding is the key word. Both the positive mathematical results (theorems, propositions) and the negative ones (counterexamples) help advance our understanding and therefore may be equally valuable. Constructing a good counterexample may require as much ingenuity as proving an important theorem.

THE CONTRADICTORY WITHIN MATHEMATICAL CONCEPTS: THE CASE OF ZERO

When one thinks about mathematics, one sees it as a domain characterized by the absence of the contradictory. But is this really the case? It is true that formal mathematics has no contradictions; that is, from the same set of hypotheses it is not possible simultaneously to deduce the truth of a statement *and* its negation. However, there are other, subtler ways in which contradictions appear in mathematics. In the following we consider a completely different way in which contradiction finds its way into mathematics. We shall see how a contradiction, or more precisely, what appears to be a contradiction at a certain moment in time, may be transformed into a powerful mathematical concept. Thus in the manner that a process can be made into an object (cf. Chapter 1) *a contradiction can be reified*—it can be embedded within a mathematical concept. Thus, even though the rules of logic ban the contradictory from formal mathematics, the contradictory finds its way into the concepts of mathematics.

The mathematical concept "zero" is related to the more basic idea of "nothing." "Nothing" is simultaneously the most elementary and yet the subtlest of concepts. On the one hand, "nothing" is the negation of "something," so it could be said that if you have the idea of "something" then you must implicitly have the idea of "nothing." On the other hand, conceptualizing "nothing" as "something" is an act of quite subtle mental gymnastics. Does "nothing" exist, and if not how can we talk of it?

"Nothing" has been seen as perplexing and mysterious to thinkers throughout the ages. "Nothingness" is the central preoccupation of "existential" authors like Sartre, Camus, Kafka, Bellow, and Nietzsche. It was even the basis for the hit television comedy *Seinfeld*. Death is the ultimate "nothing," and the fear of death may well be the fear of "nothingness." "Nothing" is feared, it is the "unknown country from whose bourne no traveler returns," but, as Shakespeare asks in this famous soliloquy of Hamlet's, why is it feared if the content of that fear is unknown? The rational mind comes up blank when confronted with "nothingness." It is a kind of ultimate contradiction. This contradictory nature of "nothing" carries over to the symbol that humankind has developed to represent it—zero.

The Difference between "Two" and "Zero"

What is the nature of mathematical objects? What kind of thing is the number "two"? Normally most people assume that there is a one-to-one correspondence between words and the objects they correspond to. "Apple" stands for the fruit; "tree" stands for the object in the back yard with a trunk, branches, and leaves. Elementary mathematical objects retain this correspondence— only it is abstracted a little. Thus "two" stands for that property that is common to all pairs of objects. "Two" is an abstract idea, yet it retains much of the concreteness of the nouns "tree" and "apple."

Every known human culture has a word for the number two. It is an extraordinary fact that human infants from a very early age can recognize, immediately and without counting, a collection of two objects. Some precursor to the concept "two" appears to be hard-wired into our brains. And not only human brains: many animals also have the ability to recognize "twoness."

Now what about the number "zero"? At first glance it is a number not unlike the number "two." For the mathematician it is just another integer, albeit one with unique properties. For example, it is a neutral element under addition, that is, $n + 0 = n$ for any integer n. Under multiplication it reduces any integer to zero: $n \times 0 = 0$. This is all at the formal level.

In fact the number zero differs from the number two in ways that have much to teach us about the nature of Greek mathematics and culture, but also about modern mathematics and aspects of contemporary culture that are our legacy from Greek civilization. Parmenides, the Greek philosopher, held that "you can only speak about what is: what is not cannot be thought of and what cannot be thought of cannot be." Thus in his view the natural world is composed of distinct objects, and the proper and harmonious use of language consists in providing a one-to-one correspondence between words, the objects of thought, and the objects of the natural world. This assumption would today be called "naive realism," and it is the "natural" point of view of most scientists and, indeed, of most ordinary people. It is the way we establish our relation to the natural world—another legacy from the Greeks. This hypothetical correspondence has the consequence of transferring the property of consistency in theoretical constructs into a property of the natural world itself. Thus, for example, the consistency of Euclidean geometry or of parts of Euclidean geometry would imply the consistency of the corresponding part of the world of physical objects.

Is it reasonable to approach modern mathematics from the point of view of naive realism? We have already seen the problem that the irrationality of the square root of two posed for the Greeks. It was a challenge that was met only incompletely. The case of zero is a yet more compelling example of the consequences of naive realism and the principle of noncontradiction. To Parmenides and his followers the idea of "nothing" could not properly be entertained, much less used in mathematics. Now it is true that "zero" is a fundamentally contradictory notion—it is a word or a symbol that stands for numerical nothing. It is a presence that implies an absence. Anyone who held strongly to the principles of noncontradiction and the correspondence between concepts and objects might be expected to reject such a notion as inherently problematic. For is it not the function of human thought to pursue and evoke clarity? How can this goal be advanced by introducing concepts that are intrinsically "flawed" by an intrinsic self-contradiction and that break the assumption of clear correspondence between concepts and natural objects? Thus it is not surprising that the Greeks, with (or because of) all their mathematical sophistication, never managed to isolate the number zero as a mathematical concept.[9]

The concept of zero is so familiar that it takes a great deal of effort to recapture how mysterious, subtle, and contradictory the idea really is. Again, zero is a name for something that does not exist. Think of a concrete instance of zero—0 apples, for example. Suppose you have a bowl of fruit that contains 5 apples and 5 oranges and then you remove all the apples. There is still something left, but it is not an apple. It would not be correct to say there is nothing there, but you could say, "There are no apples in the bowl." More completely, what you mean is, "Speaking of apples, there are none or there are zero." Thus, the expression "zero apples" evokes the idea of apples only to deny that there are any. This is a little strange. When you say "zero apples," the apples are both present and absent. This same point may be made by asking the question, "Is '0 oranges' the same as '0 apples'?" The abstract numerical zero is more subtle still, since it refers to something that all concrete instances of zero (zero apples, zero oranges, zero pears, and so on) have in common.

Since our culture inherits the metaphysical assumptions of the ancient Greeks, we can appreciate, to a certain extent at least, the resistance the Greeks would have had to the isolation of a concept of zero. The Indians of the fifth century, on the other hand, had no such problems and did manage to develop the concept of zero. Why did one culture succeed and the other did not?

The earliest recorded use of 0 in human culture was in India in the year 458 C.E. The Indians used a place-value system as early as the year 594. In a place-value system like an abacus you can have empty columns, so you need notation to mark an empty column. "0" may have started out as a simple marker, but it soon assumed the status of a numeral like 1, 2, or 3. That is, by 628 the astronomer Brahmagupta had even spelled out the rules for the arithmetic of 0. These are, for example,

> When sunya [zero] is added to a number or subtracted from a number, the number remains unchanged; and a number multiplied by sunya becomes sunya.

He also defines infinity as the number that results from dividing another number by 0. This makes some sense, since $1/1 = 1$, $1/1/2 = 2$, $1/(1/3) = 3$ (dividing 1 into thirds requires 3 thirds). So, as you divide 1 into smaller and smaller parts, it requires more and more of them.

Why did the Indians succeed in isolating the concept of 0? John Barrow in *The Book of Nothing* says that "The Indian introduction of the zero symbol owes much to their ready accommodation of a variety of concepts of nothingness and emptiness. . . . The Indian mind saw it [zero] as part of a wider philosophical spectrum of meanings for nothingness and the void." The Indians had a very rich array of meaning for nothingness, as can be seen in the vast array of Sanskrit words in which different aspects of absence were seen to be something requiring verbal isolation. These include words for atmosphere, the immensity of space, a point, a sea voyage, space, complete, hole, void, and sky. Barrow, talking about the Indian concept of zero as *sunya*, says,

> It possesses a nexus of complexity from which unpredictable associations could emerge without having to be subjected to a searching logical analysis to ascertain their coherence within a formal logical structure. In this sense the Indian development looks almost modern in its liberal free associations. At its heart is a specific numerical and notational function that it performs without seeking to constrain the other ways in which the idea can be used and extended. This is what we expect to find in modern art and literature.

Barrow goes on to say, "The hierarchy of Indian concepts of 'Nothing' forms a coherent whole." "At the top level are words including those which are associated with the sky and the great beyond. They are joined by *bindu* (literally 'a point' but more generally the 'Nothing from which everything could flow') reflecting its representation of the latent Universe." These are meanings with positive connotations. At lower levels "we encounter a host of different terms for the absence of all sorts of properties: non-being, not created, not formed, etc." These are meanings with negative connotations.

"These two separate threads of meaning merged in the abstract concept of 0 so that the concept of 'nothing' began to reflect all the facets of the early Indian nexus of Nothings, from the prosaic empty vessel to the mystics' states of non-being."

A number of comments on the Indian conception of zero are relevant to our discussion. In the first place, the Indian "nothing" is both something and nothing, both presence and absence, positive and negative, even though for us these ideas are mutually contradictory. We might *define* absence as nonpresence and vice versa, so that *by definition* the ideas are mutually exclusive. On the other hand, the Indian notion is ambiguous. *Sunya* is one idea that has two self-consistent but mutually incompatible meanings. For other civilizations (such as the ancient Greek), the idea of "nothing" has only the negative connotation of absence. It is a triumph of Indian civilization that it manages to look at nothing as this self-contradictory concept, give it a coherent meaning and even symbolic representation.

Zero is indeed a contradictory notion. Does that mean that we cannot or should not talk about it? Of course we can and we do. And having this idea has literally changed the world and our idea of the world. It is one of the "Great Ideas" of humankind. The power and importance of a concept such as "zero" may be proportional, not to its properties of harmony and consistency, but to the inner contradiction to which it gives form and by so doing resolves in some way. Thus the important thing about "zero" is precisely its inner contradiction. This is what makes it so powerful. It was this power that was harnessed and used by science. The conjecture that the power of an idea is related to the contradictory tendencies that it embraces will force us to revise our thinking about mathematics radically. The essence of the subject would not be merely the avoidance of contradiction. Mathematics is rich enough and complex enough to "contain" contradictions and use them in a productive way. The problem then will be to unravel that complexity that consists of a mixture of logical necessity together with the contingent elements of ambiguity and contradiction.

How is the notion of "zero" basic to mathematics? The most obvious use of zero is in place-value notation. Place-value notation is the way in which we write integers. For example, in the notation for the number "123," the one stands for "1 hundred," the two for "2 tens," and the three for "3 units." This is a very economical notation, and it allows for the usual algorithms for addition and multiplication that we learn at school. Compare, for example, how cumbersome it is to add with Roman numerals, where multiplication is just about impossible. In fact it is possible to develop a perfectly good notational system for arithmetic using only "1" and "0." This is of course the basis of computers' machine language. Thus commerce, science, technology, information, and computers all use the language of place-value notation for the integers. It is no exaggeration to say that the world of science and technology would never have developed had it not been for the invention of zero. All of this from a contradictory concept that from the point of view of strict logic should never have been allowed to see the light of day.

Zero and the Calculus

The introduction of the calculus by Newton and Leibniz was followed by attempts by many mathematicians to explain the concepts and justify the procedures.[10] It was evident that the calculus worked—it provided the right answers to important questions—but why did it work? "Almost every mathematician of the eighteenth century made some effort or at least a pronouncement on the logic of the calculus, and though one or two were on the right track, all the efforts were abortive." Ultimately to move the calculus from an empirical subject to one with secure foundations required a revolution in mathematics. This amounted to the invention of analysis as we know it today, what is sometimes called the arithmetization of analysis.

The situation with respect to the validity of the calculus was not so different from the situation physicists found themselves in with respect to the foundations of quantum mechanics. I remember a lecture at McGill University when Nobel-prize winner Richard Feynman was asked why quantum mechanics worked as well as it did. Feynman replied that his job, as he saw it, was to make accurate predictions, and in this regard quantum mechanics was an extremely successful theory. As for the "why," this was beyond the domain of science as he saw it. The mathematicians of the eighteenth century felt the need to answer the "why" of calculus by supplying a convincing foundational theory.

There were many reasons for the confusion among mathematicians, but we do not have the space to go into all of them. Suffice it to say that it was clear that something profound was going on in the calculus, but what it was exactly was far from obvious. "The French mathematician Michel Rolle at one point taught that the calculus was a collection of ingenious fallacies." Many other "justifications" by mathematicians were clearly fallacious.

This situation led to attacks on the calculus by people who were essentially opposed to the entire scientific revolution that was occurring at the time. The language of that revolution was mathematics, and its main tool was the calculus. Perhaps the most eloquent of the attacks on the calculus was penned by Bishop George Berkeley (1685–1753). He felt that he was defending religion against the threat posed by the new science. In

1734 he published *The Analyst, or A Discourse Addressed to an Infidel Mathematician. Wherein it is examined whether the Object, Principles, and Inferences of the modern Analysis are more distinctly conceived, or more evidently deduced, than Religious Mysteries and Points of Faith.* Berkeley pointed out that mathematicians were proceeding inductively rather than deductively and, in particular, really had no logical reasons for their procedures. To Berkeley the calculus involved as large a leap of faith as did his own religious beliefs. There is no question but that the advent of the calculus precipitated a crisis in mathematics that lasted the better part of a century.

The main conceptual hurdle was the conception of limit, and in particular the ambiguity we have encountered before—the identification of the *process* of a limit with the *object or number* that is the result of this process. All this revolves around the proper use of "0," that is, it revolves around the meaning of "0/0." Now if $n \neq 0$ then the expression $m/n = k$ is equivalent to $m = nk$. Thus "6/3 = 2" and "6 = 3 × 2." If we look at this expression for $n = 0$, then, since $0 = n \times 0$, then, as Euler said, $n = 0/0$ for any value of n. Thus 0/0 is an ambiguous animal that schoolchildren are warned to stay away from because, they are told, it is meaningless or undefined. However, it is impossible to stay away from 0/0 in the calculus. In fact, looked at in a certain way, the whole of differential calculus depends on giving a precise meaning to the expression 0/0.

The situation can be described by asking, "What is (instantaneous) speed?" Suppose that a car is accelerating from rest according to the formula $d = t^2$, where d stands for distance and t for time. Thus the car starts off at position 0 at time 0; it is at position 1 at time $t = 1$, at position 4 at time $t = 2$, at position 9 at time $t = 3$, and so on.

What is the car's instantaneous speed at time $t = 1$? We calculate the average speed from over various intervals of time around $t = 1$ and see what happens:

$t = 1/2$ to $t = 1$ average speed = distance/time = 3/2,
$t = 3/4$ to $t = 1$ average speed = 7/4,
$t = 7/8$ to $t = 1$ average speed = 15/8.

In general the average speed from time $t = 1 - 1/n$ to $t = 1$ will be $2 - 1/n$. Also the average speed from $t = 1 + 1/n$ to $t = 1$

will be $2 + 1/n$. As n gets larger and larger, $1/n$ gets closer and closer to zero, so the shorter the time interval around $t = 1$ the closer the average speed gets to 2. Thus we say that the speed at time 1 (what you would see on your speedometer) is 2.

What we have been doing is the following:

$$\text{average speed} = \frac{\text{distance}}{\text{time}} = \frac{(\text{position at time } t) - 1}{t - 1}$$

As the interval gets smaller and t gets closer to 1, the numerator of the fraction gets close to 0 and so does the denominator. So in a sense as the result should be $0/0$, yet actually the fraction gets closer and closer to 2. Thus what we want to look at in the differential calculus is the limit of a quotient which, in this case, is *not* equal to the quotient of the limits (which is equal to $0/0$).

The value of the derivative of a function (a mathematical generalization of the idea of "instantaneous speed" that we have just been considering) is therefore obtained by considering the results of an infinite sequence of approximations. In the next chapter we shall see that the identification of an infinite sequence with a single finite number requires a major leap in our understanding of the real number system. Whether we conceive of the calculus as involving such an infinite limiting process, or we think of it as involving infinitesimal quantities that are nonzero but smaller than every positive real number; in either case the very subtle concept of infinity must inevitably arise. Thus the concepts of "zero" and "infinity" are closely tied together. For example, one might say that $1/0 = \infty$ and $1/\infty = 0$. To make sense of this relationship, one could invoke the idea of limiting values just as we did for the idea of the derivative above. Thus the latter equation could be given the meaning of "the value of the expression of $1/x$ tends to the value 0 as x gets very large.[11]

Today, of course, it is well known that the calculus is amenable to an axiomatic development in the spirit of Euclid's development of geometry. However, this should not blind us to certain elements in the calculus that were legitimately problematic. From the point of view that is being taken here, these problems—the ambiguities inherent in the calculus—were extremely important. The resolution of these ambiguities required a century of work by many talented mathematicians and this work brought about the birth of modern analysis. Many of the basic elements of modern mathematics, including the concepts of real number, function, and continuity, as well as the derivative and the integral, were ultimately stabilized in the wake of the discovery of the calculus. Thus when we use the word "resolution" we do not mean merely developing a logically rigorous theory, for it is conceivable that the calculus could be developed starting from a different mathematical foundation. For example, what is meant by the term "infinitesimal" can be made precise within a mathematical structure such as the one I describe in Chapter 4. This is a different "resolution" of the calculus. In general, when I talk of a "resolution," I am referring to a stable structure that contains the original problematic situation, but I am also referring to the burst of creativity that emerges from that situation.

Concluding Commentary on "Zero"

There are various ways of characterizing our discussion of zero. One way is to say that our culture has tamed the concept of zero by removing its contradictory aspects and incorporating it into a logically coherent mathematical system. The other is that a concept that is inherently contradictory has been made ambiguous by creating out of zero "a single idea." The power of the concept is related to the way the "single idea" retains some of the incompatibility of the initial contradiction between something and nothing. In this view, then, far from excising the contradiction from our thought, we have learned to use it in a constructive and creative manner. Modern mathematics, not to speak of the world of science and technology, is a consequence of this use of the concept of zero.

For future reference, this discussion throws some light on the question that was posed earlier as to whether logic and consistency were built into the universe in some fundamental way. A concept, like zero, is ambiguous: it is a "single idea" which has two incompatible frames of reference. Focusing on the single idea or on the individual frames of reference places us in the realm of coherence and consistency. Focusing on the incompatibility of the frames of reference places us in the realm of contradiction. This is another ambiguity. Both coherence and contradiction are fundamental components of ambiguity—neither can be omitted. The conclusion that this pushes us toward is that it is ambiguity and not logical coherence that is fundamental. We cannot successfully create a perfect world of noncontradiction. Such a world would be static and devoid of depth and creativity. Logic is part of the world, but only a part. Reality in general and mathematics in particular are greater than logical consistency.

The idea of zero is a case history, a generic idea. Such a mathematical concept arises out of a vast area of human experience. One might say that the precise mathematical meaning of zero is the center of a much larger cloud of perceptions and cognitions. This larger, ill-defined cloud is usually taken to represent the *connoted* meaning as opposed to the precise, *denoted* meaning. The connoted meanings are not logical; only the denoted meaning can fit into a precise, logical scheme. Thus the connotations may include things that are ambiguous or contradictory. In the case of zero we have seen that a number of contradictory elements form part of the cloud of connotations associated with the concept.

Paradoxes and Mathematics: Infinity and the Real Numbers

PARADOX

Live at the empty heart of paradox; I'll dance there
with you, cheek to cheek.
—Rumi

As with all apparent paradoxes arising from
special relativity, under close examination these logical
dilemmas resolve to reveal new insights into the
working of the universe.
—Brian Greene, *The Elegant Universe*

In this chapter we pursue our investigation of ambiguity in mathematics by beginning a discussion of the paradoxes of infinity. What is a paradox? According to the *Encarta World English Dictionary* a paradox "is a situation or proposition that seems to be absurd or contradictory but is or may be true." It is interesting that the word "paradox" has the contrary meaning to "orthodox." Orthodox means "following established or traditional rules." Orthodoxy operates within acceptable limits. Paradox, on the other hand, breaks the limits of the orthodox. It appears "absurd or contradictory" from the perspective of the orthodox, but it "may be true" in another, larger, context. In fact there appears to be no consensus about the meaning of "paradox," so I shall explain the sense in which I shall use the word. Just as the earlier discussion of the contradictory was meant to highlight a certain aspect of ambiguity, namely, the aspect of incompatibility that exists in every situation of ambiguity, so the present discussion of paradox is also intended to illuminate and extend our understanding of ambiguity. I stressed in Chapter 1 the close relationship between ambiguity and creativity—every ambiguous situation contains a creative potential, namely, the possible resolution of the "incompatibility" through the emergence of a new idea or point of view. Our discussion of paradox is meant to emphasize the potential emergence of such a novel paradigm. Thus I shall only discuss paradoxes that have such a positive potential—every paradox that is discussed will have something to teach us about the nature and the development of some mathematical concept.

Like an ambiguity, a paradox has positive and negative aspects. The negative is, of course, the "absurd or the contradictory"; the positive is that "it may be true" from some new and different point of view. Creative situations also have a positive and a negative side. In order for something new to emerge, something old must often be broken down. As long as one is trapped within habitual patterns of thought, it is difficult for a truly original idea to break through. Thus in my discussion of ambiguity I highlighted the *incompatibility* that must exist between the two frames of reference. In a paradox this incompatibility is raised to the level of the "absurd or contradictory." A paradox is a heightened ambiguity. At least this is true for the paradoxical situations I shall discuss in this chapter. Of course we are interested, not so much in paradoxes for their own sake, but in the paradoxical as a driving force in the development of mathematics.

It is true that the difference between ambiguity, contradiction, and paradox is one of nuance, since in essence they are describing one or more aspects of a situation or event that can be viewed from multiple perspectives. There may even be times where the terms seem to be used interchangeably. In essence I shall always be discussing situations of ambiguity. Such situations may contain elements of contradiction, or they may involve the paradoxical, but the same elements will always be found—incompatibility, multiple perspectives, and a point of view that changes in time or has the potential to do so.

Now a paradox is an absurdity; a paradox is something whose very existence is unacceptable. A pure, logical contradiction is a closed case—it is wrong; it cannot be. In the situation of paradox the contradiction exists but we cannot accept the closure of the case. The situation has two aspects that are irreconcilable but we cannot leave it at that—things *must* be reconciled. Thus a paradox is a contradiction that is open, not closed.

This is a familiar situation to anyone who has grappled with a personal or artistic dilemma. If one takes such a situation seriously one frequently reaches an impasse—a moment when there seems to be no way in which to resolve the dilemma. At this stage it is very tempting to give up and quit. The situation seems to be impossible. To face this impossibility and yet continue requires great strength and great courage. Yet it is an ordinary kind of courage, manifested every day by people confronting serious illness or other personal problems. It is the precise nature of the existential dilemma faced by every one of us: Life in the full realization of death is, in a sense, impossible;[1] yet we live, we persevere, we do our work, we are kind and considerate to those around us. Paradox is not something alien to life; it is the basic fabric out of which life is composed.

It is the human ability to persevere in the face of the intractability of such situations that leads us to propose an unusual approach to "paradox," one that has the potential to radically change the way in which we look at life in general and mathematics in particular. In a few words it is that paradox cannot be eliminated from life because it is part of the basic fabric of things. Moreover, paradox has great value. Thus paradox should be seen as a generating force within the domain of mathematical practice. There has been much discussion among mathematicians and scientists about the "unreasonable effectiveness of mathematics in the physical sciences."[2] Mathematics has power! Moreover, mathematics is not a static body of knowledge. On the contrary, it is dynamic, ever changing. Where do that power and dynamism come from? Well, they come from ambiguity, contradiction, and paradox. These things are therefore of great value. They need to be unraveled, explored, developed, and not excised.

So in what sense can a paradox be true? If a paradox is true in *any* sense, then the contradiction it embodies cannot be absolute. It is neither absolutely false nor absolutely incomprehensible. What makes something into a paradox? Is it the logical structure that it carries within itself or within which it is embedded? If so, then that logical structure may be changed and what seemed an insurmountable barrier may become the gateway to deeper understanding. But it might not be the logical structure itself; it might be that the impossibility is connected to the larger cultural context in which the paradox is embedded. When I discussed "zero" in the previous chapter I emphasized that it contained a contradictory aspect, "the nothing that is." In fact, the story of "zero" does not end there; that is just the negative part. The positive part of the story is that "zero" is a seminal concept in all the ways that were discussed. Thus, if I were creating a classification scheme (which I am not doing; I am merely digging into the ramifications of ambiguity in mathematics), I would say that "zero" is paradoxical—it contains a contradictory aspect and yet is a fundamental mathematical concept.

Thus the two parts of the definition of paradox should be taken seriously. There is a contradiction, but there may also be a truth. This is illustrated beautifully in a discussion of the paradoxes associated with infinity. Many of the paradoxes that I shall discuss carry a deep mathematical idea. The object is to use the paradox as an entry into that idea. In so doing I shall be making the larger point that we have to learn to think about paradox in a different way.

INFINITY

> Mathematics is the science of the infinite, its goal
> the symbolic comprehension of the infinite with
> human, that is, finite means.
> —Hermann Weyl

To pursue the point that mathematics is not built on logically pristine concepts but includes the ambiguous and the paradoxical, I shall now turn my attention to the case history of the infinite. The infinite is an extraordinarily subtle topic. "Zero" was discussed in the last chapter, and, in a way, "zero" and "infinity" are the opposite sides of the same coin—less than anything, on the one hand, and more than anything, on the other. However, there is a major difference in the degree to which these two concepts have been assimilated into our culture. We take the concept of zero completely for granted. It appears to us to be an obvious or trivial idea. We have no doubt at all about what it means and how to calculate with it in an algorithmic, mechanical way. Though the last chapter pointed out some of the mystery that attends to the idea of "nothing," nevertheless, for the most part "zero" has lost the quality of open-endedness and has become a prosaic, well-defined, and precise mathematical object. Infinity, on the other hand, has not yet been totally assimilated. More than almost any other concept, "infinity" retains its poetic qualities—its mystery and wonder.

In *The Mathematical Experience*, Philip Davis and Reuben Hersh speculate about the origins of the idea of the infinite.

> The perception of great stretches of time? The perception of great distances, such as the vast deserts of Mesopotamia or the straight line to the stars? Or could it be the striving of the soul toward realization and perception, or the striving toward ultimate but unrealizable explanations?
>
> The infinite is that which is without end. It is the eternal, the immortal, the self-renewable, the *apeiron* of the Greeks, the *ein-sof* of the Kabbalah, the cosmic eye of the mystics which observes us and energizes us from the Godhead.[3]

When most people think about infinity, they think about the very large. I can recall an occasion when my wife and I were taking a walk and chanced to overhear an argument between two boys who must have been eight or nine years old. There seemed to be a competition going on to produce the "biggest" number. The contest went on for a while in a conventional manner: "A million, a billion," and so on. Then one of the boys played his trump card and said, "Infinity!" Not to be outdone the other boy replied in what I thought was a very interesting comment, "Infinity, infinity!" There the competition ended. We were left with a number of intriguing questions. What did the first boy mean by "infinity"? More interestingly still, what was behind the second boy's response? Did we have here a budding Cantor or was he just repeating what he had heard elsewhere? I'm afraid these questions will never be answered, but it was clear to us that "infinity" for these boys had something to do with the very large. And so it is for most people: the origins of infinity seem to be tied up with the idea of very large quantities or distances.

There is another possible source for the idea of infinity, and that is bound up with the notions of self-reference and iteration. For example, suppose we have a picture that contains a smaller

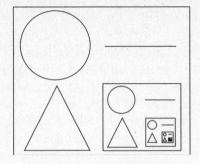

Figure 3.1. Self-reference

but otherwise identical copy of itself (figure 3.1). Then the smaller picture will have a copy of the original that is smaller yet, which will have another copy, and so on. Thus we have an infinite cascade of images, each a faithful duplicate of the original. Escher is famous for such paintings. Of course, this is an elementary version of "self-similarity" such as one finds in "fractals." The subjects of fractals and cellular automata may be popular because they capture something basic in the human experience. Remember the myth of Narcissus, who drowned after becoming entranced with his image in a pool of water. Self-image and self-reference strike a deep chord in us all. How entrancing it is to stand between two mirrors and observe a set of images that appears or recede into infinity.

Every (discrete) dynamical system is generated by a rule (a function) that can be iterated, a function that starts with a number between 0 and 1, for example, as input and produces another number, also between 0 and 1, as output. The output can then become a new input, and so on (see figure 1.3b). We are interested in the "and so on"—the asymptotic behavior of the function with a given input—and this "and so on" might be described as the behavior of the function "at infinity." Any process that can be repeated in this way, that can be "iterated," evokes the notion of infinity. With the advent of the computer this sort of thing has become very familiar. One of the basic elements of a computer program is a "loop" whereby a single computation or a sequence of computations is iterated over and over again. Thus in mathematics, in computer science, and in science in general, the infinite is explicitly or implicitly invoked.

Does Infinity Exist?

In a sense infinity is similar to every other concept in that it is an insight into some aspect of reality. Some people might object that there is no aspect of reality that is captured by the idea of infinity. They might claim that the infinite does not exist in the natural world. It may, in fact, be true that there are only a finite number of atoms in the universe, or that space is a finite four-dimensional manifold. Even if these things were true, would that imply that the infinite is not a valid human concept? What aspect of reality is the conception of infinity pointing to?

Like all other concepts, the infinite is not identical to the aspect of reality that it illuminates; that is, "the map is not the territory." It evokes the territory, yet it is not the territory that is evoked. There is something mysterious about the human use of language, and especially so in the case of "infinity." Infinity evokes something real—behind the notion of infinity there is real human experience, real human intuition. In the use of the concept of infinity we are trying to pin down this intuition, to give it some concrete form. Maybe it is impossible to definitively pin down an intuition like "infinity" in the same way that one cannot pin down what precisely one means by the "beautiful," but the very attempt reveals something admirable about the human spirit.

Infinity is a concept like other concepts, only more so. Other concepts are content to point to confined aspects of reality—infinity goes after something enormous. Thus "infinity" is an extreme example of the general process of conceptualization. After all, the infinite is held to be a defining characteristic of divinity. Perhaps because it is so extreme, a certain rawness still remains associated with the concept, at least for many people. We can all sympathize to some extent with Carl Friedrich Gauss's opinion that "the actual infinite" is not a legitimate object of thought. Maybe certain things are better not put into words; maybe some things cannot be pinned down in an analytic manner. But if this is true one could object to all words, to all ideas. However, if it is true that all words somehow fall short of the reality they would describe, then this is especially true of infinity.

Infinity reaches explicitly for that aspect of reality that is beyond words. It attempts to articulate that which is beyond articulation. It seems to be attempting to go beyond itself in the same way that a great work of art—a Beethoven symphony, let us say—does. It astounds us because it expresses something that we thought could not be expressed. We are deeply moved but we cannot say why.

Because "infinity" is a topic that is so obviously reaching beyond itself, its essential incompleteness never disappears from view, leaving a tension that never completely dissipates. Perhaps there was a time in the evolution of humanity when all words had this sense of open-ended incompleteness. However, today, most words have lost their magic for us by dint of endless repetition and commercial exploitation. Perhaps once "beauty" was an idea that had the power to stir the soul, but today everything and everyone is "beautiful." So "beauty" has been debased and its magic has been lost. "Infinity" has not yet suffered that fate.

INFINITY IS AMBIGUOUS

For my purposes in this book, infinity is an excellent topic because a discussion of the infinite inevitably brings out questions that are always present but are difficult to bring to conscious awareness. It is easy to see, for example, that the concept of infinity is inherently ambiguous. In discussing most concepts it is appropriate to begin with a definition. However, in the case of infinity, one of the basic questions is whether a definition is even possible. Many people have held that infinity cannot be defined. For, they reasoned, if we had such a definition then this very definition would transform infinity into something limited, which, of course, would make it finite and not infinite.

According to philosopher A. W. Moore, historically there have been two clusters of concepts that dominate the discussion of the infinite:

> Within the first cluster we find: boundlessness; endlessness; unlimitedness; immeasurability; eternity; that which is such that, given any determinate part of it, there is always more to come; that which is greater than any assignable quantity. Within the second cluster we find: completeness; wholeness; unity; universality; absoluteness; perfection; self-

sufficiency; autonomy. The concepts in the first cluster are more negative and convey a sense of potentiality. . . . The concepts in the second cluster are more positive and convey a sense of actuality.[4]

The first cluster of meanings is negative. It says that the infinite is *non*-finite, has no bound, no end, and so on. The second cluster is positive. It says that the infinite is *all*. This, if you think about it, is a problem in itself. If it is all, how do you talk about it, for talking about it implies standing outside of it and if you can stand outside of it then it is not all. Nevertheless, we can at least say that there is an obvious tension between these two ways of looking at infinity. To say that something is not finite is not the same as to say that it is whole and complete.

Here then is the ambiguity of infinity. It is something that is incomplete, even something that cannot be completed in principle. Yet it is complete, since we talk about it. We even aspire toward it as the highest of goals, the highest of values. In aesthetics, it is the beautiful, in ethics it is the good. It is the defining characteristic of divinity. From one perspective it is beyond all words and concepts. From the other it is precisely that—a word and a concept. Here then is the ultimate contradiction!

In *The Mathematical Experience* the authors attempt to capture this very dichotomy in the mathematical example that I used when discussing ambiguity.

Observe the equation

$$\frac{1}{2} + \frac{1}{4} + \frac{1}{8} + \frac{1}{16} + \cdots = 1.$$

On the left-hand side we seem to have incompleteness, infinite striving. On the right-hand side we have finitude, completion. There is a tension between the two sides which is a source of power and paradox. There is an overwhelming mathematical desire to bridge the gap between the finite and the infinite. We want to complete the incomplete, to catch it, to cage it, to tame it.[5]

This equation is a paradox, yet we resolve it by the equation itself, through the use of the equal sign. This equation for the summation of an infinite series is an ordinary mathematical expression no more complicated than the infinite decimals that were discussed in Chapter 1. We tend to take the mathematics that stands behind such expressions for granted. We forget what an extraordinary leap is required in order to "get it." Yet this insight is being called upon whenever we write down a real number as an infinite decimal.

Think, for example, of the decimal representation of π,

$$\pi = 3.14159\ldots$$

Again, here is π as a finite, named object versus π as an incredibly complex infinite decimal. In the former sense we know it well—the ratio between the circumference and the diameter of any circle. In the latter sense it remains mysterious to this day. We cannot answer even "simple" questions about it. One such question is, "Is π normal?"[6] Normal means that its digits are truly random, at least in the sense of being uniformly distributed. That is, for a number to be normal (in normal base 10 notation) each digit from 0 to 9 would appear $1/10$ of the time; each sequence of two digits from 00 to 99 would appear $1/100$ of the time; and so on. The answer to the question of whether π is normal seems beyond our grasp at the present time. The best we can do is to investigate the question experimentally. This has been done and, for example, for the first 6 billion decimal places, there are no anomalies, that is, each digit occurs approximately 600 million times as we might expect were the conjecture of the normality of π to be correct.

In the quotation at the beginning of this chapter, Weyl says that "mathematics is the science of the infinite"; that is, the defining characteristic of mathematics is, in some sense, the way in which it deals with the infinite. Moore gave us two general dimensions in which the infinite is conceptualized. The first comes from the word infinite itself, that is, in- or non-finite. From this point of view no finite system of thought is complete. Gauss and Aristotle, as we shall see, took this point of view when they asserted that there was something intrinsically unacceptable in the notion of actual, as opposed to, potential, infinity.

At the present time our culture is dominated by the complementary philosophies of "scientific realism" and "postmodernism." Both of these would dispense with the entire history of human thought that is represented by the clusters of "positive" concepts about infinity. Be that as it may, the infinite appears to be a valid and quite common human intuition. To take a contemporary example, consider the following quotation by the philosopher Abraham Heschel:

> What characterizes man is not only his ability to develop words and symbols, but also his being compelled to draw a distinction between what is utterable and the unutterable, to be stunned by what is but what cannot be put into words.[7]

"What is but cannot be put into words" could well be a description of the infinite. Thus it is, in a way, the ultimate ambiguity. Paradoxically, what is commonly done with "what cannot be put into words" is precisely to attempt to express "it." This may be done in words or in other ways—through music, the fine arts, or mathematics. *There exists that which cannot be expressed yet we must express it.* This is an expression of the ultimate ambiguity that characterizes the human condition itself. I shall refer to this kind of ultimate activity, in the paradoxical manner in which such an activity can only be described, as putting the ineffable into words, describing the indescribable, or expressing the inexpressible. One of the most far-reaching but least-noticed activities of this kind consists of the use of the infinite in mathematics.

If the description of mathematics as the science of the infinite rings true, it is because of the noble and audacious attempt by mathematicians to capture the infinite with finite means. Weyl continues, "It is the great achievement of the Greeks to have made the contrast between the finite and the infinite fruitful for the cognition of reality. . . . This tension between the finite and the infinite and its conciliation now become the driving motive of Greek investigation."[8] Thus the tension between the finite and the infinite lies at the heart of mathematics.

If the concept of the infinite in mathematics is really an attempt to grasp the ineffable, to conceptualize that which cannot in principle be conceptualized, then such an attempt must inevitably lead to some sort of breakdown. The breakdown occurs in the logical structure of mathematics, and what marks this type of breakdown is the appearance of paradox. Since the attempt to mathematize the infinite is a matter of pushing the rational mind just about as far as it can be pushed, then we should expect that the occurrence of any particular aspect of the infinite in mathematics is associated with the appearance of one or more types of paradox. Thus in studying various aspects of the infinite, the paradoxes they give rise to, and the way mathematics uses and resolves these paradoxes, I shall be exploring some of mankind's most profound attempts to penetrate the mystery of mysteries—the relationship between reality and the human mind.

MATHEMATICS AND THE INFINITE

In mathematics, infinity is not one concept among many. In some ways it is central to mathematics. One might say that it is precisely the way that mathematics deals with the infinite that characterizes mathematics and separates mathematical thought from that of every other discipline.

It is the use of infinity that marks the emergence of mathematics from the domain of the empirical to the domain of the theoretical. The use of infinity in any specific manner requires considerable mental flexibility. It requires a new way of using the intellect, a certain subtlety of thought, an ease with complex, contradictory notions. In the use of the concept of infinity there is always the danger that things will get out of control and slip into the realm of the purely subjective. That is, there is the danger that we will not be doing mathematics anymore.

On the other hand, the rewards are great. Davis and Hersh in the above quotation pointed to the "tension . . . which is a source of power and paradox." The infinite is itself a source of power and paradox. The usual response to such situations is to want the power but not the paradox. Thus we normally attempt to attain the power in mathematics and science through the elimination of the paradoxical. This is not my point of view. In my view the power and the paradox are inextricably linked. If we wish to gain the power, we must confront the paradox. More than almost any other mathematician, Cantor pursued the power and paid the price in both his professional and personal life.[9] He left mathematics a different subject, and mathematics is the richer for his gifts. As David Hilbert (1862–1943), one of the greatest mathematicians of his time, said, "No one shall drive us from the paradise Cantor created for us."

The Finite and the Infinite

The tension between the finite and the infinite is the great theme that resounds throughout mathematics. Human beings are finite creatures yet they aspire to the infinite. According to existentialist philosophers like Søren Kierkegaard, for example, the fundamental human dilemma resides in human beings' dual nature, on the one hand the finite body and on the other the infinite mind and spirit. Our inquiry is into how this fundamental tension plays itself out in the realm of mathematics. Mathematics has discovered its own ways to reconcile the gap between the finite and the infinite.

In modern mathematics this question of finite and infinite is still present. In a recent interview the well-known mathematician Paul Halmos was asked about his statement, "I am inclined to believe that at the root of all deep mathematics there is a combinatorial [read "finite"] insight." He went on to explain his statement as follows:

> I asserted that the deep problems of operator theory could all be solved if we knew the answer to every finite-dimensional matrix question. I still have this religion that if you knew the answer to every matrix question, somehow you could answer every operator question. But the "somehow" would require genius. The problem is not, given an operator question, to ask the same question in finite dimensions—that's silly. The problem is—the genius is—given an infinite question to think of the right finite questions to ask. Once you thought of the right answer, then you would know the right answer to the infinite question.
>
> Combinatorics, the finite case, is where the genuine, deep insight is. Generalizing, and making it infinite, is sometimes intricate and sometimes difficult, and I might even be willing to say that it's sometimes deep, but it is nowhere as fundamental as seeing the finite structure.[10]

This comment seems to me to state very well the situation as it is lived by the mathematician. In a sense, everything in mathematics is a dialogue between the finite and the infinite. The most trivial observation, such as "The sum of two odd integers always gives an even integer," applies to an infinite number of cases. We cannot check the veracity of this statement by checking every case, so we need a stratagem for dealing with every case at once. This stratagem is contained in the observation that an integer is odd if and only if it can be written as $2k + 1$ for some (other) integer k. This observation enables us to reduce the argument from an infinite number of concrete cases to one abstract case. What we have done is to *abstract* the situation by the introduction of a variable. The variable is the means of reducing the infinite to the finite. However, as I pointed out in the discussion of ambiguity, the notion of variable is ambiguous. Thus the reduction from the infinite to the finite is accomplished through the introduction of the ambiguous.

Mathematical Induction

The subtle use of variables and their use in making the infinite concrete is illustrated by the form of mathematical argument called "mathematical induction." Induction in science and philosophy refers to a situation in which one discerns a pattern or regularity in nature and infers that the regularity will continue for all time. Thus the sun rises every morning and so we infer inductively that the sun will rise tomorrow. This form of induction is never certain, for after all the sun may not rise tomorrow. Thus induction is never absolutely certain in comparison with deduction, which is usually taken to be absolutely certain. Mathematical induction however is a form of reasoning that is deductively certain even though it retains some of the form of scientific induction.

A typical situation of mathematical induction is the following:

$$1 = 1 = 1 \times 1,$$
$$1 + 3 = 4 = 2 \times 2,$$
$$1 + 3 + 5 = 9 = 3 \times 3,$$
$$1 + 3 + 5 + 7 = 16 = 4 \times 4.$$

There seems to be a pattern here. This pattern could be expressed by saying that the sum of the first n odd integers is equal to n times n. Is this pattern true for all values of n? That is, is it possible that for some very large value of n the pattern ceases to be true? How can we assert that this formula is always true even though the number of cases is infinite? The Principle of Mathematical Induction allows us to do this. It asserts that, if there is a family of formulas $P(1)$, $P(2)$, $P(3)$, and so on as above, then every occurrence of the formula is assured to be valid if the following two conditions hold:

C1: $P(1)$ is valid.

C2: For every positive integer k, if $P(k)$ is valid then so is $P(k + 1)$.

In our example $P(1)$ says that $1 = 1 \times 1$, which is certainly valid. The verification of condition C2 is slightly more involved. In the first place, it requires writing our conjecture in a more "mathematical" form. That is, we must introduce a variable and write out the formula that we are trying to verify. Thus without writing out the infinite number of cases: $P(1)$, $P(2)$, and so on as one general case $P(n)$ we cannot proceed. In this case the general formula could be written out as

$$P(n): 1 + 3 + 5 + \cdots + (2n - 1) = n^2.$$

Now we are in shape to work on condition C2. We assume that $P(k)$ is valid for some k, that is, $1 + 3 + 5 + \cdots + (2k - 1) = k^2$. Now we work on the validity of $P(k + 1)$:

$$1 + 3 + 5 + \cdots + (2k - 1) + 2(k + 1) - 1$$
$$= 1 + 3 + 5 + \cdots + (2k - 1) + (2k + 1)$$
$$= k^2 + (2k + 1) \text{ (since we know that } P(k) \text{ is valid)}$$
$$= (k + 1)^2.$$

Thus $P(k + 1)$ is valid (if $P(k)$ is) and by the principle of mathematical induction the formula $P(n)$ is valid for all $n = 1, 2, 3, \ldots$

Notice that the only calculation involved is a simple one. One has the nagging feeling that one is getting off with too little effort. All these infinitely many formulas are now established as true, when the only formula that has been verified directly is $P(1)$. All the rest of the work is accomplished by the condition C2. Thus mathematical induction also involves the reduction of the infinite to the finite. It demonstrates the great power that resides in the ambiguous notion of variable and in algebraic manipulation.

Why is the principle of mathematical induction valid? Can one prove that it is true using the arithmetical properties of the integers, say? The answer is that, just as the parallel postulate of Euclidean geometry cannot be derived from Euclid's other axioms and postulates, the principle of induction cannot be derived from the field or arithmetic axioms. It must be taken as an additional assumption. It can be shown to be equivalent to other principles such as the well-ordering principle, which states that every subset of the positive integers has a first (least) element. However this doesn't put us any farther ahead. The fact is that the principle of mathematical induction is an assumption that we make about the number system. We make it because it seems reasonable to do so and because it is an invaluable tool. It enables us to prove a vast array of interesting mathematical statements that we feel *must* be true.

It is interesting that the last sentence implicitly differentiates between what is true and what can be proved. Of course this is the content of the famous incompleteness theorem of Gödel. Is the truth of a mathematical statement dependent on our being able to prove it? Could something be true even if we could not *in principle* prove that it is true. In the case of our above example we could prove the formula directly. Suppose

$$S = 1 + 3 + 5 + \cdots + (2n - 1)$$

Then

$$S = (2n - 1) + \cdots + 5 + 3 + 1.$$

Adding up the columns, we see that $1 + (2n - 1) = 2n$, $3 + (2n - 3) = 2n$, etc. Thus the sum $2S = n \times 2n = 2n^2$, and so $S = n^2$ as required.

However, suppose that we had no direct way to obtain this result and the use of induction was the only way to prove it. Are there such formulas, and is it reasonable to claim that they are valid? In one sense we *know* that they are true. In another the *proof* of their validity depends on an assumption about infinity, on an axiom that enables us to work with infinite sets in a systematic manner. Thus a wedge has been inserted between proof and truth. I shall return to this discussion later on.

Paradoxes of Infinity

The notion of infinity is inherently paradoxical. Thus the use of infinity in mathematical or philosophical arguments naturally gave rise to many of the famous paradoxes of the past. I shall investigate a number of these paradoxes in some detail. Each case history will certainly teach us something both about the idea of infinity and about the process whereby human thought continually reaches beyond the constraints inherent in any particular cultural epoch. The notion of infinity can be considered to be a diamond with many different facets. Different paradoxes will illuminate different facets of the diamond.

Paradox #1: Infinity Is Not an Object

Every "it" is bounded by others. "It" exists only by
being bounded by others; But when "Thou" is spoken,
there is no thing. "Thou" has no bounds.
—Martin Buber

I protest above all against the use of an infinite quantity
as a completed one, which in mathematics is never
allowed. The Infinite is only a manner of speaking.
—Carl Friedrich Gauss[11]

As soon as you mention the word infinity, you run into the problem of whether it is a unique idea or a concept like other concepts. This is related to the fundamental paradox of infinity. Can infinity be conceptualized? This is a true paradox. On the one hand, it is impossible. To conceptualize infinity is to make it a concept *commes les autres*, to remove precisely the quality that makes infinity infinite. On the other hand, it *must* be conceptualized. If ever there was something that was worth talking about, it is infinity. This is a situation that goes beyond mathematics. It is restatement of a basic human dilemma: the imperative to go beyond our finite limits, to reach for the impossible, to transcend death itself. It will be interesting to look at how the Greeks managed the dilemma of infinity. In the end they managed to come up with a very rational compromise: potential infinity, yes; actual infinity, no. In other words infinite processes were acceptable, infinity as an object was unacceptable.

In my earlier discussion of the number zero I mentioned Greek culture's failure to come up with the abstract notion of zero. I attributed that failure to the Greek idea that concepts should correspond to existent objects and to their worship of harmony and logical consistency. If "zero" was not deemed to be a suitable object of thought, then "infinity" should have been even more problematic. Yet the story of infinity in Greek mathematics is a complex one. On the one hand, Aristotle rejects the notion of dealing with infinity as a completed object of thought. On the other hand, there is a wide-ranging and ingenious use of infinite processes by Greek mathematicians, the use of "potential infinity."

Potential infinity arises in association with the "Method of Exhaustion." This method was introduced into Greek mathematics by Eudoxus of Cnidos. The method of exhaustion is used by Euclid in the *Elements*, but it is used most systematically and brilliantly by Archimedes to arrive at his amazing results about the areas and volumes of geometric figures. In essence the method involves creating an infinite sequence of approximations to a geometric figure in which you are interested. For example, suppose we wished to calculate the area of a circle C, $A(C)$. We would inscribe a series of polygons inside the circle. Let us say that $P(1)$ was a square, $P(2)$ a polygon of 8 sides, $P(3)$ a polygon of 16 sides, and so on. Notice that each of these figures has an area that approximates the area of the circle more and more closely. Outside the circle we would circumscribe another series of polygons, which we could call $Q(1)$, $Q(2)$, $Q(3)$, and so on. Given that the areas of these polygons were well known, we could now pin down the area of the circle by noting that

$$A(P(1)) < A(P(2)) < A(P(3)) < \cdots < A(\text{circle}) < \cdots <$$
$$A(Q(3)) < A(Q(2)) < A(Q(1)).$$

Of course for the argument to work we had to be sure that the area of the circle was approximated *arbitrarily closely* by the areas of the inscribed and circumscribed polygons. That is given any (small) number $\varepsilon > 0$, there was an integer n such that

127

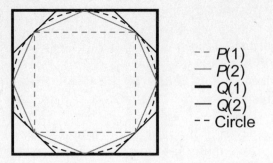

P(1)
P(2)
Q(1)
Q(2)
Circle

Figure 3.2: Approximating the circle by polygons

$$A\ (\text{circle}) - A\ (P\ (n)\ ) < \varepsilon$$

and ·

$$A\ (Q\ (n)\ ) - A\ (\text{circle}) < \varepsilon.$$

Euclid uses this method in Book XII, when he proves:

> **Proposition**: *The area, A, of a circle with circumference C and radius r is equal to the area, T, of the triangle with base C and height r, that is, $rC/2$.*

(This is essentially the usual formula for the area of the circle $A = \pi r^2 = (r)(\pi r) = (r)(C/2)$.

> **Proof**: The proof involves assuming that $A > T$ and arriving at a contradiction, then assuming $A < T$ and arriving at another. This was called the method of "double reductio ad absurdum." (It is interesting that once again infinity—here in the guise of a sequence of approximations—is associated with a sophisticated proof by contradiction.)
>   Let us just go over the first part of the proof. Assume that $A > T$. Using the method of exhaustion there must be some interior polygon, $P\ (n)$, whose area is so close to the area of the circle that $A\ (P\ (n)\ ) > T$. Now Euclid had already established a formula for the areas of regular polygons, it was $A\ (P\ (n)\ ) = sD/2$, where s was the length of a perpendicular drawn from the center to the middle of the opposite side and D was the length of the circumference of $P\ (n)$. Thus $1/2\ sD > T = rC/2$. However, it is clear from figure 3.3 that

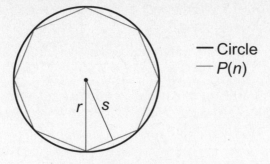

Figure 3.3: $A > A(P(n)) > T$

$s < r$ and $D < C$. This is a contradiction and establishes that $A \leq T$. In the second part of the argument we assume that $A < T$, use a circumscribed polygon and get a similar contradiction. Thus we are able to conclude that $A = T$ as required. ∎

The method of exhaustion as it is used above is really a method of successive approximation. Archimedes used this method to arrive at the pinnacle of Greek mathematical thought. For example, he used this method to find numerical approximations to that enigmatic and elusive quantity π. It also brought Archimedes very close to the invention of the calculus, since integration is in a sense a systematic application of this method.

In all of this work one was skirting on the edge of paradox and contradiction. This was noted by many Greek thinkers (notably Aristotle), who went to great lengths to keep themselves on what they would consider the rational side of the infinite. Let us look again at the sort of approximation that was used in the method of exhaustion.

The method of exhaustion consisted in constructing an approximation, K, to an unknown quantity, U, that one wished to investigate. Now a physical approximation is a finite thing, that is, given a margin of error, $\varepsilon > 0$ (which could be $1/10$ or $1/1000$) we would construct the known quantity U so that it would satisfy the inequality

$$|K - U| < \varepsilon.$$

However, our argument involved constructing a *mathematical* approximation. A mathematical approximation is really a sequence of approximations that get better and better: $K_1, K_2, K_3,$

$K_4, \ldots$ This means that for *any* positive number ε you can find an approximation K_n, such that

$$|U - K_n| < \varepsilon.$$

You might say that the value of the unknown quantity, U, is completely determined by the sequence of approximations. Thus, for example, the value of 1/3 is completely determined by the series of approximations

$$\frac{3}{10}, \frac{33}{100}, \frac{333}{1000}, \ldots$$

The Greeks believed that their reasoning was on solid ground as long as they did not consider the sequence of approximations as constituting one completed mathematical object. You will notice that in my description of a mathematical approximation above it was not necessary to use the word infinity or infinite. The quantities that were dealt with were all finite, and in any situation one would only need to go through a finite number of steps in order to arrive at the needed approximate value. The fear of the paradoxical nature of infinity was so great that mathematicians for another two thousand years would maintain that infinity was merely a way of speaking and therefore not an intrinsic part of the mathematics being discussed (just read the statement of Gauss with which this section was introduced).

It is perhaps due in part to the residual hesitation of dealing with the notion of infinity that accounts for the eventual victory of the modern definition of limit. Suppose for example we wish to say that the limit of the sequence of numbers; 1, 1/2, 1/3, 1/4, ... is 0. That is we wish to say that this sequence of numbers *"tends to zero as n tends to infinity."* What we have italicized is the way we talk about this limit to this day. We *talk* about infinity and we use infinity in our notation:

$$\lim_{n \to \infty} \frac{1}{n} = 0$$

However, when we rigorously define what we mean it is in the following way:

For every number $\varepsilon > 0$ there is some integer N such that $1/n < \varepsilon$ whenever $n > N$.

Notice that the formal definition has excised all reference to infinity. This is what Gauss meant when he said that, "The Infinite is only a manner of speaking," Thus the Greek ambivalence about infinity has left a considerable mark on mathematics.

It is also interesting to point out again how deep and significant results in mathematics occur, not by keeping as far away as possible from the contradictory and paradoxical, but, on the contrary, by seeming to work right up to its edge. This is a phenomenon that we shall observe over and over again.

Remember that at this stage we are still very far from being able to write (and think of) $1/3 = .3333. . . .$ This seemingly small step from a series of approximations to the notion of the infinite series would have to wait for a conceptual breakthrough that involves seeing through the profound taboo that had grown up around the paradox of infinity. Before we get to that breakthrough, I will discuss a related Greek paradox.

PARADOX #2: CAN AN INFINITE PROCESS GIVE A FINITE RESULT? ZENO'S PARADOX AND INFINITE SERIES OF POSITIVE TERMS

Zeno of Elea proposed some of the most influential paradoxes of antiquity. Bertrand Russell claimed that "Zeno's arguments, in some form, have afforded grounds for almost all the theories of space and time and infinity which have been constructed from his day to our own."[12] We actually know little of Zeno except what we read in Plato's *Parmenides*. Evidently Zeno proposed a series of paradoxes in order to support the philosophical positions of Parmenides. "Parmenides rejected pluralism and the reality of any kind of change: for him all was one indivisible unchanging reality, and any appearances to the contrary were illusions, to be dispelled by reason and revelation."[13] Zeno attempted to defend Parmenides' counter-intuitive thesis against his critics by showing that the denial of his views led to absurdities. In the case of Achilles and the Tortoise, which I shall discuss below, he attempted to demonstrate that if motion was infinitely divisible then it followed that nothing moved. Thus Zeno appears to be attacking commonsense notions of plurality and motion.

In light of our discussions about contradiction and paradox, it is interesting that Zeno's arguments have the form of *reductio ad absurdum*, an argument that is very close in spirit to the mathematical "proof by contradiction." Even though we shall give a "resolution" of the paradox in terms of modern mathematics, it would be wrong to underestimate the importance of these paradoxes of Zeno. In the first place he is credited with being the inventor of this kind of argument in philosophy, in which one argued against a point of view by showing that its logical consequences were unacceptable. Of course this is an example of how reasoning processes that were used in mathematics later carried over to other subjects. It also shows how the notion of paradox is central to Greek thought. Of course the Greek position was that paradoxes must be avoided at all costs, but underlying that position is the historical fact that these paradoxes had an enormous effect on Greek and subsequent thought. Zeno was pointing to a real problem, one that is very close to problems in mathematics that took a great deal of time and effort to resolve. Again the paradox has the enormous value of highlighting a fertile area of thought.

ACHILLES AND THE TORTOISE

Achilles runs faster than the tortoise, so the tortoise starts the hundred-yard dash fifty yards ahead of Achilles. When Achilles has gotten to the tortoise's initial position, $T1$, the tortoise will have moved ahead to its second position $T2$. When Achilles gets to $T2$, the tortoise will have moved to $T3$. This will continue forever, so Achilles will never actually catch up to the Tortoise. On the other hand, we can calculate exactly how much time it takes each to run the race and it is clear that Achilles gets to the finish line first.

"Resolution"

Suppose Achilles (A) and the Tortoise (T) are running a 100-yard race. A starts at the beginning and T starts at the 50-yard line. A runs at a speed of 10 yards/second, T at 1 yard/second. Here is a table of their positions at various times:

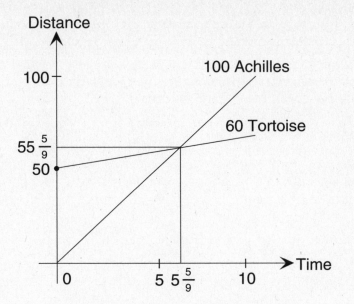

Figure 3.4. Achilles and the tortoise

Time	A's Position	T's Position
1	10	51
5	50	55
5.5	55	55.5
5.55	55.5	55.55
5.555	55.55	55.555...
5.555...	55.555...	55.555...

Thus Achilles will catch up to the Tortoise after 5.555... seconds, at which time they will both be at a distance of 55.555... yards from the start. After that time Achilles will be ahead. Now what precise time is represented by that infinite decimal? A little calculation will show us that if $x = .555$ then $10x = 5.555...$ Subtracting gives $9x = 5$ and $x = 5/9$. Thus Achilles and the Tortoise meet at time 5 5/9 seconds at position 55 5/9 yards.

With this resolution it is difficult to see what all the fuss was about. It does not involve the calculation of the time when they meet, for that is in general a simple algebraic calculation (in this case $50/(10 - 1)$). What it may involve is the assumption that time and space are both infinitely divisible. If so, then there is a problem here that is worth thinking about. What are we dividing? Is it some physical substance? Is it a mathematical substratum? Either of these creates a problem, the first with the quantum nature of the physical world, the second with the nature of the number system. Perhaps the problem lies in the relationship between the physical and the mathematical. That, too, is a deep question that is not resolved to this day.

At the very least there is the problem of an infinite process that purports to end up with a finite result. Achilles' race with the Tortoise can be broken down, as we have seen, into an infinite number of sub-races. Yet the end result is a definite time and position. We have seen that the Greeks accepted implicit infinite processes in the method of exhaustion, but resisted accepting these infinite processes as completed objects. Zeno's paradox demonstrates that it is not so easy to keep these categories distinct. When we write

$$\frac{5}{9} = .555 \ldots$$

we are mixing up an infinite process with the finite result of this process.

The ambiguity of finite objects with infinite representations is already a problem for rational numbers. This problem is compounded when we consider irrational numbers. In the case of 1/3 we may look at the infinite decimal as a series of approximations to the relatively concrete object 1/3. But in the case of irrational numbers what is it that we are approximating? The root of two exists as a length, a geometric number, by virtue of Pythagoras's Theorem. π exists in a slightly less concrete way as the ratio of the circumference of a circle to its diameter. In both cases we may imagine that our decimal representation is approximating *something*. However, most irrational numbers as we understand them today, something like .10100100010000. . . , have no name or symbolic representation that is distinct from their decimal representation. They *are* their representation. Thus the infinite decimal now is both the approximation and that which is approximated. There is a discontinuity here in our understanding; what is required is a creative insight.

Thus Zeno's paradox leads to a systematic consideration of infinite series which, in a special case, includes the theory of infinite decimals and therefore of the real number system. Thus there can be no modern mathematics and science without a resolution of this paradox. Of course, by resolution I don't mean "the solution" to the paradox. Rather, I mean discovering a way of working with the situation that is brought to the fore by Zeno's paradox. This requires opening up the paradox to discover its vast mathematical implications.

What are these implications? I shall begin examining the treasure chest of the real numbers by discussing the problem of summing infinite collections of positive numbers. Take, for example, the series mentioned earlier,

$$\frac{1}{2} + \frac{1}{4} + \frac{1}{8} + \cdots = 1.$$

What is the problem here? Well, we know how to add a finite collection of numbers. We know how to add two numbers, so we can add three by the stratagem of adding the first two and then adding the result to the third:

$$x + y + z = (x + y) + z.$$

Continuing in this way we can add any finite collection of numbers. So if we have a finite collection of numbers contained in a collection, B, we know what we mean by the sum of the elements in B, which we could write as ΣB. The question becomes what meaning can be given to the sum of an *infinite* collection, $I, \Sigma I$? Does it make sense to talk of the sum of an infinite collection as a finite number? It is a strange idea when you come to think of it. It is possible to think of summing a finite collection in a finite time (if there are 10 numbers and it takes one second to do a sum, then it should take 9 seconds to sum all 10 numbers). For an infinite collection, if each sum took the same finite time. then of course you would never get to the end of the calculation. Thus in a sense an infinite sum can never be accomplished in practice. The sum must be inferred. For example, let

$$S = \frac{1}{2} + \frac{1}{4} + \frac{1}{8} + \cdots$$

Then

$$\frac{S}{2} = \frac{1}{4} + \frac{1}{8} + \cdots$$

Subtracting gives

$$\frac{S}{2} = \frac{1}{2}.$$

So $S = 1$.

Of course one might object that this reasoning begs the question, since in the first line it implicitly assumes that the series does add up to a finite number, namely S, and only then finds the value of S. If we wish to develop a more self-contained argument, then we would actually have to start adding:

First term = 1/2
Sum of two terms = 3/4
Sum of three terms = 7/8
Etc.

We would then note that these (partial) sums are getting closer and closer to 1. Thus they form a sequence of approximations to 1 (and to no other number). In situations like this we say that the series *converges*. We are irresistibly led to the conclusion that if the sum exists then the only reasonable value for it to assume is the value 1. Nevertheless, the proof of this fact requires one to assume a property of the real number system, which can be expressed as follows:

Every increasing sequence of (real) numbers that is bounded above must approximate (converge to) a (unique) real number.

That is to say, every increasing sequence of positive terms (and in particular the partial sums of an infinite series of positive numbers) either gets unboundedly large (gets larger than any number you can mention) or converges to (approximates) some finite positive number. This is really an elementary axiom of the real number system and cannot be derived from the arithmetic or algebraic properties of numbers with which the Greeks were no doubt familiar (the so-called field properties).

Thus the paradox of the infinite series of positive terms points to an actual mathematical difficulty, one that the Greeks could not have solved for a number of reasons. They had no concept of the real number system as a whole; in fact, the real number system is vastly larger than the number systems with which they worked. The other problem was of course the potential/actual infinity paradox that was discussed above.

In fact, by identifying this question as problematic the Greeks were displaying a legitimate mathematical intuition. There *was* a problem here. The solution to the problem was not mathematical so much as conceptual. The problem is not so much solved as defined away by *assuming* a new axiom, something like "every convergent sequence of rational numbers *defines* a real number." In other words, the problem is not solved in the conventional way; rather, somehow a new mathematics is built on top of the old by taking the process/object paradox as the new *definition* of number, so that now a number is entirely defined by an approximating sequence.[14] The side effect of this definition is that you open the door for a vastly enriched concept of number. This makes possible the calculus, analysis, and, in a way, the modern world.

Thus in retrospect one can see that the problem the Greeks faced was not merely whether individual geometric numbers were legitimate numbers. It was to construct a new number system in which these new irrational numbers would be on a par with the old rational numbers. As we shall see later on, the construction of the required number system entailed the acceptance of a vast collection of these new irrational numbers. As I said above, the great majority of these numbers had no obvious geometric or analytic meaning. What we know of these numbers is that they are given by infinite decimals. Luckily one of the strengths of the decimal system is that it provides a way to distinguish between rational and irrational numbers. A number is rational if and only if its decimal representation is terminating or eventually repeating; for example $.25 = 1/4$ or $.1333\ldots = 2/15$. Thus we have a way to write down the decimal representation of irrational numbers—the decimal merely has to be nonrepeating, like $.01001000100001\ldots$, for example.

Does the Harmonic Series Converge?

Infinite series are very subtle. Even if all the terms are positive it is sometimes difficult to tell whether the series converges or diverges.[15] A case in point is the harmonic series

$$1 + \frac{1}{2} + \frac{1}{3} + \frac{1}{3} + \cdots$$

All the calculations in the world will not serve to determine whether this series converges or diverges. In fact it diverges, but it does so very slowly, that is, it takes a very large number of terms for this series to add up to a large number. Nevertheless it is possible to understand the series with a simple conceptual argument that is due to the French scholar Nicole Oresme (1323–1382). We only need to write the series in the following way:

$$1 + \left(\frac{1}{2}\right) + \left(\frac{1}{3} + \frac{1}{4}\right) + \left(\frac{1}{5} + \frac{1}{6} + \frac{1}{7} + \frac{1}{8}\right) + \left(\frac{1}{9} + \cdots + \frac{1}{16}\right) + \cdots.$$

Thus after the first bracket the sum is 1.5. The second bracket is larger than $1/4 + 1/4 = 1/2$, thus the sum at this point is larger than 2. The third bracket is larger than $1/8 + 1/8 + 1/8 + 1/8 = 1/2$, thus the sum is now larger than 2.5. We continue in this way, with each bracket contributing at least $1/2$ to the eventual sum. Thus, if we want the sum to add up to at least 10, we must add about 18 brackets. Of course this is an enormous number of terms. Since each bracket ends with the number 1 divided by some power of 2, we would need approximately 2 to the power 18 or 262,144 terms. And things get much worse very quickly. This makes the series almost impossible to calculate directly, yet the argument that it diverges is simple and convincing.

The harmonic series is itself paradoxical, but not in as dramatic a way as other paradoxes we have been considering. Here the message is that even for series of positive terms the idea of convergence is a subtle one. Series can appear to converge and yet diverge. In fact the harmonic series is on the boundary between convergence and divergence. This can be made precise. Look at the family of series of the form

$$x + \frac{x^2}{2} + \frac{x^3}{3} + \cdots,$$

where x can take on different values (this is called a power series). Then the harmonic series is obtained by setting $x = 1$. It
turns out that the series converges when x is between -1 and 1
and diverges when x is greater than 1 or less than -1. Thus the
harmonic series is close to converging but just misses.

The message here is that the behavior of the harmonic series
may appear to be paradoxical as long as it is not fully understood. Bernoulli proposed a proof[16] for the divergence of the harmonic series which amounted to assuming that the sum is A and
then showing that $A = 1 + A$. If A were finite this would lead to
$1 = 0$. Thus, by contradiction, A is infinite. This proof by contradiction is valid if we assume that series of positive terms either
converges or, as we say, diverges to infinity. In a sense Bernoulli
is working with infinity as a number. This is legitimate and
problems are avoided as long as all the terms are positive. If the
latter assumption is not satisfied then there are further complica-
tions, as we shall see below.

COMPLETION

Our discussion of infinite series of positive terms leads us to the
question of what it means for a number system to be complete.
What does completeness mean? It is a word that refers not to
individual mathematical objects but to a system, a number system, for example. It is related to the notion of "closure." A system is said to be "closed" under a certain operation if that operation does not take you out of the system. For example, the
counting numbers, 1, 2, 3, . . . are closed with respect to addition
but not to subtraction. The complex numbers are algebraically
closed. That is, if you write down a polynomial with complex
coefficients then its roots will be complex. Completion is a kind
of closure with respect to convergence. This means that every
infinite sequence of numbers that looks as though it should converge, actually does converge.

To be precise, a sequence $r_1, r_2, r_3, r_4, \ldots$ that looks like it
should converge is called a *Cauchy sequence*. This means that the
distance between the terms get closer and closer to 0 as you go
farther out in the series.[17] Does the sequence actually have a
limit? Well, that depends. If you are working in the rational

number system, then the decimal representation for any irrational number will give you a Cauchy sequence of rational numbers that does not have a rational limit. For example the irrational number .01001000100001... is really the infinite sequence of rational numbers 0/10, 1/100, 10/1000, 100/10000, 1001/100000, etc. All of these numbers are within 1/10 of each other, all except the first are within 1/100, all terms past the first two are within 1/1000, and so on. Thus they form a Cauchy sequence, but the rational number system does not contain the limit, namely, the irrational number that is referred to by the entire infinite decimal.

Thus one can pose the question: starting with the rational numbers can one create a system that contains all the numbers that correspond to the limits of Cauchy sequences? This is called "completing the rational numbers." This is indeed possible, and what you end up with is the real number system. It is interesting that you cannot go through this procedure a second time and get anything new. If you take Cauchy sequences of real numbers, then the limits exist and are all real numbers. In this sense the real number system is complete but the rational number system is not. The subject that we call mathematical analysis is best done in the context of a complete system.

Thus the Greeks' problems with infinity would not be resolved until the rational number system was successfully completed. This entailed creating a new system of numbers and considering it as one set or object. This of course would have brought them up against the notion of "actual infinity." This again involved working with the paradoxes of infinity. As long as one is considering the nature of individual irrational numbers like the various roots, the golden mean, and π, the notion of completion does not really come up. The construction of the real number system required facing up to a number of different paradoxes related to numbers and infinity.

Paradox #3: Series with Different Sums

As infinite series go, the simplest to understand are series all of whose sums are positive. After the initial epistemological obstacle is overcome, namely, that an infinite series could converge at all, then the only problem with series of positive sums that distinguishes them from ordinary sums is the fact that the sum of such a series, one like $1 + 1 + 1 + \ldots$, could get infinitely large. In such a case we shall say that the series diverges. But even then things work out the way you imagine they should. For example, if you add a convergent series to a divergent series then the resulting series diverges, as it should if we imagine that the rules of extended arithmetic should work: $N + \infty = \infty$.

What happens if we mix up positive and negative terms in our series? Suppose $A = 1 - 1 + 1 - 1 + 1 - 1 + \cdots$. Then, grouping the terms two by two, we have $A = (1 - 1) + (1 - 1) + \cdots = 0 + 0 + 0 + \cdots = 0$. On the other hand grouping them in a slightly different way gives us $A = 1 + (-1 + 1) + (-1 + 1) + \cdots = 1 + 0 + 0 + \cdots = 1$. So have we shown that $1 = 0$? How is this possible?

The problem here is that the infinite series $1 - 1 + 1 - 1 + \ldots$ cannot be said to have a sum. So writing the sum of the series as the letter A is misleading. Using the letter A assumes that the series has some fixed value, that you can assign a precise meaning to the sum of the series. This is in line with the ambiguous nature of the notation for infinite series, where the same symbol stands both for the formal series and for the sum of that series. However, the ambiguity is only fruitful when the sum exists. Thus there is another implicit ambiguity in the notation: the notation for the series stands for the sum *only when the sum is meaningful*. When is the sum meaningful? Well, that is what we usually mean by the series converging. However, there may still be meaning in the series even when it diverges. Thus the question of when the notation is clear is a subtle one involving no less than the entire theory of infinite series. Saying this in other words, the entire theory of infinite series has as its aim the investigation of the ambiguity contained in the notation $A = \Sigma \, a_n$.

Now certain series of alternating positive and negative terms do indeed add up to something concrete. For example,

$$1 - \frac{1}{2} + \frac{1}{3} - \frac{1}{4} + \frac{1}{5} + \cdots = \ln 2 \text{ (the natural logarithm of 2). (*)}$$

Nevertheless, the situation is tricky. The above series is actually made of two series: $1 + 1/3 + 1/5 + 1/7 + \cdots$ and $1/2 + 1/4 + 1/6 + \cdots$. Both of these series are very close to the harmonic series and for that reason both diverge to infinity. The question is, "Can we make sense out of the difference of these two series?" Well if we subtract them in the way that we have above by alternating the terms of the two series then the sum is ln 2.

However, it turns out that what is vital is the order in which the terms are taken. If we multiply both sides of the equation (*) by 1/2 we get

$$\frac{1}{2} - \frac{1}{4} + \frac{1}{6} - \frac{1}{8} + \cdots = \frac{1}{2} \ln 2. \ (**)$$

Let's add (*) and (**) term by term. We get

$$1 + \left(-\frac{1}{2} + \frac{1}{2}\right) + \frac{1}{3} + \left(-\frac{1}{4} - \frac{1}{4}\right) + \frac{1}{5} + \left(-\frac{1}{6} + \frac{1}{6}\right) + \frac{1}{7} + \cdots = \frac{3}{2} \ln 2.$$

Notice that the terms with denominators 2, 6, 10... cancel out, whereas the terms with denominators 4, 8, 12... double to give terms −1/2, −1/4, −1/6, ... Thus the new series is

$$1 + \frac{1}{3} - \frac{1}{2} + \frac{1}{5} + \frac{1}{7} - \frac{1}{4} + \cdots = \frac{3}{2} \ln 2.$$

The terms of this new series are identical to the terms of the original series (*) but in a different order. Now mathematicians of the eighteenth century were used to ignoring the order of terms and only considering their value. This is indeed valid for series of positive numbers, but here it leads to trouble! For if the series were identical to its rearrangement it would follow that their sums would be equal, that is, ln 2 = 3/2 ln 2 or 3/2 = 1! Richard Courant[18] says, "it is easy to imagine the effect that the discovery of this apparent paradox must have had on the mathematicians of the eighteenth century, who were accustomed to operate with infinite series without regard to their convergence." This is indeed a paradox and, as Courant says, the paradox points to the need to better understand infinite series with positive and negative terms.

In fact, matters are even worse than this. Not only can you rearrange the series to add up to two different sums, but this series (and others like it) can be rearranged to add up to any number at all! The argument is easy to understand when you see that these series are made up of two series that each diverge to infinity. Suppose you wish to make the rearranged series add up to 10. Start by adding on enough of the first series that the

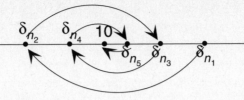

$$\delta_{n_1} = 1 + \frac{1}{3} + \frac{1}{5} + \ldots + \frac{1}{k_1} > 10$$

$$\delta_{n_2} = \delta_{n_1} - \frac{1}{2} - \frac{1}{4} - \ldots - \frac{1}{t_1} < 10$$

$$\delta_{n_3} = \delta_{n_2} + \frac{1}{k_1 + 2} + \frac{1}{k_1 + 4} + \ldots + \frac{1}{k_2} > 10$$

$$\delta_{n_4} = \delta_{n_3} - \frac{1}{t_1 + 2} - \frac{1}{t_1 + 4} - \ldots - \frac{1}{t_2} < 10$$

Figure 3.5. $1 - 1/2 + 1/3 - 1/4 + 1/5 - \ldots$ adds up to 10

sum gets to be larger than 10 (you can do this because the sums of the first series are unbounded). Then subtract enough of the second series that the combined sum is less than 10. Then add new terms of the first until it gets larger than 10 again. Continue in this way, alternating between the first and the second series so that the sums alternate between being larger and smaller than 10. If the terms of both series get closer and closer to 0 as you add more and more terms, this implies that the sum is indeed getting closer and closer to 10. So 10 must be the sum.

But then we could have picked any other number instead of 10, say 20. The same argument would show that some re-arrangement of the initial series would add up to 20. Thus we could make the series add up to whatever number we choose. Thus we have proved:

Theorem: *An alternating series that is composed of two divergent series whose terms tend to zero can be rearranged so that its sum is any given number.*

Thus the paradox is explained. The problem was in imagining that ΣB always represents a definite number. In the discussion about series of positive terms we got into trouble by ignoring the fact that ΣB could be infinite, but this was a small mathematical problem because in some ways an infinite sum of positive terms can be treated the same way as a finite sum. For example, series of positive terms (finite or infinite) have the pleasant property that any rearrangement will add up to the same number, that is, ΣB has only one value. For series of positive and negative terms this isn't true any more. When we write down an infinite series ΣB we must specify both the value of the terms and their order.

CONCLUSION

The development of the modern concept of the real number is accompanied at every stage by the paradoxes of infinity. A real number is a convergent infinite series of rational numbers and, as such, is an extremely subtle and complex idea. To begin with, every infinite series is both process and number, and this ambiguity creates both danger and possibility. As soon as the infinite is introduced the threat of incoherence arises. Then why, one might ask, is the infinite introduced at all? The answer is that the infinite is *forced* on the mathematical mind as it grapples with its attempts to describe the world. The power of theoretical explanation is somehow tied up with the concept of infinity. However, this power is accompanied by a threat, a threat that is controlled by erecting barriers of the type "infinity is not an object," whose objective is to harness the power inherent in the concept of infinity without being destroyed by the threat of chaos implicit in the situation. Yet whenever one set of barriers is erected some genius eventually figures out a way to circumvent the restrictions and get closer to the black hole of infinity without having their mathematics destroyed. Thus the threat of the irrationality of the square root of two is eventually transcended by the invention of the decimal representation for real numbers, where the process/object ambiguity of the notation becomes the modality through which the power of infinity is harnessed and put to constructive use.

A concept like infinity is never completely domesticated. Rather, different aspects of infinity are conceptualized, as we have seen in the case of the real numbers. This might lead us to conclude that the story is over, that we have arrived at the ultimate theory in the form of the real numbers as a complete ordered field. Within that theoretical framework we may convince ourselves that we understand infinity in some sort of definitive way. That would be a mistake. There are many ways to think about infinity that have not been discussed in this chapter. For example, it is possible to create a number system that is larger than the real numbers—that contains so-called "infinitesimals" or infinitely small quantities as well as infinitely large quantities. This justifies an intuition of Leibniz, for example, and his way of looking at the calculus, that was overlooked in the now conventional development of the real numbers that I have been describing. Infinity always retains its mystery—the question "What is infinity?" will always be with us in one way or another. The mystery of infinity resides in the impossibility of ever completely reducing it to a well-defined concept that fits into a rigorous theory. As long as it maintains this mystery it will be capable of producing further conceptual riches, further revolutions of thought—new paradoxes giving rise to increasingly sophisticated resolutions.

More Paradoxes of Infinity: Geometry, Cardinality, and Beyond

IN THE LAST CHAPTER I discussed the manner in which the Greeks achieved a stable notion of infinity, how that equilibrium broke down as a result of developments in mathematics such as the invention of the calculus, and the way a new equilibrium was forged. This new way of thinking —the system of real numbers—remains definitive for most mathematicians to this day. Nevertheless, as we shall see, there are other legitimate intuitions about infinity that can provide the basis for further mathematical development.

This chapter continues the discussion of infinity by considering a series of paradoxes that are associated with more revolutions in the history of Western intellectual thought. The first area to be considered is geometry, and includes the nineteenth-century assault on that holy of holies, Euclidean geometry. Most people are not aware of how problems that arose within mathematics forged not only our contemporary understanding of the world but, more important, our intuitive sense of the nature of reality itself. The contemporary movement from modernism to postmodernism is paralleled by the movement from a stable Euclidean world to a "relativistic" non-Euclidean world. In the former, the space of Euclidean geometry is identified with the natural world; while in the latter, space is something distinct from geometry, which now only "models" the natural world. This is a revolutionary shift in perception.

The second area we shall examine is set theory and more particularly the revolution that arose through the work of the mathematician Georg Cantor. Cantor's work arises directly from a new understanding of the actual infinite. Like many radical changes in the world of ideas, Cantor's work engendered virulent opposition in its time, yet its conclusions are today accepted without question in the world of mathematics. They involve the very nature of mathematical truth and its relationship with reality.

Finally we shall briefly discuss an alternate model for the foundations of analysis—a system that contains not only the usual real numbers but also quantities that are infinitely large and infinitely small. Within this system many of the considerations of the previous chapter are seen in a new light, notably, the ideas of "limits" and therefore of "derivative." In considering this "nonstandard" theory we shall see how the infinite is always capable of supplying new mathematical ideas. In many cases the development of these seminal ideas was accompanied or preceded by the emergence of the paradoxical. Each of these paradoxes flowered into a new way of understanding mathematics and the world.

PARADOX #4: INFINITY IS A POINT

Recall the postulates with which Euclid began his axiomatic development of geometry. The second was

A line segment can be extended indefinitely in either direction.

The fifth was the notorious "parallel postulate" discussed in Chapter 2, pp. 94–96.

Most authors point to the clear difference between the first four postulates, on the one hand, and the fifth, on the other. The statement of the "parallel postulate" is more complex than the statements of the others. However these earlier postulates are not themselves without their problematic aspects. In particular, consider the second postulate.

What does it mean that a segment can be "extended indefinitely"? Does it mean that given any line segment we can extend it a little further? As we extend it, can it repeat the same territory over and over again like a circle, or does the extension have to be into new territory, so to speak? Or are we thinking of an "infinite" line as a whole? Now the open interval $(0,1)$ consisting of all points between 0 and 1 has the first property of extendibility: if we take a subinterval (a,b) of $(0,1)$ then we can always extend it to the larger interval $(a/2, (1 + b)/2)$. But this kind of extendibility will not do for some of Euclid's constructions.

In a sense the problem here is the old problem of "potential" versus "actual" infinity. Is the straight line a completed (infinite) object like the decimal $.333\dots$, or is it a potentially infinite object, one for which it is possible to add a few more decimal points, so to speak? There are a number of problems with the seemingly obvious notion of the straight line—problems that really come to the fore with the advent of the Riemannian model for non-Euclidean geometry. But before discussing non-Euclidean geometries consider first of all an earlier development: parallel lines and projective geometry.

In Euclidean geometry parallel lines were defined to be straight lines that did not meet if they were extended indefinitely in either direction. Our everyday experience tells us that we can have very long pairs of lines that remain equidistant from one another (railroad tracks, for example). The Euclidean definition of parallelism captures the feeling that the essential quality here is "never meeting." The problem is, of course, with the "never" in "never meeting." It opens the door to conceptualizing the straight line as an infinite object that in turn gives rise to various stratagems for managing the problematic aspects of introducing an infinite object into a geometry that ostensibly deals with finite objects and processes.

For centuries the Euclidean definition was the one way to think about parallel lines. Yet there are no (infinite) parallel lines in the natural world, and our visual experience contradicts this definition. We "see" that equidistant lines, like railroad tracks, get closer to one another in the distance. In fact, they seem to converge toward, but never quite hit, some distant point on the horizon. In other words, the Euclidean definition refers to an intellectual reconstruction of what appears to be "really" going on and not to the primary visual experience itself.

A new way of looking at parallel lines and infinity arose in the fifteenth century, not from mathematics but from art. Artists desired to draw a more faithful representation of nature than their predecessors in the Middle Ages. The problem is, of course, how to produce the illusion of space and distance on a flat two-dimensional painting. The art of linear perspective was developed by the artist and architect Filippo Brunelleschi and per-

Figure 4.1. Piero della Francesca: architectural view of a city.
Kaiser Friedrich Museum, Berlin.

fected by such notable artists as Albrecht Dürer and Leonardo da Vinci. Paintings using linear perspective contain the following elements:

1. A *horizon line* that crosses the canvas at the viewer's eye level. Of course it represents where the sky appears to meet the ground.

2. A *vanishing point* that is located near the center of the horizon line. This point is where all the parallel lines (called "orthogonals") that run toward the horizon appear to meet like train tracks in the distance.

3. *Orthogonal lines* are parallel lines that help the viewer's eye to connect points on the edges of the canvas to the vanishing point. The artist uses them to align the edges of walls and other elements of the painting.

Of course when we say that the "orthogonal lines" are parallel we mean that they are parallel in the real scene that the painter is working from, but that in the painting these parallel lines are represented by a family of lines that are not parallel but meet at the vanishing point.

149

Naturally we are interested in the "orthogonal lines" and the "vanishing point." What the painter has done, using the artifice of this one-point perspective, is to represent a family of parallel lines by a family of lines that meet at a "point at infinity," the vanishing point. In so doing infinity has now become a tangible object; that is, the concept of infinity has been reified. It is no longer "potential infinity," a process that never terminates but merely "points to" infinity. It is now "actual infinity," an actual point.

Now one could claim that there is no "actual infinity" in the situation at all, merely a finite representation of actual infinity. In supporting the position of "no actual infinity" one would carefully distinguish between the "actual" scene where the point at infinity does not exist and the "represented scene" with its vanishing point. The "represented scene" involves portraying how the scene looks to the observer, be it the artist or the viewer of the painting. To paint using linear perspective is implicitly to include the subjectivity of the observer in the painting. The painting is not merely the objectively rendered scene but the scene painted from a certain viewpoint—the painted scene now includes the eye of the observer. Thus the price of creating an accurate rendition of the "scene as seen" requires one to include the observing eye, and it is to this eye that the "vanishing point" and the "orthogonal lines" make reference.

Even if one could still claim that "actual infinity does not exist," one has made an enormous stride toward working with geometrical infinity as something concrete. It is not so surprising, therefore, that mathematics quickly came up with a geometry that describes this new situation. This new geometry is called *Projective geometry*, because it studies the properties of geometric figures that remain unchanged under projection. "Projection" refers to those hypothetical rays of light that would join a point in the actual scene that is being painted to the corresponding point on the two-dimensional canvas. It follows from what was said above that parallel lines are *not* preserved under projection. But what about that line that we called the "horizon" where parallel lines seemed to converge?

Projective geometry takes the step of considering the horizon as an ordinary line, a "line at infinity." Every family of parallel lines now meet in a point, a "point at infinity" on this line.

Whereas in Euclidean geometry nonparallel lines meet in one point but parallel lines do not meet, in projective geometry *every* pair of lines meet at one point.[1] Thus all pairs of lines are now on an equal footing, so to speak. And not only do all pairs of lines determine one unique point but there is a unique line that joins any two points. As far as projective geometry is concerned, points and lines are *dual* to one another: in any theorem the words "point" and "line" can be interchanged to give you another valid theorem. This *principle of duality* was used to give rather simple proofs of theorems that were previously quite difficult.[2]

As Eli Maor points out,

The notion that a line is made up of points is so deeply rooted in our geometric intuition that no one seemed to have ever questioned it. Projective geometry, through the principle of duality, dispelled this notion: it put point and line on an equal footing, so that either could be regarded as the fundamental building block from which the rest of geometry is built.[3] Thus by breaking with tradition, projective geometry. . .[opened] the door for a flux of new discoveries which greatly enriched mathematics and affected its subsequent development. But all of this could not have happened had we not introduced, right from the beginning, the points and line at infinity.[4]

Projective geometry is not the only place where points at infinity enter into mathematics. The real number system can be extended to include either one or two points at infinity. The "two infinity" situation includes what we might call $+\infty$ and $-\infty$ and makes the real line into a closed interval that is equivalent in some sense to $[0,1]$. In the "one infinity" situation the same infinite point is obtained by going either right or left along the real line. It amounts to adding one point to the real line, which makes it into a circle (imagine taking a line and identifying the two ends, as in figure 4.2). Within the extended system with one infinite value, the statement that the function $f(x) = x^2$ tends toward infinity as x tends toward infinity is no more mysterious than saying that the function approaches the value 4 as x gets close to 2.

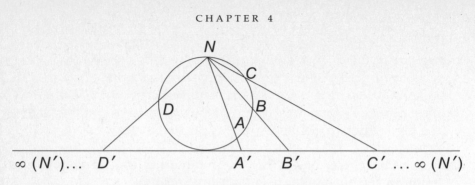

Figure 4.2. One point of infinity

The complex numbers have a geometric interpretation as points on the two-dimensional plane, where the complex number $2 + 3i$ is identified with the point in 2-space with coordinates $(2, 3)$. It is conventional to add one complex point at infinity to this system. In this manner we create what is topologically a sphere—the so-called Riemann sphere (Figure 4.3). The way this works is that one places an ordinary unit sphere on the plane in such a way that the equator lies along the unit circle, the northern hemisphere is above the plane, and the southern hemisphere is below. One then draws lines from the North Pole to points on the sphere. Each such line hits the plane in exactly one point, and each point on the sphere is identified with the point in the complex plane that is on the same ray emanating from the North Pole. Notice that under this mapping, which is called "stereographic projection," points in the southern hemisphere will project to points on the complex plane inside the unit circle, whereas points in the northern hemisphere will project to points outside the unit circle. If we follow a ray in the complex plane starting from the origin, we can see that the corresponding curve on the sphere will approach the North Pole. Looking at things on the sphere, the phrase "going to infinity" now has the concrete interpretation of "going toward the North Pole." In this way "infinity" becomes just another complex number and can be handled pretty much in the same way as any other complex number. Therefore, a complex function can be "continuous or differentiable at infinity." It can have a "singularity at infinity" or a "pole at infinity" and so on.

The process of adding "ideal points" representing infinity is well established in mathematics. There are usually various ways

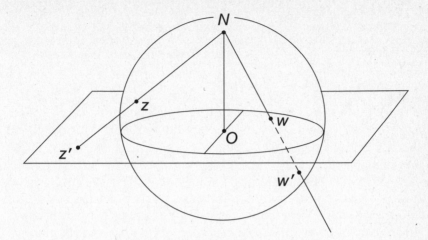

Figure 4.3. Stereographic projection

of constructing these "compactifications" of spaces. As we saw above, the real line can be compactified by adding either one or two infinite points. The complex plane has one infinite point—the so-called one-point compactification but the plane itself can be compactified in other ways—by adding a "circle at infinity," for example. Different compactifications are useful in different theories, but all involve the geometric reification of infinity, making that vague and mystical idea of infinity into one or many concrete points in a geometric, analytic, or topological space.

Paradox #5: Is Geometry Real?

For thousands of years Euclidean geometry was the definitive mathematical theory—the theory of space. The expression "the geometry of space" contains the reasonable assumption that there is something real out there called "space" and that Euclidean geometry is a way of discovering and codifying the preexisting properties of that space. However, there is another position that one might take: that the Greek geometers *created* space, that they created what we now call Euclidean space. Recall the Greeks' reluctance to accept concepts like zero or infinity that did not correspond to objects that they considered real. So it is reasonable to imagine that the Greeks did not think of their geometry as creating space, nor was their geometry a theory of space; for them Euclidean geometry *was* space. This identification between space and Euclidean geometry was broken with the advent of non-Euclidean geometries. This break was one of the most significant intellectual revolutions of all time. In many ways it marks the beginning of the modern era.

It is worthwhile to go back and spend some brief time looking at some of the problematic aspects of Euclidean geometry. These involved the seemingly straightforward idea of "straight line" and, of course, "parallel lines." These ideas contain unsuspected subtleties that gave rise to paradoxes and inconsistencies that ultimately led to the creative leap to new geometries, new ways of looking at the relationship between geometry and the natural world, and ultimately new ways of thinking about the nature of mathematics itself.

I said earlier that Euclidean geometry was responsible for our notion of space. It was not so much that this geometry described space, as it was that the nature of space itself is inseparable from Euclidean geometry. For millennia after the Greeks no distinction was made between Euclidean geometry and the real world ostensibly described by that geometry. The theorems of Euclidean geometry described the very nature of physical reality. You could say Euclidean geometry *was* the geometric structure of the natural world. Today we might say that the "was" in the previous sentence refers to the metaphoric nature of this relationship. There are those who would say that the world is not an objective entity that comes into a relationship with a subjective entity, namely our minds, but rather our metaphoric constructions determine the structure of the world that we experience.[5] If this is the case, then we can understand the power of the Euclidean model. Newton, for example, believed that the world existed within an absolute three-dimensional Euclidean space and all the events of the universe happened within that container.[6] Even today we "feel" that we live in a Euclidean three-dimensional space even though the theory of relativity places us in a curved four-dimensional space-time continuum.

This belief in the identity between Euclidean geometry and the structure of the physical world seems strange to us today, especially considering what has been said about straight lines and the parallel postulate. Indeed it was in the context of the parallel postulate of Euclid that the absolute nature of this structure came tumbling down. The reader will recall that there had been a long history of doubting the status of the parallel postulate. There was a general feeling that it was of a different nature from the other postulates. It seemed more like a theorem than an axiom, and the feeling was that it should be possible to derive it from the other Euclidean axioms. An incredible amount of mathematical ingenuity was devoted to this attempt. Mathematicians succeeded in showing that there were many equivalent formulations of the parallel postulate. One, which goes under the name Playfair's Postulate, stated that, "Given a straight line, L, and a point, P, that is not on L, it is possible to draw a unique line through P that is parallel to L." This not only seems reasonable, but given a piece of paper and a pencil any student in high school can actually draw the required line.

What, then, was the problem? Remember that Euclid claimed that all of his results followed from the axioms and the application of standard logical reasoning; the results should not depend on the particular pictures that were drawn. The theorems of Euclid should be true for any system, for example, any geometric system that contained classes of objects that satisfied the axioms. In particular, straight lines did not have to be "straight" but could be any collection of curves that satisfied the axioms. Looked at in this abstract light Gauss, Nikolai Lobachevsky, and Janos Bolyai produced geometries that satisfied all the Euclidean axioms except one—the parallel postulate. These were the famous hyperbolic and elliptic non-Euclidean geometries. For example, in the elliptical example due to Riemann, which takes place on the surface of a sphere, "straight lines" are replaced by great circles. Great circles are the routes taken by airplanes on long voyages. They are formed by cutting the sphere by a plane that goes through the center of the sphere. They are geodesics, that is, routes that minimize distance. Any two such circles must meet in two points, and so such a geometry has no parallel lines.

It is difficult to go back in time and appreciate the consternation that was evoked by the realization that Euclidean geometry was merely one geometry in a whole family of possible geometries. Today we have no difficulty accepting the "relativity" of knowledge in the sense that scientific theories are models of reality that are valid for a certain range of phenomena but may always be changed for better and more exact models. We are in the post-Euclidean world that arose precisely out of the crisis of non-Euclidean geometry. However, before this revolution, Euclidean geometry *was* the natural world. The philosopher Immanuel Kant even claimed that we are born with an a priori knowledge of the Euclidean world—it is hard-wired in, so to speak. There could be lacunae in Euclidean geometry, but these did not matter very much because Euclidean geometry was backed up by the nature of physical reality itself. They were the same thing.[7] However, with the advent of non-Euclidean geometries, a huge and unsettling question naturally arises. If these geometries are models of the reality of physical space, how can we come to know this reality? Is it only available to us through the filter of some geometric model and so essentially unknowable in itself? Or, more radically, is there no reality at all and we merely create the idea of space and thus our experience of it through our adherence to some particular model?

In the light of non-Euclidean geometry the world became much more complicated. Doubt had entered the mathematical universe. This was the beginning of the loss of the certainty that mathematical theories had seemed to provide to human beings' investigation of the world.[8] There were now mathematical theories that applied to certain situations and not to others. The axiom systems had to be tight enough to ensure validity on strictly logical grounds and not through reference to a particular model.

In an attempt to get mathematics back on firm ground, there arose a new philosophy of mathematics. This was formalism, the belief that mathematics consisted of deriving logically necessary results from sets of consistent axioms. From the point of view of the strict formalist, mathematics was a kind of game with no meaning. The crucial thing was to feel secure that no errors were being introduced into the reasoning process. Perhaps it would one day be possible to program computers to reason in this way and thus to produce mathematical theorems. Perhaps mathematics was algorithmic. This was a powerful dream and it dominated the thinking about mathematics for many years.

It is now believed that the great mathematician Gauss was the first person to realize that non-Euclidean geometries were possible. However, he refused to publish his results, perhaps realizing what a furor this would cause. From the current state of mathematical knowledge it is difficult to understand why anyone would be bothered by yet another mathematical theory. To understand what the fuss was about, we must see the paradox inherent in non-Euclidean geometries—not in the results but in the very existence of these geometries. It was as though a whole culture was built around the idea that Euclidean geometry was the way things were and there could be no other way of looking at the natural world. Then, all of a sudden, there was another way of seeing. Faced with this paradox, with this impossibility, something fundamental has to give way. What dies is a particular worldview and what is born in a rush of creativity is a whole new world, a new and richer understanding of mathematics and its relationship to the natural world.

Cantor's Revolution: "Counting" Infinity

Now I shall move on to consider a completely different way in which infinity came to be used in mathematics. I stated in Chapter 3 that there is an inevitable incompleteness about the concept of the infinite—"infinity" is open-ended. The infinite has been incorporated into mathematics in various ways; each such attempt is important and has something to teach us but none of them are definitive.

Recall that by the nineteenth century a consensus had been attained with respect to the "correct" way to understand and use infinity in mathematics. This equilibrium position was initiated by the Greeks and lasted for more than two millennia. The consensus, you will recall, involved distinguishing between process and product. Infinity as process, the so-called potential infinity, was acceptable to the Greeks, but infinity as object, "actual infinity," was anathema. The notion of actual infinity, as I pointed out in the last chapter, is implicit in the foundations of analysis, in particular in the decimal representation of real numbers. This really comes to the fore with the question of the nature of irrational numbers, especially those that have no familiar representation other than as decimal numbers. If you don't believe in actual infinity it is hard to believe in the reality of irrational numbers.[9] These questions came to a head in the work of Georg Cantor, born in St. Petersburg, Russia in 1845. Cantor gave the world a new way in which to work with the notion of actual infinity based on the theory of sets. In his theory "infinity as object," that is, infinite numbers became as legitimate parts of the mathematical firmament as the fractions, for example.

Cantor's work is permeated with a mathematical richness that, in most cases, is associated with counterintuitive, not to say paradoxical elements. The underlying tension that generates many of these paradoxes is that Cantor's approach arises from the experience of "counting," and therefore his definitions are often incompatible with other intuitions about infinity that arise from different areas of experience—measuring, for example. However, the many paradoxes that arise out of Cantor's approach will give us a point of entry into the mathematical theory.

PARADOX #6: CAN THE PART EQUAL THE WHOLE?

The story begins not with Cantor but many centuries earlier with one of the mathematical insights of Galileo. Galileo believed that it was given to man to understand the secret workings and regularities of nature. However, he believed that these secrets were not visible for all to see. They were encoded, so to speak, and the code with which nature guarded her secrets was mathematics. He said, "Philosophy [science] is written in that great book which lies before our eyes; but we cannot understand it if we do not learn the language and characters in which it is written. This language is mathematics, and the characters are triangles, circles, and other geometric figures."[10] Thus it is interesting that Galileo should be the author of the following paradox. What he made of the fact that mathematics, the language of nature, was capable of producing such anomalies we cannot be certain. However, he certainly managed to put his finger on a problem that would not be fully resolved for hundreds of years.

Galileo's paradox involves the sequence of square numbers:

$1, 4 = 2 \times 2, 9 = 3 \times 3, 16 = 4 \times 4, 25 = 5 \times 5, 36 = 6 \times 6$, and so on.

Of course the collection of such numbers is infinite. Now compare the collection of square numbers to the collection of counting numbers as shown in figure 4.4a, b.

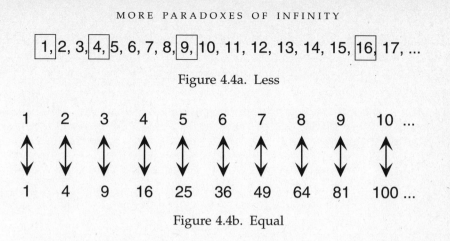

1, 2, 3, 4, 5, 6, 7, 8, 9, 10, 11, 12, 13, 14, 15, 16, 17, ...

Figure 4.4a. Less

1	2	3	4	5	6	7	8	9	10 ...
1	4	9	16	25	36	49	64	81	100 ...

Figure 4.4b. Equal

On the one hand, the collection of squares is "smaller" than the collection of counting numbers since there are many numbers that are integers but are not squares. In fact the collection of squares is clearly a "part" (or subset) of the collection of all of the natural numbers. On the other hand, every square can be matched up with the number of which it is the square; 1 with 1, 4 with 2, 9 with 3, 16 with 4, 25 with 5, 36 with 6, and so on. In this matching all the square numbers as well as all the natural numbers are accounted for, and each is counted exactly once. We are led to the conclusion that there are exactly as many squares as there are natural numbers. This is the paradox. Which is it to be? Is the number of squares "equal" to the number of counting numbers, or is the number of squares "smaller" than the number of counting numbers?

The problem here lies with the idea of "equal" and the related ideas of "larger" and "smaller." The paradox is rooted in the ambiguity of the notion of "equality." There are two contexts at play. The first is set theoretical: two sets are equal if and only if they have identical elements. Thus the set of square numbers is not equal to the set of counting numbers. The second context involves "matching": two sets are "equal" from this point of view if their elements can be perfectly matched up in what we call a one-to-one correspondence (figure 4.4b).

The paradox can be stated as "the whole (the counting numbers) is equal to the part (the square numbers)." Now remember that the fifth "common notion" of Euclid was precisely that "the whole is greater than the part." This statement seems so obvious that it appears to require no explanation; it seems to be true be-

159

cause of the very definitions of the words. "Part" means smaller than "whole" if they are referring to the same object, does it not? Galileo's paradox puts this self-evident notion into question.

Of course there is a nonscientific, mystical tradition that appears to say something else about parts and whole, and what it says is related to the idea of the infinite. Recall the famous lines of William Blake:

> To see the world in a grain of sand,
> And heaven in a wild flower,
> Hold infinity in the palm of your hand
> And eternity in an hour.[11]

This seems to say that the part is equal to the whole in some way. The modern sophisticated person might scoff at the ancient notion that the whole of reality is somehow enfolded in every segment of reality, no matter how small. However, this idea has emerged in many ways in modern science, from the holograph to the mathematical theory of fractals. In the latter, for example, the notion of self-similarity is introduced. This means that the pattern of the fractal is replicated exactly by arbitrarily small pieces of the fractal. The whole here is indeed identical to the part, albeit on a smaller scale. To use a more mathematical term one might say that the part is *isomorphic* to the whole, where, for now, isomorphism is to be understood in the sense that some crucial structural features of the situation that we are interested is common to both situations. This notion of isomorphism is very powerful and basic and I shall return to it in Chapter 5.[12]

Galileo's paradox is ultimately resolved by a breakthrough—the invention of a new way to think about infinity. This new idea is interesting because it involves going backward, so to speak, in the direction of more elementary ideas in a manner that is reminiscent of Halmos's comment that the questions of the infinite are resolved by solving "finite" problems. In this case it involved thinking carefully about a very basic idea—the idea of "counting." What is counting? It involves "matching" the objects you are counting and the "counting numbers," namely, 1, 2, 3, 4, etc. The last number in the sequence is the number of objects in the collection that you are counting. Counting therefore involves matching the collection that you intend to count against a standard reference collection, the natural numbers.

Figure 4.5. Counting as a one-to-one correspondence

Counting is the informal term for certain kinds of one-to-one correspondences. (A "one-to-one correspondence" is the technical name that we give to the normal idea of "matching.")

Now what is more elementary "matching" or "counting"? We have seen that counting involves matching, but is it possible to match without counting? The answer is, yes, of course it is possible. Suppose that two children both have large collections of marbles but cannot count past three, say. Can they decide whether they have the same number of marbles? Certainly they can. They just need to match the marbles in one collection with the marbles in the other. If all the marbles in both collections get matched up with none left over, that is, if there exists a one-to-one correspondence between the two collections, then the children have the same number of marbles.

It follows that it is possible to determine the "cardinality" of sets using only the notion of matching. Two sets of objects have the same cardinality if their elements can be matched with one another in such a way that nothing in either collection is left out. The stroke of genius was to apply this idea—one that is self-evident for finite sets—to infinite collections. Cantor's genius also involved sticking to his intuition that this was the correct way of thinking about the size of sets in the face of the seemingly bizarre results that followed from it. And one of the bizarre results was that the whole could now be "equal" to the part. Of course "equal" now took on the precise meaning of "can be matched up with" or "has the same cardinality."

The same idea of the equivalence between the whole and the part forms the basis for the paradoxical tale that is due to David Hilbert (1862–1943). It is the story of an "infinite motel" (actually Hilbert talked about a hotel, not a motel). This story is constructed in such a way as to bring out strongly the manner in

161

which an infinite set can be said to have the same size as a proper subset of itself. It goes as follows.

There is a motel that has an infinite number of rooms that are numbered in order: room 1, room 2, room 3, and so on. Suppose that each of the rooms has somebody in it; that is, there are no vacancies. Now suppose that a new customer arrives at the desk and asks for a room. "No problem," says the ingenious clerk. "We can move the person in room 1 to room 2, the person in room 2 to room 3, and so on. Every person will move to the next room and this will leave room 1 vacant and the new person can move into this room." So the question is, was the motel originally full or not? If it was originally less than full, then just where were the empty rooms? If it was full, then how was there room for another person? These are the considerations that make the story of the infinite motel paradoxical. (Of course if we can make one room vacant in this way then we can make two rooms vacant by repeating the process twice. In fact we could make a million rooms vacant, and even better we could empty out an infinite number of rooms—all the odd numbered rooms, for example. Do you see how?)

The paradox of the infinite motel is also a paradox of whole and part. Somehow the same configuration of motel rooms will accommodate both the initial group of people and a larger group. Thus a smaller group would seem to be "equal" to a larger group that contains it, that is, the whole is equal to the part.

In the paradox of the infinite motel what is going on is that we are matching up the set of rooms, which we designate by the counting numbers: $R1$ (for Room 1), $R2$, $R3$, . . . with the set of guests which we designate by $G1$, $G2$, $G3$, . . . Initially the rooms and the guests match up in the obvious way: $R1 \leftrightarrow G1$, $R2 \leftrightarrow G2$, $R3 \leftrightarrow G3$, etc. Then there is the problem of the additional guest, let us call her $G0$. We create a new matching in the following way: $R1 \leftrightarrow G0$, $R2 \leftrightarrow G1$, $R3 \leftrightarrow G2$, etc. In this way every room is matched with a person with no rooms or people left out. Since both collections $\{G1, G2, G3, . . .\}$ and $\{G0, G1, G2, G3, . . .\}$ are matched with the set $\{R1, R2, R3, . . .\}$ they have the same cardinality. In a sense this is saying that as far as cardinality is concerned infinity is equal to infinity plus one.

Set equality is not the same as cardinality equality. Two unequal sets can indeed be matched. It turns out that *every* infinite set has the property that it can be matched perfectly with some (proper) subset. In fact this property could be taken to be the defining property of infinite (as opposed to finite) sets. A set is infinite if and only if it has the same cardinality as one of its subsets. What has happened here is that our paradox, "the whole equals the part" has now become the defining property of the infinite.

COUNTABLE INFINITY: INFINITE SETS THAT CAN BE LISTED

Countable infinity is the cardinality of the counting numbers. It is usually represented by the symbol $\aleph_0$, aleph null. It is the infinity with which we are most familiar. In fact, a set is countably infinite if its elements can be arranged in an infinite list. The defining characteristic of a list is that it has an order; there is a first element, a second, a third, and so on. For example, the set of even integers is countable because they can be listed: 2, 4, 6, 8, ... So is the set of odd numbers. The set of all integers is countable because it can be written in the following way: 0, 1, −1, 2, −2, 3, −3, ... Now this isn't the usual way of writing the integers. The usual way is ... , −3, −2, −1, 0, 1, 2, 3, ... However, the fact that the integers *can* be written in a list makes them countable.

Notice that from the usual point of view everything we are saying is problematic, not to say paradoxical. Since the positive integers and the negative integers have cardinality $\aleph_0$, the statement that all of the integers are countable could be written as $\aleph_0 + \aleph_0 = \aleph_0$ (*). That is, if you put together two countably infinite sets (take their union), the resulting set is also countably infinite. So the arithmetic of infinite cardinal numbers does not follow the same rules as ordinary arithmetic. In fact, Cantor was able to prove that the union of a countably infinite collection of infinite sets is also countably infinite. In symbols we would write $\aleph_0 + \aleph_0 + \aleph_0 + \ldots = \aleph_0$ (**). When you stop to think about it, both of the seemingly paradoxical equations (*) and (**) are statements of "the whole is equal to the part."

163

The Rational Numbers Are Countable

Now a naive approach to the question of the cardinality of the set of all fractions would lead a reasonable person to conclude that there are vastly more fractions than there are natural numbers. A fraction is determined by a *pair* of integers and, in a sense, pairs are two-dimensional whereas single numbers are one-dimensional. I shall return to the question of dimensionality later on, but for now let's just say that it seems reasonable to think that a higher-dimensional set is "larger" and paradoxical to assert that it is not. In the case of rational numbers there are certainly many proper subsets that are copies of the natural numbers, for example, the set {(1/1), (2/1), (3/1), (4/1), . . .} but also {1/2, 2/2, 3/2, 4/2, . . .} and so on.

When one sets out to answer the question about the cardinality of the rational numbers, one begins by attempting to create a possible list. Try it for a moment. What you are doing is creating an order for the rationals with a first, a second, a third, and so on. The problem is that the rationals come equipped with a perfectly good order and that order is size, that the rational number r is less than s if $s - r > 0$. In practice we know that $10/11 < 123/131$ because 1310 (10 times 131) is less than 1353 (11 times 123). Most people would proceed to try to list the rationals using the regular order according to size. This, of course, does not work (what fraction would be first?) and might lead to the erroneous conclusion that the rationals cannot be listed, or that there are more rationals than integers.

The genius of Cantor is that he ignores the normal order and constructs a completely new way of ordering the rationals. Again we encounter the phenomenon that the insight involves focusing on the new definition of cardinality as matching and ignoring the additional structure that happens to be present.

Once discovered, the argument is very simple. There is really nothing more to it than drawing the correct picture. Consider the following array:

$$
\begin{array}{llll}
1/1 & 1/2 & 1/3 & 1/4\ldots\ldots \\
2/1 & 2/2 & 2/3 & 2/4\ldots\ldots \\
3/1 & 3/2 & 3/3 & 3/4\ldots\ldots \\
4/1 & 4/2 & 4/3 & 4/4\ldots\ldots \\
\ldots\ldots
\end{array}
$$

Order it in the following way:

$$(1/1), (1/2, 2/1), (1/3, 2/2, 3/1), (1/4, 2/3, 3/2, 4/1), \ldots.$$

Notice that in the first bracket numerator and denominator add up to 2, in the second they add to 3, in the third to 4, and so on. Notice also that every (positive) fraction appears in the list. Thus we have listed all the positive rationals. We can similarly list all the negative rationals and combine the two lists to get one master list with all of the rational numbers.

Simple but brilliant!

THE SET OF ALL "KNOWABLE" NUMBERS IS COUNTABLE[13]

The method of argument that was used to show that the set of rational numbers is countable can be used to show that the union of countably many countable sets is countable. That is, if A_k is a countable set for $k = 1, 2, 3, \ldots$, then $\cup_{k=1}^{\infty} = \{a : a \in A_k$ for some $k\}$ is also countable. If even putting together an infinite collection of infinitely countable sets only results in another countable set, one might be tempted to say that there are no other kinds of sets, that all sets are countable. This is a powerful intuition; that is, the notion of countable infinity captures something quite fundamental about the nature and limits of what human beings can know.

Suppose that we are given a finite set of symbols such as the English alphabet augmented by punctuation marks, spaces, and such additional symbols that may be needed to write down a phrase or sentence in English. Call a number "knowable" if it can be described or named by some phrase or sentence. For example, "twenty-two" is knowable, and so is every integer and fraction. The number π is knowable, since it can be described as "the ratio of the circumference to the diameter of a circle." The square root of two is knowable, since it is described by "the positive number that is a solution to the equation x squared minus two equals zero." Similarly the solution to any polynomial equation whose coefficients are integers (the so-called algebraic numbers) is knowable. Thus every number we could describe now or in the future is knowable.

So how many knowable numbers are there? Well one could consider all possible sentences in the letters and symbols of our language. Let B_1 be the set of all expressions that can be written using one letter or symbol from our language. Let B_2 be the set of expressions that can be written with two. and, in general, let B_n be the set of expressions that can be written with n letters and symbols. Now each one of these sets is countable, in fact finite, and we could put it into a list by first ordering the original set B_1 of letters and symbols and then ordering the elements of B_2 in the way they would be ordered in a dictionary—first by the first symbol and for those with the same first symbol by the second symbol, and so on until we have ordered all of the sets B_n. Then the set B of all possible expressions is equal to the union of all these finite lists. This set, by the above reasoning for the rational numbers, must also be a countable set. Now the set of "knowable" numbers arises from a subcollection of B and so must also be countable.

Paradox #7: Multiple Infinities

In the light of the previous discussion it would be entirely reasonable to doubt that there could possibly be infinite collections of numbers whose cardinality is not countable. Yet there are indeed other infinite cardinal numbers that arise in mathematics. They are right under our nose, so to speak, but to identify them and demonstrate the existence of other infinities requires a bit of genius.

If infinity is a problematic notion, how much more so is the idea of multiple infinities. After all, infinity is a defining characteristic of divinity, is it not, and the mathematics we are talking about was developed in a monotheistic religious context. There is something vaguely pantheistic, not to say "primitive," about the notion of two infinites. Isn't infinity something absolute, like beauty or excellence? There seems to be something wrong about the phrase "more excellent." Something is either excellent or not excellent. Maybe this is one of the problems that Aristotle and all those who came after him were trying to save us from by banning the use of "actual infinities." It is bad enough to deal with infinity as a well-defined mathematical concept. It is worse to have a situation of multiple infinities of various sizes. Isn't infinity by definition absolute and incomparable? Having multiple infinities seems to contradict those very properties that we would like the infinite to possess.

Nevertheless Cantor, who was now armed with a precise definition of cardinality, could legitimately investigate the cardinality of other infinite sets. Since all the sets he had so far encountered were countable, he could ask whether in fact *all* sets were countable. In fact he was able to show that:

Theorem: *The cardinality of the real numbers is not countable.*

(This new cardinal number is usually denoted by "*c*" for continuum. So the theorem proves that $c \neq \aleph_0$.)

Proof: Notice that what is required here is to demonstrate that it is *impossible* to put all the real numbers into one infinite list. As usual the only possible approach is to argue by contradiction. Assume that the real numbers are countable and therefore that it *is* possible to list all real numbers. Let us write down this list:

$$r_1, r_2, r_3, r_4, \ldots .$$

Now each r_n is a real number, so it can be written as an infinite decimal (if the decimal is finite, then just add a tail of 0s). Now our list looks like

$$r_1 = .n_{11}\, n_{12}\, n_{13}\, n_{14} \ldots ,$$
$$r_2 = .n_{21}\, n_{22}\, n_{23}\, n_{24} \ldots ,$$
$$r_3 = .n_{31}\, n_{32}\, n_{33}\, n_{34} \ldots ,$$
$$r_4 = .n_{41}\, n_{42}\, n_{43}\, n_{44} \ldots ,$$
$$\ldots$$

Remember that by assumption this list contains *all* real numbers. Of course it doesn't really—that is only our assumption. What we really want is to produce a contradiction. This contradiction will be that in fact the list *cannot* contain all real numbers. Why not? Because we shall now find a number that cannot possibly be on this list. Call this new number r and we can give a formula for determining its decimal representation. Write r as an infinite decimal using the notation

$$r = .n_1\, n_2\, n_3\, n_4 \ldots ,$$

167

so we need to give a rule for determining the digits n_1, n_2, n_3, ... It's easy; all we have to do is change all the diagonal terms (the terms in **bold**) in the infinite array above:

$$n_1 \neq n_{11},$$
$$n_2 \neq n_{22}.$$
$$n_3 \neq n_{33},$$

and so on. We could do this in many ways, but to be definite,

Let $n_1 = 5$ if $n_{11} \neq 5$ and $n_1 = 7$ if $n_{11} = 5$.
Let $n_2 = 5$ if $n_{22} \neq 5$ and $n_2 = 7$ if $n_{22} = 5$.
Let $n_3 = 5$ if $n_{33} \neq 5$ and $n_3 = 7$ if $n_{33} = 5$.

And so on. Then the number r cannot equal r_1 since its decimal representation differs from it in the first place. It cannot equal r_2 since its decimal representation differs from it in the second place. And, in general, it cannot be equal to any of the members of the list. Thus r is a new number that is not on the list. This contradicts the assumption that you could put all real numbers on the list. It follows that no list is possible, and so the real numbers are uncountable. ∎

Now this is a seminal result in the recent history of mathematics, and it is equally important for the analysis of mathematics that I am putting together. Let us start with the argument. As usual, it is an argument by contradiction. The contradiction is internal, one that depends only on the terms and ideas contained within the argument itself. In fact the argument depends on only two ideas: countability and the decimal representation of real numbers. Consider the latter. Even though Cantor is given credit for being the first person to deal with infinity as an object, in fact the proof depends on the fact that real numbers are objects with a representation as *infinite* decimals. This representation, as we have seen, amounts to reifying an infinite process. By the time of Cantor, it was acceptable to think of an infinite series as a number. In a sense, the argument for the uncountability of the real numbers builds on that accomplishment.

Why was there such resistance to Cantor's ideas and methods? I have already mentioned the long-standing prohibition against treating infinity as an object. But in addition there was the feeling that somehow Cantor was cheating. He was achieving too much, too easily. To see this, we must take a moment to look at a consequence of noncountability of the reals. Recall that the reals are made up of two kinds of numbers: the fractions or rational numbers, and the others, the so-called irrational numbers. That is, an irrational number is merely one that cannot be written as a fraction; think of the square root of two, for example. Suppose that the set of all irrational numbers were countable. Since we have shown that the rational numbers are countable and we know that the union of two countable sets is countable it would follow that the set of all real numbers was countable. However, Cantor has shown us that this is not true. This small argument by contradiction shows us that:

Corollary 1: *The set of all irrational numbers is uncountable.*

This is a result that bothered many people. You may recall from an earlier chapter that it took a considerable argument to prove that one number, the square root of two, is irrational. In the same way one can show that other roots are irrational. There are also a few other numbers that one can show to be irrational, but altogether all the numbers that one can directly show to be irrational form a countable set. Here was Cantor with one relatively simple argument proving that there were a vast number of irrational numbers, that there were, in fact, many more irrationals than rationals. This is very strange. There are all these irrationals but we do not *know* most of them in the sense that we know the rationals, the roots, or even the numbers that the Greeks could obtain through geometry, the so-called constructible numbers.

Things are even worse than this. An *algebraic number* is one that is the solution to a polynomial equation with integer coefficients. All the rationals are algebraic numbers since m/n is the solution to the equation $nx - m = 0$. The square root of two is the solution to $x^2 - 2 = 0$. Similarly all the square roots, cube roots, and other roots are algebraic. Now the set of all algebraic numbers is countable. The reason for this is that there are a countable number of equations of order n for $n = 1, 2, 3, \ldots ,$ Since the countable union of countable sets is countable, this implies that there are a countable number of such equations. Each equation has only a finite number of roots, and so there are a countable number of roots, that is, of algebraic numbers. The numbers that are left over are called the *transcendental* numbers.

How many transcendental numbers are there? That is, what is the cardinality of the set of transcendental numbers? It follows from the same reasoning we used above for the irrationals that

Corollary 2: *The set of all transcendental numbers is uncountable.*

Now this is an even stranger result. Euler was the first to speculate that transcendental numbers even existed, and Joseph Liouville in 1844 was the first to produce an example of a specific transcendental number. The most familiar transcendental number is the number π, but the proof that π was transcendental did not come until after Cantor had constructed his theory. Thus we have the paradoxical situation that most real numbers are transcendental but we are familiar with almost none of them. In fact things are even "worse" than this. Remember that we showed that the set of all "knowable" numbers was countable. Using the same reasoning, it follows that

Corollary 3: *The set of all "unknowable" numbers is uncountable.*

This is indeed a strange and paradoxical result. The set of individual numbers that can be "known" (in the sense of being described or constructed in some finite language) is countable. It must follow that the remaining numbers are "not knowable," yet we can and do "know" them as a set by proving that they collectively have properties such as those described in Corollaries 1, 2, and 3. We know these remaining numbers in one sense but do not know them in another. This is the paradoxical situation that was forced on mathematics from the time of the introduction of irrational numbers but was made more explicit through Cantor's investigations.

We have seen how the real number system—this vast expansion of the concept of number—imposes itself on mathematics. Our vaunted real number system, the basis for calculus, differential equations, and almost all of physics, is composed of a relatively small set of numbers with which we are familiar together with a much larger set of numbers with which we have no experience and of which we know little or nothing. And as Corollary

3 says, in some sense we *can* know nothing about "most" real numbers. Thus the price of constructing the real number system, and in this way resolving some of the ambiguities of Greek mathematics, was the introduction of a vast collection of "new" numbers that must inevitably retain a certain mystery and "unknowability."

Not all mathematicians were willing to make that particular bargain. They concluded that any way of thinking that could produce such bizarre results must be fundamentally flawed. But where was the mistake? There were two possible areas to attack. Perhaps the entire theory of Cantor was intrinsically illegitimate. This would be the critique of "actual infinity" as mathematical object, which we have already discussed. The second was to look at the argument itself and note that the key arguments are arguments by contradiction. Maybe there was something wrong with the kind of argument that could demonstrate the existence of an infinity of numbers of a certain type (irrational, transcendental, or "unknowable") without being able to produce a single instance of such a number. Talk about a paradox!

This paradox exists by virtue of proof by contradiction. After all, how do you say something directly about a number about which you know very little? All you know about irrational and transcendental numbers is negative. An irrational number is *not* rational, a transcendental number is *not* algebraic, and, of course an "unknowable" number is not "knowable." The only way to reason about them is negatively, that is, by contradiction. If one were to disallow such reasoning, one could avoid the seemingly paradoxical results that followed.

In retrospect, although one can sympathize with this reaction, mathematics followed the path of development. As was the case with the parallel axiom of geometry, mathematics grew to include the new theories and ways of thinking. It absorbed the aspects of the new developments that were contradictory or paradoxical at the time of their introduction. However, the infinite cardinal and ordinal numbers of Cantor are just mathematical concepts, no different in kind from other mathematical concepts. Each deep concept—I am thinking of zero, for example—changed the mathematical and therefore the human landscape. Many appeared paradoxical, or, rather, did indeed have a paradoxical aspect to them. When they became part of normal mathematics, mathematicians learned to take them for granted. Mathematics then moved on to the next frontier and forgot about the paradoxical elements of past mathematical development. It is only when mathematics is taught and teachers see the difficulties that students experience in grasping these concepts, that one becomes aware that these difficulties are real. The paradoxes of the past have become the "epistemological obstacles"[14] that students face in the present.

(0, 1) has the same cardinality as (0, 2)

Cardinal numbers give one way of measuring the "size" of a set. Using this measure, the set of all rational numbers is "smaller" than the irrationals, the transcendentals, or the reals. Of course there are other measures of size. One that was mentioned above was set inclusion—a set is "larger" than any of its proper subsets. As we have seen, one of the difficulties of working with infinite collections is that a concept like "size," which has a clear meaning in the context of finite sets, can take on multiple meanings in the context of infinite sets.

When it comes to measuring the size of line segments, the most intuitive measure is the length of the segment; the set of real numbers between 0 and 1, (0, 1), has length one whereas (0, 2) has length two. (0, 1) is thus "smaller" than (0, 2) but, surprisingly, it has the same cardinality! This is really another variation on the story of the whole and the part and could be considered as a paradox in its own right, one that is often attributed to Bernhard Bolzano (1781–1848). Bolzano was a priest, philosopher, and mathematician who did fundamental work in the foundations of analysis. He was inspired to extend Galileo's thinking concerning the one-to-one correspondence between infinite sets to the situation of intervals of numbers. Consider the function $y = 2x$. To every number x between 0 and 1 there corresponds precisely one number $y = 2x$ between 0 and 2. Looking at this correspondence from the other direction, that is, starting with y, every number y between 0 and 2 corresponds to the number $x = y/2$ between 0 and 1 (figure 4.6). Therefore, the set (0, 1) matches up perfectly with the set (0,2), that is, they have the same cardinality.

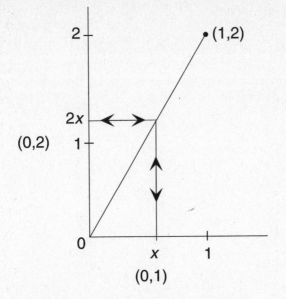

Figure 4.6. Matching (0,1) and (0,2)

The argument may seem to be very simple, but remember that Bolzano is creating a correspondence between sets both of which have an uncountable infinity of elements. Not only does 1/3 match up with 2/3 but $\sqrt{2}/2$ matches up with $\sqrt{2}$, $\pi/4$ with $\pi/2$, and so on. We are making a definitive statement about a vast collection of mathematical objects most of which are transcendental numbers and thus poorly understood.

In fact there is nothing special about these two particular intervals; any two intervals of real numbers have the same cardinality. For example, if we wish to match up (0,1) with (0,10) we need only multiply by 10 instead of 2. In general, we have:

Proposition: *All finite intervals regardless of length have the same cardinality.*

In fact the situation is even more counterintuitive. Cardinality, which was introduced to measure the "size" of infinite sets, does not distinguish between the length of finite and infinite intervals!

A Finite Interval has the Same "Size" as an Infinite Interval

In particular the unit interval (0,1) has the same cardinality as the set of *all* real numbers. To establish this fact, all that is necessary is to produce a matching between these two sets. This is easily established by the function

$$f(x) = \frac{2x - 1}{x\,(x - 1)}\,.$$

The graph of this function clearly indicates that this function matches the set {x: 0 < x < 1} on the x-axis with all the real numbers on the y-axis (figure 4.7).

In the same way, it is not difficult to show that *all* intervals, finite or infinite, have the same cardinality, that is, the same number of points. The last result is interesting because it juxtaposes two different ways of thinking of infinity. The first considers infinity to be a cardinal number, the second a length. One's "feeling" or intuition of infinity in one situation doesn't carry over to the other. It is this inconsistency which when placed against the idea that infinity is unique accounts for the feeling of paradox.

THE CANTOR SET

One of Cantor's most ingenious ideas was the construction of the set that now bears his name. The Cantor set is one of the most beautiful mathematical objects. Its properties are subtle and appear paradoxical in various ways, one of which again involves the equating of a part to a whole. The unraveling of these paradoxes gives us a deeper insight into the nature of the real number system and sets the stage for many far-reaching developments in mathematics.

The previous example showed that our intuition of size in terms of length does not correspond to size as measured by cardinality. But the situation is even more complex. It is possible to assign a "length" to sets of numbers that are not intervals. The simplest such set would be the union of two intervals, say {x: either 0 < x < 1 or 2 < x < 4}, that is, (0, 1) ∪ (2, 4). It

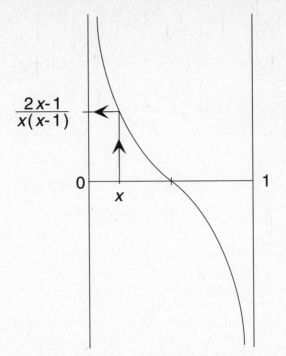

$$\frac{2x-1}{x(x-1)}$$

Figure 4.7. Matching (0,1) and the real numbers

would seem reasonable to give this set the length $1 + 2 = 3$. On the other hand, if a closed (including its end-points) interval $[a, b]$ has length $b - a$, it would be consistent to assign a point the length 0 since it could be considered the length of the trivial interval $[a, a]$.

Now lengths are additive, so any countable set would also have length zero ($0 + 0 + 0 + 0 + \ldots = 0$). Thus the rationals would have length 0 and the set of irrational numbers between 0 and 1 would have length equal to the length of $[0, 1]$ minus the length of the rationals in $[0, 1]$, that is 1. This accords with the cardinality measure of size that would also have the irrationals "larger" than the rational. However, Cantor has a surprise in store for us—the construction of a set which has "length" 0 but cardinality equal to that of all of the real numbers, that is, a set that is small in one measure but large in the other. This beautiful example is really another paradox of infinity.

175

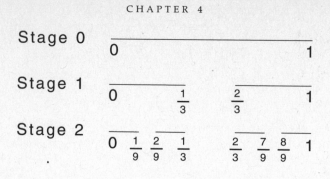

Figure 4.8. Construction of the Cantor set

CONSTRUCTION OF THE CANTOR SET

The Cantor set is constructed in stages:

Stage 0: Start with the interval, $C_0 = [0,1]$.
Stage 1: Remove the middle third interval $(1/3, 2/3)$, leaving $C_1 = [0, 1/3] \cup [2/3, 1]$.
Stage 2: Remove the middle third of all intervals in C_1, leaving $C_2 = [0, 1/9] \cup [2/9, 1/3] \cup [2/3, 7/9] \cup [8/9, 1]$.
. . .
Stage n: Remove the middle third of all intervals in C_{n-1}, leaving $C_n = 2_n$ intervals, each of length $1/3_n$.

The Cantor set, **C**, consists of all the points, if any, that are left after this infinite process is completed, that is the points of **C** would be members of C_n for all values of n. Are there any such points? Well certainly there are, since 0, 1, 1/3, 2/3, 1/9, 2/9, 7/9, 8/9, . . . and all the endpoints of the intervals will never be removed. However, there will be no intervals in **C** since we have removed the middle-third of all intervals. So what kind of a set is **C**, and how big is it?

C has length zero

Consider the length of the intervals that are removed form [0,1] to form the Cantor set:

$$\frac{1}{3} + 2\left(\frac{1}{9}\right) + 4\left(\frac{1}{27}\right) + \cdots.$$

This is a geometric series and its sum is known to be $a/(1-r)$, where $a = 1/3$ is the first term and $r = 2/3$ is the common ratio. Thus the series adds up to 1. Since what is removed has length 1, the set that remains can only have length equal to the length of [0,1] minus the length of the removed set, that is, $1 - 1 = 0$. *We have shown that* **C** *has length* 0.

C has the cardinality of the real numbers

If we write the numbers between 0 and 1 as decimals to base 3 instead of using the usual base 10 notation, it is easy to pick out the points that remain in the Cantor set. Recall that the notation for decimals to base 3 is identical to decimals to base 10 except that all powers of 10 are replaced by powers of 3. Thus .12 (base 3) stands for $1/3 + 2/9$ instead of the usual $1/10 + 2/100 = 12/100$. There are infinite decimals in base 3 notation such as .111... (base 3), which stands for

$$\frac{1}{3} + \frac{1}{3^2} + \frac{1}{3^3} + \cdots = \frac{(1/3)}{1 - (1/3)} = \frac{1}{2}.$$

Now base 3 notation has a simple geometric interpretation. It involves breaking the interval [0, 1] into three equal intervals (of length one-third), subdividing these into three (now of length one-ninth), and continuing in this way forever, each time subdividing every interval into three of length one-third of the original. The base 3 notation is an address that tells you the location of the point with respect to these various intervals. If it begins with .1... (base 3), the point will belong to the middle-third interval. If it begins with .12... (base 3), it will belong to the middle-third interval and to the right-hand or largest interval when that middle interval is divided into three. The first two digits after the decimal point allow us to locate the point up to an accuracy of one-ninth. The point $x = .12...$ (base 3) must lie between 5/9 and 6/9. Thus the decimal representation describes the position of a number with respect to precisely those intervals that we used to construct the Cantor set.

177

All the points that are removed at Stage 1 above have a 1 as their first digit after the decimal point. Recall that 1 in the first digit (base 3) stands for 1/3. The points that remain after stage 1 can be represented (to base 3) as either .0. . . . or .2. . . Stage 2 will remove the points with a 1 in the second place (base 3), leaving points that look like .00. . . or .02. . . or .20. . . or .22. . . . Continuing in this way we can see that the set **C** consists of exactly those numbers that can be written to base 3 without any 1's. For example, .2 or .022 or .020202. . . are all points of this type. How many such points are there?

Now there are the same number of decimals with 0's and 2's as there are of decimals with 0's and 1's since, for example, .020202. . . would be uniquely associated with .010101. . . . But *every* number between 0 and 1 can be written as a decimal with 0's and 1's, since this is the *base* 2 representation of a number. (The base 2 representation of a number follows exactly the same principles as the base 3 representation except that the only digits allowed are 0's and 1's and all powers are powers of 2 instead of 3.) Therefore, there is a one-to-one correspondence between **C** and [0,1]. For example, the real number 1/4 has base 2 representation .01 (= 0/2 + 1/4) so it would match up to the number with base 3 representation .02, which is in **C** and is equal to 2/9. Thus **C** has the cardinality of the real numbers, *c*.

In summary, we have proved the counterintuitive proposition that a set can be simultaneously large (cardinality *c*) and small (length 0).

The Cantor set is really an extraordinary mathematical object with many properties that have not been mentioned yet. One of these properties is self-similarity. For example, if you look at the portion of the Cantor set that lies between 0 and 1/3 it is identical to the whole Cantor set except for size. That is, if you blow up the first third of the Cantor set by a factor of 3 you will get the whole thing. This is true for any part of the Cantor set that lies in one of the intervals that we used to construct **C**. Equipped with our base 3 notation we can easily see why this is so. Since the Cantor set consists of all points whose decimal base 3 consists only of 0's and 2's, the first third consists of all such points that begin with a 0. These points match up perfectly with *all* of **C** by simply matching every point .0*abcd*. . . in this interval with the point .*abcd*. . . in the whole of **C**.

In other words, the Cantor set has the property of being self-similar—any segment, no matter how small, is essentially identical to the whole set. In this way "the whole is equal to the part" or "the part contains the whole." The Cantor set is in a sense the most elementary fractal and is thus the beginning of that fascinating story in the recent history of mathematics.

The square has the same "size" as the line

The paradox of the whole being equal to the part is brought out even more dramatically by the paradoxical result that the square has the same cardinality as the line. Not only does cardinality not respect length, but it does not even respect dimensionality. This is again counterintuitive, another paradox. Dimension is one of the most elementary characteristics of the natural world. Lines and circles are one-dimensional (it requires one number to locate a point on these figures); surfaces are two-dimensional; solids are three-dimensional.

In a sense dimension is another possible way of determining size. Intuitively speaking, a two-dimensional set is "larger" than a one-dimensional one, that is, there are more points in two dimensions than in one. Surely the plane is "bigger" than a line. For one thing, two-dimensional figures have positive area whereas one-dimensional figures have area zero.

Cantor himself initially believed that a higher-dimensional figure would have a larger cardinality than a lower-dimensional one. Even after he had found the argument that demonstrated that cardinality did not respect dimensions: that one-, two-, three-, even n-dimensional sets all had the same cardinality, he said, "I see it, but I don't believe it."[15] He was aware of how upsetting this result would be to other mathematicians, many of whom would see this as indicating that there was some problem with the definition of cardinality. What good was this definition if it could not even distinguish between dimensions?

Nevertheless what is one to make of the following argument which purports to match the points of the unit interval $I = \{x: 0 < x < 1\}$ with those of the unit square

$$S = \{(x, y): 0 < x, y < 1\}.$$

Cantor's argument is again simple but ingenious. In the manner of Descartes, every point in the square is determined by its two coordinates x and y, both of which are real numbers and can therefore be written as infinite decimals:

$$x = .x_1 \, x_2 \, x_3 \, x_4 \ldots,$$
$$y = .y_1 \, y_2 \, y_3 \, y_4 \ldots.$$

We make these two real numbers correspond to the single number whose decimal representation is given by

$$z = .x_1 \, y_1 \, x_2 \, y_2 \, x_3 \, y_3 \ldots.$$

For example, the point in the square with coordinates (1/2, 1/3) or (.5000, .333. . .) would correspond to the single number .53030303. . ., which turns out to be 1/66. Conversely, the number $12/99 = .121212. . .$ would correspond to the point in the square with coordinates (.111. . ., .222. . .) = (1/9, 2/9). This is not just any correspondence: it is a one-to-one correspondence since every pair of numbers gives rise to precisely one number and conversely every single decimal number can be "decoded" to give a pair of infinite decimal numbers. Thus the square and the interval have the same cardinality. As a generalization it can be shown that the cube in three dimensions, the hypercube in four dimensions, and even the higher-dimensional analogues of the cube in n-dimensional space for any positive integer n all have the same cardinality. Notice how this uses the fact that real numbers have a representation as *infinite* decimals.

It follows from our discussion that there are (at least) three different ways of determining the size of infinite sets: the counting size or cardinality and two geometric measures, the length of intervals and the dimension. Cantor's paradoxes involved the juxtaposition of these notions. As usual, the paradoxes indicate that the "size" of a set is a complicated mathematical notion that can be considered from various perspectives. The notion of length leads in the direction of measure and probability theory. The notion of dimension itself can be formulated in various ways and further investigated in topology and in the study of fractals, those generalizations of the Cantor set that were spoken of above. Even the difference in dimension between the interval and the square is a subtle matter. It is not a matter of the cardinality of the underlying sets. However, there is an interesting difference: if you remove a point from the interval you have divided it into two connected pieces, but if you remove a point from the square it remains connected. This is a topological distinction between these two geometric figures—what is called a "topological invariant." A topological invariant is not preserved under ordinary one-to-one correspondences but is preserved under one-to-one correspondences that are continuous and whose inverses are continuous, so-called "homeomorphisms."

Infinitesimals[16]

In this chapter, I have discussed various ways in which the notion of "infinity" is used in mathematics. We saw how many problems about "infinity" were resolved by the adoption of what has become the standard approach to the real number system. This standard manner of looking at the foundations of analysis is to a large extent due to the work of two men—Karl Weierstrass (1815–1897) and Cantor. In my attempt to demonstrate how the paradoxes and problems of mathematics are often resolved by moving to a "higher" or more general point of view, I may have inadvertently given the impression that this "standard" viewpoint is definitive—that the theory of the real numbers that we have created is the last word, the Truth, the only way to look at analysis. This is not my position; indeed it would not be consistent with my claim that "infinity" can never be completely pinned down—there will always be new ways to think about infinity, new ways to mine its riches to the benefit of mathematics and science. The infinite cardinal numbers arise, as we have seen, from generalizing the mundane activities of counting and matching. However there are other legitimate intuitions about infinity that can be turned into fruitful mathematical concepts—measuring numbers, for example.[17]

To make this point I will now turn to another tradition, one that goes back to Archimedes and was used by many mathematicians including both Newton and Leibniz in their development of the calculus. This tradition uses the notion of an "infinitesimal" number. An infinitesimal is a number that is infinitely small yet greater than zero. The discussion of infinitesimals highlights many of the themes I have been discussing—here is another intuition about infinity that is paradoxical from a certain point of view yet leads to a resolution that is mathematically significant. Today, the subject of infinitesimals is often associated with nonstandard analysis as described by the eminent logician Abraham Robinson[18] in the context of mathematical logic. However, I shall discuss a much more elementary approach that only uses the idea of an "ordered field," a mathematical system of "numbers" within which one can do arithmetic and which has an "order" in the sense I shall describe below.

PARADOX #6: DO INFINITESIMALS EXIST?

Do there exist numbers x such that $x > 0$ but $x < \varepsilon$ for every real number $\varepsilon > 0$? In the real number system as I have developed it, such numbers cannot exist, that is, their existence would be paradoxical. Yet many brilliant mathematicians, especially those of the eighteenth century, used infinitesimal methods freely—seemingly content to ignore the logical difficulties that arose—because of the intrinsic interest of the mathematics that could be obtained in this way. It was these logical difficulties that were the basis for the devastating criticism of Bishop Berkeley that was discussed in Chapter 2. Eventually these logical difficulties were resolved and a rigorous theory developed at the cost of excising these infinitesimal quantities from mathematics. In doing so, something was gained and something lost. What was gained was a theory that secured the foundations of mathematical analysis. What was lost (and felt to be mistaken) was the intuition that there was mathematical substance in the notion of infinitesimals.

Remember how we thought of the real numbers as a "completion" of the rational number system. Developing the system of real numbers from the system of rationals involved creating a vastly larger system of numbers. It was an extension of the rationals in the sense that it contained them as a subsystem—that is, every rational number is also a real number. Well, it turns out that it is possible to construct number systems that contain the reals as a subsystem. Now such a system cannot have *all* the properties of the real numbers—in particular it is not complete, that is, it does not have the property that every bounded increasing sequence defines a unique number. The reason for this is that completeness is a property that *characterizes* the real numbers.

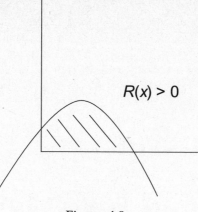

$R(x) > 0$

Figure 4.9.

Nevertheless, you can still do arithmetic in the larger system that is, you can add, subtract, multiply, and divide. Moreover, the notion of "order" can be defined. We do this below in the manner of Tall (2001).

Consider the set of all rational functions of the form

$$R(x) = \frac{a_0 + a_1 x + \ldots + a_n x^n}{b_0 + b_1 x + \ldots + b_m x^m},$$

where x is a variable and the coefficients are all real numbers with $b_m \neq 0$. These are the usual functions of high school algebra. They form an extension of the real numbers since every real number a can be thought of as a rational function $R(x)$, where $a_0 = a$, $b_0 = 1$, and $a_i = b_j = 0$ for i and $j > 0$. If $R(x)$ is not identically zero then its graph is either strictly positive or strictly negative in some interval of the form $0 < x < k$ (figure 4.9).

We shall say that such a rational function $R(x)$, is *positive* if $R(x) > 0$ for such an interval. For example, the function $f(x) = x$ is *positive*. Also, for any real number $a > 0$ the corresponding constant function $f(x) = a$ is *positive* by our new definition of what it means for a function to be positive. Thus the idea of "positivity" carries over from the reals to this new situation of rational functions. Now any reasonable notion of what it means to be positive such as the one that we have given for rational functions leads directly to a corresponding notion of order in the following way. We define $R(x)$ to be *greater than* $T(x)$ if and only if $R(x) - T(x)$

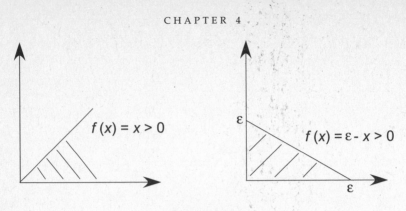

Figure 4.10.

is *positive*. The entire collection of rational functions can be ordered in this way. For example, if $\varepsilon > 0$ is some positive real number, then the function $f(x) = \varepsilon - x$ is positive (figure 4.10). Therefore (as functions), $0 < x < \varepsilon$ for every positive real number $\varepsilon > 0$ and so the function $f(x) = x$ is an infinitesimal. We have thus created a legitimate mathematical number system that contains the reals and elements that play the roles of infinitesimals. It also has corresponding "infinite" elements such as $R(x) = 1/x$.

Do such systems have mathematical value? Are they artificial and arbitrary constructs, or do they correspond to a valid mathematical intuition? Another way of putting this is to ask whether the real number system as we understand it today is more "real," more anchored in reality, than a system such as the above that is larger and includes infinitesimal quantities. There have been many mathematicians who have taken the position that only the positive integers are "given by God," that is, are real, and that all other systems including the rationals, the reals, and the system we are now discussing are human creations. If this is so, then their value is to be determined by what they have to contribute to mathematics.

Let us take a moment to look at a simple mathematical example from the "infinitesimal" point of view. We have seen how the real numbers can be considered as functions in this larger number system. "Infinitesimals," as we have seen, are rational functions, R, that are positive and are less than any positive real number. Now it can be shown that any finite function is of the form $R = a + \varepsilon$, where a is a real number and ε is an infinitesimal. (Infinite elements are of the form $1/\varepsilon$ for an infinitesimal ε, as we might expect.) The number a is called the *standard part* of R.

Now consider the calculus problem I discussed in Chapter 2, where the distance traveled by a moving body, s, was a function of time, $s = t^2$. We were interested in determining the speed of the body at time $t = 1$. In Chapter 2, we thought of the instantaneous speed as being the limit of an infinite sequence of average speeds. However, we could think of this calculation in another way.[19] Let dt (one symbol) stand for an infinitesimal increment in time and ds for the corresponding increment in distance. We want to calculate the ratio ds/dt and we wish this to be a finite number. Now $s = 1$ when $t = 1$. When $s = 1 + dt$, $ds = (1 + dt)^2 - 1 = 2dt + (dt)^2$. Thus $ds/dt = 2 + dt$. The standard part of this number is 2; that is, the derivative is the standard part of the extended number ds/dt. Notice that we have calculated this derivative without recourse to any limiting process. This is a conceptual and pedagogical advantage. The disadvantage of this way of looking at things is that it depends on introducing a new and larger number system. This balancing of advantages and disadvantages is reminiscent of considerations that arose in the construction of the real number system itself.

There are those who claim that human beings have some sort of special intuition for the real "number line," and therefore that systems that contain infinitesimals are unintuitive. Others (like Tall 1980, 2001) claim that they arise from an equally valid but different intuition. Cantor's definition of (infinite) cardinality arose, as we saw, from thinking about the nature of counting and asking what it would mean to "count" an infinite set. Tall (1980) discusses extrapolating the *measuring* properties of numbers instead of the *counting* properties. He says, "By posing an alternative schema of infinite measuring numbers, we may at least see that our interpretation of infinity is *relative* to our schema of interpretation rather than an *absolute* form of truth." This conclusion applies not only to "infinity" but to a whole host of seminal mathematical concepts.

Conclusion

After spending two chapters discussing the infinite, it may be appropriate to spend a few more paragraphs reflecting on this marvelous journey we have taken. Let us assume that human beings have always had a sense of the infinite, but that it originally was an aspect of the numinous. It was, as I said earlier, ineffable. It belonged to another realm, a realm of awe and mystery, a realm of the divinely inspired, so to speak, certainly not the domain of the rational and the empirical. Nevertheless there exists an intuition of something that we call the infinite and therefore an irresistible desire to capture its essence, to define it and understand it.

It is clear from the discussion that to define the infinite is a contradiction in terms. By definition the infinite cannot be defined. Even to say that there is something that stands outside the infinite will not accord with many people's intuition of the infinite. Yet how are we to see a good part of intellectual history except as an attempt to get at the idea of the infinite, to separate out the aspects of the infinite that could be discussed from those that could not be? As time went on, the infinite was first of all conceptualized, and then that conceptualization was increasingly refined. Thus the infinite is implicit in the definition of the straight line or in the "method of exhaustion." Then it is there explicitly in the "point at infinity" or in infinite decimals. Finally it is further refined through the notion of an infinite cardinal number, and this leads to the existence of multiple infinities. At every stage in this process we are moving away from the infinite as the singular and the incomparable toward the infinite as a well-defined but multifaceted mathematical concept.

Every step on this journey is accompanied by the paradoxical. This is fair enough since we now understand why the idea of the infinite *must* induce the paradoxical. New stages of mathematical development arise out of a "resolution" of a set of paradoxes. However, like the development of the real number system, there is a price to pay. We "resolve" the paradox of irrational numbers through the development of the real number system, but at the price of the introduction of an uncountable set of new numbers, numbers that can never be "known" in the way that the rationals are known. In other words each step toward simplicity in one sense is accompanied by a complexification in another.

Finally, this discussion of infinity is not something that is taking place on the boundaries of mathematics; it is central to mathematics, and because mathematics is central to science and technology, it is central to contemporary culture as a whole.

Mathematics is the science of the infinite—the infinite more than anything else is what characterizes mathematics and defines its essence. If the infinite is impossible in the sense that it is inevitably accompanied by the paradoxical, then the same holds for mathematics. This does not detract from the importance of mathematics. On the contrary, it is what makes mathematics great! To grapple with infinity is one of the bravest and extraordinary endeavors that human beings have ever undertaken.

SECTION II
THE LIGHT AS IDEA

*

A MBIGUITY IN MATHEMATICS, the theme of the book, was introduced as a way to discuss the creative aspects of mathematics, what was metaphorically called the "light," as opposed to the logical surface structure. The creative in mathematics is expressed through the birth of new ideas. These ideas may consist of a new way of thinking about a familiar concept or they may involve the development of an entirely novel concept. An idea is usually at the heart of a mathematical argument but an idea may even entail a new way of looking at a whole area of mathematics. Creativity in mathematics is inseparable from ideas, but what is the nature of these mathematical ideas and where do they come from? Mathematical ideas form the most exciting and yet the most mysterious aspect of mathematics. Thus this section begins with a discussion of ideas in mathematics.

The connection between "ideas" and "ambiguity" arises from a consideration of the origins of ideas. I am especially interested in ideas that arise out of situations of ambiguity. It is not situations of the greatest logical coherence that give rise to profound ideas. Counter-intuitively, great ideas often arise out of the kind of problematic situations we have been discussing—great ides come out of situations of great ambiguity. Some people imagine that creativity in mathematics arises out of a novel rearrangement of facts that are already well understood. Ideas that are the result of such superficial activity are rarely profound. Deeper ideas often arise out of situations that are murkier than this—situations that are experienced as in flux, situations that contain elements that do not seem to fit well together, where there is conflict, in short, the kind of ambiguous situations I have been discussing.

What is there about situations of ambiguity that is propitious to the birth of new ideas? An idea emerges in response to the tension that results from the conflict inherent in ambiguity. We generally avoid situations of ambiguity because we find this tension disagreeable. With the sudden appearance of the idea the conflict is reconciled and the tension disappears. Think about the following metaphor. Suppose a bar magnet is introduced into a field of iron filings that are aligned in a random order. As we all know, the filings will realign themselves into a regular order because of the magnetic field that is generated by the magnet. Now an ambiguous situation should also be seen as a field

191

of force, a force that is generated by the two conflicting frames of reference that characterize the ambiguity. This is analogous to the poles of a magnet and may, in certain situations, precipitate the birth of an idea that will give the mathematical situation a coherent organization in the way that the iron filings are "organized" by the magnet. Thus we are led to think of a mathematical idea as a principle that organizes a given mathematical domain. This section begins with an introduction to ideas in mathematics and then goes on to discuss the relationship between ideas and ambiguity.

The Idea as an Organizing Principle

Something's happening here, what it is
ain't exactly clear. . .
—"For What It's Worth," written by Stephen Stills
(1966), performed by Buffalo Springfield

Inside and outside mathematics, Thom's interest
was turned to forms and ideas. His passion was
to understand geometrically the nature of things,
and for this he used mathematical proofs. But he
was not a formalist, and, for him, proofs
remained secondary to the conceptual
landscape they revealed.
—David Ruelle on René Thom[1]

INTRODUCTION: WHAT IS GOING ON HERE?

What is the core ingredient of mathematics? Is it logic or precision? Is it "number" or "function"? Is it "structure," or "pattern," or the subtlety of mathematical concepts? Perhaps it is abstraction? In our search for the inner nature of mathematics we might do well to listen to the words of mathematicians. Not just the words they use when we ask them to explain the nature of their subject. The language they use in such an artificial situation is alien to the language that they use when they are discussing mathematics among themselves.

When discussing a particular piece of mathematics, the mathematician may ask, "Now, what is really going on here?" That is, what is the core mathematical idea? A piece of mathematics, a proof for example, may go on for pages and pages and may include detailed calculations and subtle logical arguments. However, there is often a surprisingly concise mathematical idea that forms the basis for all the detailed work. Strangely enough, when a research article is published in a professional journal this underlying idea may not even be mentioned. The reader may go through the many pages of detail and end up none the wiser for the effort unless he or she manages to grasp the core idea. For the author of the paper the construction of such a complex argument would have been impossible were there not some explicit idea around which the proof was organized. Thus skillful expositors of mathematics are those who manage to lay bare the essential mathematical ideas that underlie a particular piece of mathematics. Skillful teachers are those who let their students in on the secret of "what is really going on."

Mathematical ideas are the building blocks of mathematical thought. What is an idea? An idea is a principle that organizes experience, in this case mathematical experience. Every mathematical concept is an idea; every proof is built around an idea. Without ideas to organize experience, the world would be chaotic and unmanageable. Everything that we deal with in mathematics is an idea. "One" is not a number, it is an idea. "Zero" is an idea, and, as I discussed earlier, it is a subtle and powerful idea. Axioms are ideas that imply other ideas. There is no logical hierarchy of ideas—logic itself is an idea. This chapter and section consider the implications of the statement that ideas are at the center of mathematics.

Ideas are extremely difficult things to get a handle on. The idea is formulated, made explicit. Therefore the idea itself stands behind any explicit formulation. It is extraordinary how much mathematics can be obtained from a fruitful idea. It follows that an idea is not something that is completely objective—it has both objective and subjective characteristics. Furthermore, the idea is not, strictly speaking, logical. The logic comes later—logic helps to formulate and organize ideas. Nevertheless, ideas are where the action is—they are an essential aspect of mathematics.

Ideas can seemingly come out of anywhere. Mistakes, contradictions, even paradoxes have been used to generate important mathematical ideas. Mathematics has been called "the science of pattern," but what are patterns if not ideas? Even "randomness," which could be defined as the absence of pattern, is an important mathematical idea. It is even a fairly common philosophical position today that mathematical knowledge is "constructed" by each and every practitioner. From this point of view, ideas are all there is and all there can be. The world we live in—mathematical and natural—is a world of ideas.

This is not to say that every piece of mathematical work contains a mathematical idea; certainly it may not contain a profound idea. "Trivial mathematics" is mathematics devoid of an idea. A "trivial" argument is one that follows in an automatic, algorithmic way from a given set of definitions and a few purely logical manipulations.

For the mathematician, the idea is everything. Profound ideas are hard to come by, and when they surface they are milked for every possible consequence that one can squeeze out of them. Those who describe mathematics as an exercise in pure logic are blind to the living core of mathematics—the mathematical idea—that one could call the fundamental principle of mathematics. Everything else, logical structure included, is secondary.

The mathematical idea is an answer to the question, "What is going on here?" Now the mathematician can sense the presence of an idea even when the idea has not yet emerged. This happens mainly in a research situation, but it can also happen in a learning environment. It occurs when you are looking at a certain mathematical situation and it occurs to you that "something is going on here." The data that you are observing are not random, there is some coherence, some pattern, and some reason for the pattern. Something systematic is going on, but at the time you are not aware of what it might be. This is a tangible feeling. We might call it the "something is going on here but I don't know what it is" feeling. Whereas "something is going on" is the mathematical idea, what we are talking about here is the potential, unformulated idea. Perhaps we could think of it as the germ of a possible idea. One is still not conscious of the idea, and one has to work very hard in order to bring this potential idea to consciousness.

The feeling that "something is going on here" can even be brought on by a single fact, a single number. A case in point happened in 1978, when my colleague John McKay noticed that $196884 = 196883 + 1$. What, one might ask, is so important about the fact that some specific integer is one larger than its predecessor? The answer is that these are not just any two numbers. They are significant mathematical constants that are found in two different areas of mathematics. The first arises in the context of the mathematical theory of modular forms. The second arises in the context of the irreducible representations of a finite simple group called the Monster. McKay intuitively realized that the relationship between these two constants could not be a coincidence, and his observation started a line of mathematical inquiry that led to a series of conjectures that go by the name "monstrous moonshine." The main conjecture in this theory was finally proved by Fields Medal winner Richard E. Borcherds. Thus the initial observation plus the recognition that such an unusual coincidence must have some deep mathematical significance led to the development of a whole area of significant mathematical research.

It is interesting that at a seminar called "Monstrous Moonshine and Mirror Symmetry" given by Helena Verrill at Queen's University in 1997 she made the statement, "These results have been proved by Richard Borcherds, using vertex algebras, but it's still not really understood what is happening. So, our problem is to understand what's going on."[2] This is precisely my point. McKay noticed that there was something going on. This led to mathematical activity leading to theorems and proofs. But still it is possible to say "we do not really understand what is going on." Understanding what is going on is an ongoing process—the very heart of mathematics.

Some might claim that the germ of an idea is nothing and that all that counts are those ideas that actually work out. This would be a mistake, for the germ of the idea is an essential prerequisite for the creative act. One can feel when "something is going on," and good mathematicians are very sensitive to that feeling. Of course this feeling is not unique to mathematicians. The poet Denise Levertov is alluding to this same potential idea when she says, "You can smell the poem before you can see it."[3] The poet T. S. Eliot said, "first an inert embryo or creative germ is conceived and then comes the language, the resources at the poet's command."[4] As in poetry so in mathematics everything begins with this "creative germ" to which the mathematician brings life by changing the feeling of "There must be something going on here!" to "Aha! Now I see what is going on here."

The mathematician's work can be broken down into various stages. The first involves spade work: collecting data and observations, performing calculations, or otherwise familiarizing oneself with a certain body of mathematical phenomena. Then there are the first inklings that there exists in this situation a pattern or regularity—something that is going on. This is followed by the hard work of bringing the embryo into fruition. Then, finally, when the idea has appeared, there is the stage of verification or proof. Is the idea in fact correct? However, this verification stage is really part of a larger and more elaborate process that involves the exploration of the full range of applicability of the idea. The latter stage may involve the abstraction or generalization of the initial idea. In practice, "writing things up" often sets the stage for the modification of the idea or, in fact, for the emergence of a new idea. So it is not a question of having the ideas on one side and the logic or proof on the other. The practice of mathematics involves the interaction between ideas and logical rigor. Unfortunately, since the formal structure is more obvious and therefore so much easier to get hold of, the importance of mathematical ideas is often neglected. Nevertheless, ideas are the principal actors on the stage of mathematical activity.

In addition to the interactive relationship between ideas and proofs, there is another way in which the description of mathematical activity that is sketched out in the first part of the above paragraph is incomplete. In fact, it neglects an essential ingredient, one that is reminiscent of what I called "reification" in Chapter 1. Remember that reification arises when a process "becomes" an object, such as when we write a number as an infinite decimal, for example, $1/3 = .333\ldots$, and think of this situation as one ambiguous process/object. It is now possible to work with this new entity and to embed it into a more general system or process. In this way a complex, potentially infinite situation has been transformed into something that can be conceptualized and worked with as a single mathematical unit.

Now, the mechanism of reification that has been described above also applies to mathematical ideas in general. What I mean by this is that a certain mathematical situation is captured by a mathematical idea that is then thought of as a single mathematical object and often represented by one symbol. The idea becomes a new unit of thought, what has been called a "thinkable entity," and can then be embedded into a higher order process or system. This explains how ideas, as mathematical objects, can be organized by new ideas (and so on) in situations of increasing complexity and abstraction. The successful mathematical thinker does not carry around in his head all possible detail, but only works with the ideas at an "appropriate" level. The unsuccessful thinker is unable to recognize what level is appropriate to the given situation and so often gets bogged down in irrelevant detail. For example, to be successful in arithmetic, a child must learn to deal with multiplication as something in its own right and not only as repeated addition.

In other words, what is happening here is that the "idea as organizing principle" becomes the "idea as thinkable entity." Without such a mechanism for the compression of knowledge, it would not be possible to work with the extremely complex structures that make up advanced mathematics.

The Idea as an Organizing and Generating Principle: A Simple Example

The mathematical idea is an organizing principle. It organizes a body of mathematical phenomena. It is the heart of mathematics, but it is not itself (finished) mathematics. Rather than stable, unchanging, finished mathematics, the generating idea is dynamic. It creates mathematics. In recognition of the importance of fundamental ideas, Harvard, for example, has, in recent years organized a seminar called, the "Basic Notions Seminar." In the words of the mathematician Barry Mazur, "The aim of that seminar is to survey each week some central theme of mathematics—some idea that has different manifestations as it crops up in different fields of mathematics—an idea, in short, that deserves to be contemplated by students not only in the context of its usefulness for this or that particular result but also because of its service as a unifying thread."[5]

As a first very elementary example, consider this "proof" for the sum of a geometric series:

Let $S = a + a^2 + a^3 + \cdots + a^n + \cdots$.
Then $aS = a^2 + a^3 + \cdots + a^n + \cdots$.
Subtracting gives $S - aS = a$
or $S = a/(1 - a)$.

Now this is not a valid argument. When $a = 1$ you get $S = 1/0$, which can be a problem; when $a = -1$ you get $S = -1/2$, but as we saw in the section on the paradoxes of infinity, one might equally say $S = 1$ or $S = -1$. The "problems" with this idea revolve around the first equation. Does the infinite sum $a + a^2 + a^3 + \cdots$ always converge, does it necessarily give you a specific number, that is, does S always exist? However, there *is* an idea here. You could formulate the idea in the following way: if you have a sum of powers of a single number, a, and then you multiply the sum by a, and subtract, everything cancels out except the first and possibly the last term. It is what is called a telescoping sum. So this mathematical idea may generate many questions: What do we mean by an infinite sum? What is convergence? For what values of a is the final formula valid and for what values is it invalid? In its application to the decimal representation of real numbers this idea can be used to show that every eventually repeating infinite decimal represents a fraction. It also raises the question about the converse of that statement: whether the decimal representation of every rational number is eventually repeating. This in turn can lead us to consider irrational numbers, the real number system as a whole, and a host of other questions.

Simple though it is, this example also raises the question of the relationship between ideas and proof that I mentioned briefly in the previous section. Even though I maintain that ideas, strictly speaking, are not logical, nevertheless attempting to write down a proof is a very important step. Attempting to make the idea precise leads us to penetrate the situation more fully; it allows us to become aware of new questions and perhaps sets the stage for the emergence of new ideas. Even though the idea has been expressed in a mathematical format it may not yet be mathematics. At this preliminary stage it is neither valid nor is it invalid. Thus the idea is really more basic than any proof or result that may flow from it. The organizing principle is more elementary than what we normally call mathematics.

IDEA AS PATTERN

Recall the following situation from the section on mathematical induction (pp. 123–125):

$$1 = 1^2,$$
$$1 + 3 = 4 = 2^2,$$
$$1 + 3 + 5 = 9 = 3^2,$$
$$1 + 3 + 5 + 7 = 16 = 4^2.$$

Clearly there is a pattern. We grasp the pattern. so we could say that we intuit that "something systematic is going on here." In discerning the pattern we are already engaging in a certain kind of mathematical activity. Yet we are still at a preliminary stage. We could go further by expressing that pattern explicitly. In words we might say "the sum of the first n odd integers equals n squared." We have now given some precision to our intuition. The pattern is now explicit but this is not enough. We might ask whether this pattern is valid for all values of n, even values that are so large that an actual computation is impractical if not impossible. The validity (or not) of the pattern requires an answer to the question, "Why?" Why does that pattern hold? The question demands the probing of the pattern at a deeper level. Though the pattern is itself an idea, verifying the validity of the pattern demands another idea, an idea that will convince us that we know why the pattern is valid. In this case there are many ways to validate the pattern. One that we described earlier was by mathematical induction. A more immediate and satisfying geometric idea is contained in the picture in figure 5.1.

Now the notion of pattern is so central to mathematics that mathematics has even been defined as the science of patterns. According to Keith Devlin,

> What the mathematician does is examine abstract "patterns"—numerical patterns, patterns of shape, patterns of motion, patterns of behavior, voting patterns in a population, patterns of repeating chance events, and so on. Those patterns can be either real or imagined, visual or mental, static or dynamic, qualitative or quantitative, purely utilitarian or of little more than recreational interest. They can arise from the world around us, from the depths of space and time or from the workings of the human mind. Different kinds of patterns give rise to different branches of mathematics.[6]

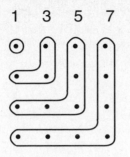

Figure 5.1.

What, then, is the essence of pattern? As we stop to consider this most elementary notion, we shall discover that the essence of pattern is elusive; "pattern" is a mysterious and subtle notion. From the dictionary definition, the everyday use of the word "pattern" is as a template. A blueprint, for example, is a pattern for a building. When one knits a sweater, one uses a "pattern." In this sense, a pattern is a model or a representation of an object. In this situation the pattern is something distinct from that which it patterns. However, when looking at the finished product one might work backward and say that it has a pattern. The "pattern" in the sense of the blueprint is implicit in the finished product.

However, there is another sense in which one uses the word "pattern." It arises when one sees something systematic in a body of data—when one sees that "something is going on here." Given a string of numbers: 2, 5, 8, 11, 14, . . . we might say that there is a pattern; each number is 3 more than the preceding number. Here we start from the other end, in this case the list of numbers, and make a jump to the principle that was used to generate the list.

Thus the notion of a generating idea is very close to the idea of a pattern. Where, we might ask, is the pattern located? Is it implicit in the data? Is it there objectively in the sense that any observer would agree on the existence of the same pattern? Or is the pattern subjective, dependent on an idea that is imposed on it through some act of intelligence? This is like asking whether a pattern is discovered or invented and leads to the question about whether mathematics is itself discovered or invented. This is an important question that is implicit in much of what has been discussed. I shall return to it in Chapter 8.

For the time being, let us just say that whether it is pattern recognition or pattern formation, producing or recognizing patterns seems to be a basic function of intelligence. Clearly the ability to recognize the regularities of seasonal variations, for example, would be an evolutionary plus. It is conceivable that natural selection favored species or individuals who developed this ability. If this "patterning" is related to what we mean by intelligence, then we can see one of the advantages of intelligence in the battle for survival. If mathematics is, as we believe, one of the areas of human activity where "patterning" of the most subtle and complex kind is the essential activity then we see why mathematics might be regarded as "intelligence in action." It is not surprising then that mathematics is so successful in all areas of human activity. If mathematics is indeed the expression of intelligence in action then it is necessary to understand mathematics if we wish to understand ourselves and the world in which we live.

"Ideas Are Ambiguous" or the Ambiguity "Is" the Idea

It is interesting that when we talked about idea and pattern our description inevitably contained a certain ambiguity. A pattern is both a kind of blueprint and the sense that there is a blueprint to be found. It is both explicit and implicit, both discovered and invented, objective and subjective. An idea is the feeling that "something is going on" but it is also the feeling "now I understand what is going on." It arises from the genius of the great mathematicians, but once it has been made explicit, it has the feeling of inevitability, the feeling that it has always been out there waiting for us in some sort of Platonic heaven.

If "the idea" in general is ambiguous, so it is also with particular mathematical ideas. The examples of ambiguity enumerated in Chapter 1 were really a list of deep mathematical ideas. The "idea" of the Fundamental Theorem of Calculus is that integration is the inverse of differentiation. I said at the time that calculus consists of one process that looks like integration if you look at it in one way and differentiation if you look at it in another.

However, to grasp this idea you have to look at things in a certain way; or, rather, grasping the idea *means* looking at things in a certain way. Thus if you make an integral into a function by letting the upper end point vary:

$$F(x) = \int_a^x f(t)\, dt,$$

then the derivative of $F(x)$ is the function $f(x)$. If one does not look at things in the right way, then the "idea" of the Fundamental Theorem is not so obvious.

It is clear from our discussion so far that there is a connection between "ambiguity" and "idea" in mathematics. An ambiguity can be a barrier, as the square root of two was a barrier to the Greek understanding of number. When this barrier is overcome, it is overcome by a mathematical idea that dissolves the barrier. In this case what is required is an expanded idea of "number" that results from reifying a limiting process, that is, identifying an infinite sequence of rational numbers with a single real number. Nevertheless, the ambiguity persists in the resolution. Yes, $\sqrt{2}$ is a constructible number, but it is also an irrational number—it is as legitimate a number as a rational number or integer but it is also different in the sense, for example, that its decimal representation is nonrepeating. On the other side of the barrier, so to speak, there remain (at least) two frames of reference. The difference is that, armed now with a more general idea of "number," one is free to think of a number as either rational or irrational, as a length or as an infinite decimal. In this sense ideas are often ambiguous. The ambiguity does not limit the idea —the ambiguity is the very thing that flowers into the idea. One could go so far as to say (in an ambiguous, metaphorical sense), "The ambiguity is the idea."

Examples of Mathematical Ideas

Mathematics contains data fields of enormous complexity that are organized by mathematical ideas. However, the data are themselves made up of other mathematical ideas. For example, in the case of Euclidean geometry, the basic geometric forms: points, lines, circles, and triangles; the postulates, say, for example, that "any two points can be joined by a line"; the elementary notions (p.88) like "the whole is greater than the part"; the propositions and theorems; all these are expressions of mathematical ideas. Mathematics is a vast self-organizing body of ideas.

In an axiomatic system like Euclidean geometry, it is not a matter of merely going from obvious truths to more complex truths via acceptable modes of reasoning. For example, the parallel postulate has a relationship to the theorem that says that the sum of the angles of any triangle is equal to two right angles. Both statements harbor complex mathematical ideas. Their relationship is complex. In one situation we might be interested in how the sum of the angles in a triangle depends on the parallel postulate. In another we might be interested in the converse relationship: how the sum of the angles of a triangle being more, less, or equal to two right angles affects the idea of parallelism. Moreover the relationship between the two is a matter of logical reasoning involving other mathematical ideas. Even the logical reasoning itself is an organizing principle that is part and parcel of the overall flux of ideas that we are calling mathematics.

For a subject that revels in the abstract, mathematical ideas are often very concrete. Think of the discussion of "variables" in Chapter 1. The domain of a variable may include an infinite range of values, but when a variable is used we think of it as having a definite and particular value. This enables us to think concretely. This mental technique is used very generally in mathematics. We want to understand some mathematical phenomenon that arises in a wide variety of circumstances. How is one to think about such a general phenomenon? Often one thinks in depth about some particular but generic example— some computation or picture. Of course the genius is in picking the right example or examples. One looks for some specific example that captures all the subtleties of the general situation. Thus "what is going on" is often revealed within the specificity of a particular example. The task of extending and abstracting that understanding is often secondary. The role of specific counterexamples in establishing the boundaries of some mathematical theory is balanced by the role of specific generic examples for which one can say that they illustrate the "general case."

Despite the contention that mathematical ideas do not fit into an inevitable hierarchy, it is still possible to pick out certain elementary mathematical ideas. By elementary I mean that these ideas are embedded in almost every mathematical situation. The most obvious candidate for an elementary idea is logic itself. And, indeed, it was the project of Russell and Whitehead and before them of Gottlob Frege to show that logic was *the* elementary idea in mathematics. Their feeling was that mathematics was inherently and definitively hierarchical—that there were ideas (axioms) that were intrinsically the most elementary and that mathematics was in fact all that and only that which followed logically from these axioms. Reality, as Gödel showed, turns out to be considerably more complex.

Nevertheless, I shall continue the discussion of mathematical ideas by considering certain key ideas that arise in a variety of areas of mathematics. Do you remember the discussion of "1 + 1 = 2" in Chapter 1? To make the point that even the most elementary mathematical objects contain profound mathematical ideas, I shall begin by considering the three ideas that arise in this most elementary mathematical statement, namely, "one," "two," and "equality."

Elementary Ideas

"One" is *the* fundamental organizing principle—an idea so basic that without it there would be neither natural nor conceptual worlds. "One" is so basic that we take it completely for granted and therefore it is difficult to bring its importance into the open. Davis and Hersh[7] might well be talking about "one" when they say, "We who are heirs to three recent centuries of scientific development can hardly imagine a state of mind in which many mathematical objects were regarded as symbols of spiritual truths or episodes in sacred history. Yet, unless we make this effort of imagination, a fraction of the history of mathematics is incomprehensible." Not only would the history of mathematics be incomprehensible without the idea of "one," but in fact there could be no mathematics whatsoever.

"One" is an idea that goes beyond mathematics. As was the case with "zero" and "nothing," the mathematical "one" is an idea that grows out of the human condition itself. It points to such a fundamental aspect of reality that it is impossible to grasp all of its ramifications and yet there is an overpowering need to do just that—understand it, use it, and control it. For the neo-Platonist Plotinus,[8] "The One is the absolutely first principle of all. It is both 'self-caused' and the cause of being for everything else in the universe." One might say that "One" represents the ultimate in simplicity. Yet Plotinus's idea of The One is (he claims) indescribable directly. It is another name for the ineffable. As such, it is basic to religious thought and can be found in the theology of all the great religions: Judaism, Christianity, Islam, Hinduism, Buddhism. Low,[9] in his books on Zen Buddhism, speaks of oneness as a force, an imperative that could be expressed as, "Let there be One!"

Before we go farther, notice that "one" and the connected idea of "oneness" are ambiguous, that is, they are used in two different and conflicting senses. In the first sense, "one" represents something that is a unit and distinct from all others—a unique individual in a world of other individuals. The second sense comes from the word "oneness" or "to be one with," which means connected or part of a larger whole. It gives the sense of a number of distinct parts merged in a larger unity. The first sense emphasizes the uniqueness and separateness of that which is designated as the "one," the second the harmonious integration of parts. Thus every human being is "one" in both senses—as an utterly unique individual and as an integrated part of the human race, equal in rights and dignity to everyone else and sharing in the common human legacy.

What is the nature of the insight that the idea of "one" is attempting to capture? It is the fact that the world presents itself to us not as an undifferentiated chaotic soup but as broken up into units or "unities."[10] This is as true for our conceptual world as it is for the natural world. This was something that was noticed and emphasized by Gestalt psychology. Imagine that the world as we know it does indeed emerge out of chaos, that is, out of a kind of universal flux. If this flux is primary then how do we come to see the world as composed of objects? There must be a mechanism that converts processes into objects—a mechanism of reification.

In a fascinating chapter in one of his books, the neurologist Oliver Sacks[11] tells the story of a man who, blind from birth, had his sight restored to him as an adult. When the bandages were removed from his eyes some weeks after the operation, what did he see? The fact is that he saw nothing except a flux of confusing and irritating sensory impulses with no stability or coherence. The man had to *learn to see*. He had to learn to convert the sensory data into stable forms that could be named and so be made more or less permanent. The idea "one" is connected to this naming of the world, at least to the way in which the world presents itself to our senses and our minds as whole objects. When I look around my study I do not see parts of objects; I see one desk, one computer, one book, one pad, one pen, and so on.

To prove that there is something substantial going on in our ability to "unitize" the world, consider the following difficult problem in the field of artificial intelligence. Suppose you wish to construct a robot that can "see." The robot has visual sensors and so can receive sense impressions from the surface of my desk, say. How is the robot to perceive that there is a book on the desk? The sensor receives visual data from all over the desk. How is it to identify part of that data as coming from one object, the book? Another way of saying this is: How is the robot to identify the boundaries of the book? The boundary of the book will separate the book from the non-book, so to speak. Evidently this is a really difficult problem in the field of artificial intelligence.[12] The difference between a field of raw visual data and the field that a human being perceives involves breaking reality down into "wholes," into a set of clearly differentiated objects. Whatever process is involved in making that enormous leap, it is part of the significance of the idea of "one."

An idea as basic as "one" must find its way into mathematics in many ways. We have already encountered this phenomenon in various places in our discussion. In Euclidean geometry and subsequently in almost every well-defined mathematical subject we encountered the drive to bring together all the known truths in one grand axiomatic system. Mathematics itself is an immensely complicated whole—mathematics is "one." On the other hand, mathematics strives ceaselessly to isolate the most primitive ingredients in any situation. In fact mathematics is the result of the isolation of some of the most basic elements of human experience—number, function, chance, geometric figure, and so on.

Consider the number one—the primordial mathematical object. One is the generator of all number systems—you iterate one to produce the positive integers; invert it to produce the negatives; take quotients to produce the rational numbers; take sequences to produce the real numbers; and take pairs of real numbers to produce the complex numbers. In that sequence "one" is the irreducibly simplest starting point. Part of the meaning of "one" is connected to simplicity. You get that feeling in the axioms of Euclidean geometry—in the definition of "point" as "that which has no part," for example. In fact the whole idea of axioms is to isolate the irreducible elements of a mathematical situation. Isolating the elementary building blocks and putting them together in structures that are complex and yet harmoniously connected—in this way we could say that mathematics is a manifestation of Plotinus's drive toward the "one."

The world of our perception and cognition is made up of "wholes" or "unities"; yet these unities possess an internal structure. The body is made up of organ systems, the organs of subsystems and so on down to cells, molecules, and atoms. The most elementary structure that a "whole" may possess is represented by the number *two*. I shall briefly take up the idea of "two" and its relation to "one." If I have two pens on my desk lying close to one another I might see them as "a pair of pens." In so doing I am grouping the pens together into one set or unit. A couple is itself a unit; it is *one* pair. Metaphorically one could say, two (the couple) is one (unit). Thus the idea of "two" is ambiguous; from one perspective, the couple has an inner structure—the two is composed of multiple units; from another, it is a unified whole.

Another way of seeing this is to think of "one" geometrically—as a line, say the interval [−1, 1] of all real numbers between −1 and 1. Then "two" could be thought of as the line with symmetry, where every number x is associated with $-x$. As mappings, "one" could be represented by the identity map on the interval $i(x) = x$. "Two" would be represented by the map $t(x) = -x$, where $t(t(x)) = x = i(x)$, that is, this symmetry has order 2. In other words, "two" is a unity with a deeper structure. Indeed we could look at all the (small) positive integers as unities with additional structure. Another way of saying this is that as ideas the small positive integers are insights into the more subtle aspects or properties of the idea "one." This is perhaps what psychologist Carl G. Jung and his followers mean when they refer to the small positive integers as archetypes: ideas that structure both the outer natural world and our interior mental world.[13] Whatever opinion one has of Jung's thought, the important point is that the small positive integers are among the most basic ideas.

The ability that enables human beings to recognize a couple of objects *without counting the elements* and to differentiate them from a single object is called "subitizing." Subitizing is the ability to recognize instantaneously the cardinality of small collections of objects. This ability appears in human infants at a young age, in fact, before they have learned to speak.[14] This "numerosity" or propensity for number may be hard-wired, so to speak, in human babies (and many animals) and not learned. Thus the ambiguity, and therefore the flexibility, of the idea of (counting) number would develop out of this, more basic, biological substratum.

• • • • •

The next elementary idea that I intend to take up is the notion of *equality*. The idea of equality, along with the various symbols by which equality of one sort or another is represented, is to be found everywhere in mathematics, yet any teacher who has looked at the way the "equal" sign is misused by students is forced to conclude that there is something in the concept of equality that is subtle and difficult to understand. Why is this? Doesn't "equal" mean "the same" or even "identical"? Can't we always use the metaphor of a balance where equal quantities balance and unequal ones do not?

The first thing to notice about the use of equality in mathematics is that there is an ambiguity present. Again, the ambiguity is not an imprecision. Rather, in the same way that a metaphor says, "*A* is *B*" when it is obvious that "*A* is not *B*, " so "1 + 1 = 2" says that "1 + 1" *is* 2 when it is clear that "1 + 1" is not identical to "2"; after all, "1 + 1" is "1 + 1," that is all. You could say that "1 + 1" is "2" with additional structure. For example, "5 =

3 + 2" but it also equals "4 + 1." So "3 + 2" actually carries more information than does "5" alone. This consideration alone dispenses with the idea that "equality" is so obvious that there is nothing to say about it.

In fact, one of the most basic intellectual tasks involves the classification of objects. Any classification depends on the notion of "sameness," which is in many ways a synonym for "equality." A noun, like "tree," is a classification; every object is either a "tree" or a "non-tree." The act of naming implies that there is a criterion for determining when we will call an object a "tree"; that is, naming determines a set of trees which is defined by criteria that allow us to determine which objects are members of that set and which objects are not. All trees are the "same" in their "treeness," that is, insofar as they belong to the set of trees. Thus naming involves abstracting some particular property of objects and using this criterion to establish the relationship of "sameness" between them. Of course, any two objects are the same with respect to certain criteria but different with respect to others. The process of establishing criteria for "sameness" is fundamental to language and no less to mathematics. The classification of objects is usually considered to be obvious and a mere matter of common sense, but, as Steven Pinker[15] points out, "there is nothing common about common sense." Classifying objects is a subtle affair, as is seen when we attempt to program computers to identify the "sameness" that human beings usually determine so effortlessly.

In mathematics we have a series of notions that are related to this idea of "sameness." Of course at its extreme to be the same means to be identical. Some authors now use the symbol ":=" to mean "defined to be," as when we change variables by writing $x := t + 3$ or defining a set A as $A := Z$. This usage was forced upon us by the need that computer languages have for extreme clarity. However most uses of "=" are not so simple nor so clear. Above and also in Chapter 1 I discussed some of the complexity associated with the use of the equality sign in equations like "2 + 3 = 5" or "2 + 3 = 4 + 1." I said "=" was understood in an operational sense. University students often preface every line of a calculation with this ubiquitous "=" sign, whether it is appropriate mathematically or not. We saw earlier that even an elementary algebraic equation such as "$x + 2 = 5$" is more of an event than an object. It evokes a certain class of objects, namely, the set of solutions to the equation, where, again, the solution set designates not only those numbers that are solutions but also those numbers that are *not* solutions to the equation. Again, the equation is usually defined relative to a definite mathematical context. However if that context is absent or inadequate the equation *evokes* a context. One could say that an equation is a mathematical idea that carries with it one or more optimal contexts that give it meaning.

Then there are uses of the equality sign that look like equations but are actually identities. Like $x^2 - 1 = (x + 1)(x - 1)$, they are valid for all values of the variable x within the understood context. Such identities may have a geometrical interpretation, as in the case of $(x + y)^2 = x^2 + y^2 + 2xy$, and, in fact, the Greeks regarded them as a kind of geometric algebra. Trigonometry contains a series of famous identities such as $\sin^2(x) + \cos^2(x) = 1$. They tell us something about the relationship between the various trigonometric functions. However identities represent quite different mathematical ideas than do equations.

Equality is normally thought of in terms of equations, but there is a logical notion of equality that is basic to the formal dimension of mathematics. Two propositions are formally "the same" if they are logically equivalent. If **P** and **Q** denote the two propositions, then logical equivalence means **P** is a consequence of **Q** and vice versa. In words we say **P** *if and only if* **Q** or **P** *is a necessary and sufficient condition for* **Q** and write **P** $\Leftrightarrow$ **Q**. I will refer to this situation as a *tautology*. Most people take logical equivalence to mean that there is no difference between the propositions **P** and **Q**. Yet this is not the case at all.

Let us look back at a couple of examples that make this point. When we talked about infinite sets we discussed the proposition, "A set is infinite if and only if it has the same cardinality as a proper subset of itself." For example, the natural numbers $\{1, 2, 3, \ldots\}$ have the same cardinality as the even numbers $\{2, 4, 6, \ldots\}$. The existence of a subset of equal cardinality is initially a surprising fact, *the* surprising fact about infinite sets, and yet saying that such a subset exists turns out to be equivalent to the nonfiniteness of the set. In other words, the proposition gives us a new way to think about infinite sets. Recall that the fact that a set could be "equal" to a subset was initially considered to be paradoxical because it broke Euclid's rule that "the whole is greater than the part." The proposition in question, therefore, reveals a new way of thinking about infinite sets. How can one say that it is "merely" tautological?

The second example comes from the domain of the real numbers. Recall that rational numbers can be thought of as ratios of whole numbers, but they may also be thought of as decimals. In this latter context we have the proposition, "A real number is rational if and only if its decimal representation is finite or eventually repeating." Thus $.25 = 1/4$ and $.313131\ldots = 31/99$. It is a surprising fact that the rationals have a representation as decimals that differentiates them so neatly from the irrationals. Of course this criterion not only tells us something about rational numbers but also allows us to easily pick out the irrationals for they will have nonrepeating decimal representations. It makes the irrationals accessible in a way that they were not before. Thus the dual representation gives us new information. There is no way one can say that looking at a fraction as the quotient of integers is identical to looking at it as a repeating decimal.

Not unexpectedly, I maintain that tautologies in the form of logical equivalences are ambiguous. They are ambiguous because what they do is to compare two frames of reference and show that they are really both referring to the same situation. Of course some tautologies are trivial and some reveal important new mathematical insights. Both of the above propositions are important because they reveal a deeper structure to the situations that are being investigated. Nevertheless, one has to differentiate between the "formal" point of view in which a tautological statement is merely restating the same fact in different language and the "ambiguous" point of view in which these equivalences may say very important things indeed. A question that has often been asked about mathematics is the following, "If mathematics is merely tautological, how do we ever do anything that is new? Why is mathematics so successful in so many ways?" The answer to this question is frankly obscure from a formalist perspective. From my perspective, on the other hand, certain tautologies are valuable precisely because they are ambiguous and so contain a multiple perspective that expands our understanding of the mathematical situation in question.

In this regard consider the normal situation in mathematics that is summed up by the statement (due, I believe to the mathematician John Kelley), "A fundamental theorem becomes a definition."[16] The theorems one has in mind here are precisely the "if and only if" propositions I am discussing above. With the proposition about infinite sets we can now *define* an infinite set to be one with the same cardinality as a proper subset. In this case it is not clear what is gained by doing so. However, there are many cases where one of the equivalent conditions can be extended more easily than the other to a more general situation. For example, there are many equivalent ways to define the continuity of a function of one variable, $y = f(x)$. One of these involves the way the function treats converging sequences and generalizes quite naturally to metric spaces, situations in which distance is defined. Another involves the way the function treats certain kinds of sets that are called "open" and generalizes to what are called "topological spaces." Thus each of the original definitions gives us another way of thinking about what continuity means. It adds new flexibility to our understanding of the concept.

● ● ● ● ●

Yet another form in which "equality" arises in mathematics is the notion of equivalence. What does equivalence mean? Think, first of all, about fractions. Two fractions are equivalent if they represent the same number: $1/2, 2/4, 3/6, \ldots$ are all equivalent fractions. All equivalent fractions have the same value, so when we refer to a fraction we may be referring either to their common value or to one specific numerical example such $3/6$. Thus there is an ambiguity here.

Another example comes from Euclidean geometry, where two triangles are equivalent (the actual word that is used is "congruent") if one fits exactly over the other. A further example comes from modular arithmetic, let us say, for instance, clock arithmetic, where two numbers are equivalent if they differ by twelve: 1, for example, is equivalent to 13 or to 25.

Now each of these notions of equivalence is a form of "sameness." $1/2$ is the same as $4/8$. Two right-angled triangles with sides 3, 4, 5 are the *same* no matter where they are located in the plane. If 12 hours have passed then the arms of a clock are in exactly the *same* position as they were before. These examples of equivalence can be abstracted into a general definition of an *equivalence relation* **R** on a set *X*. This is a relationship between elements of the set; we shall use the notation *x***R***y* to mean that *x* and *y* are in the given relationship "**R**." This relationship is subject to three very natural rules: (i) *x***R***x*, that is, every element has the relationship *R* with itself;[17] (ii) *x***R***y* implies *y***R***x*, that is, the relationship is symmetric; and (iii) if *x***R***y* and *y***R***z* then *x***R***z*, that is, the relationship is transitive. Such a relationship partitions the set into mutually disjoint subsets where the elements of any subset are in the given relation to one another. For example, suppose the set is the real numbers and the relationship is "differing by an integer," that is, *x***R***y* means that $x - y = n$ for some integer n.[18] For example, the number .5 is in the relationship **R** with the numbers $-.5$, ± 1.5, ± 2.5, and so on. **R** divides the real numbers into classes that typically look like $\{x, x \pm 1, x \pm 2, \ldots\}$, where x is any given real number.

A very basic mathematical operation involves constructing the "quotient space" obtained by "dividing out by the relation." What is meant by this? A new space is formed and designated by $X/\mathbf{R}$ whose elements consist of the equivalence classes. Notice that what is going on here is that a deliberate ambiguity is introduced—the relationship or rather a whole class of equivalent objects in X is now considered to be an object or point in $X/\mathbf{R}$. Denote the class that contains the element x by $[x]$. In the above example each class would have a unique representative between 0 and 1, so we could think of the quotient space as the interval [0, 1] with 0 and 1 identified (since they differ by an integer). Topologically this would give us a circle. This gives us a new (ambiguous) way to think about a circle—a circle consists of the real numbers "divided" by the above relationship, $X/\mathbf{R}$, or it is the unit circle, S^1, around the origin in 2-space, $\{(\cos 2\pi x, \sin 2\pi x)\}$, where $2\pi x$ is the angle that the line joining the point to the origin makes with the positive x-axis. The connection between these two representations is given by the angle $2\pi x$; more specifically, the element $[x]$ of the quotient space representation will correspond to the point with angle $2\pi x$. There is a natural mapping from the reals to the quotient space, $X/\mathbf{R}$, given by $p(x) = [x]$ for any real number x ("p" stands for "projection"). Geometrically the map $p(x)$ consists of spiraling the real line around the circle over and over again (see figure 5.2). In words

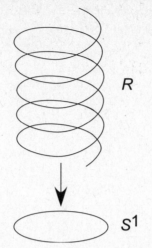

Figure 5.2. The real line "covers" the circle

we say that the real line covers the circle or the line is a "covering space" for the circle.

The fact that the real line is a covering space for the circle implies that there is a deep geometric and analytic relationship between the line and the circle. For example, there is a relationship between continuous functions from the circle to itself and functions on the line that is described by figure 5.3. This means, for example, that a function from the circle to itself, for example the function $f(\cos \Theta, \sin \Theta)) = (\cos 2\Theta, \sin 2\Theta)$ which wraps the circle around itself twice can be "lifted" to a map, $F(x)$, of the line into itself. In this case the function $F(x) = 2x$. In general, this allows us to measure how many times a given continuous map "winds" the circle around itself (here two times). The *index* of rotation and of a map of the circle is given by $F(1) - F(0)$, which must be an integer. (In our example the index $F(1) - F(0) = 2$.) It is an important concept in the theory of complex functions of one variable.

On the other hand, a function $F(x)$ defined on the real numbers would give rise to a function on the circle if $F(x)$ preserved the equivalence classes, that is, if the value of the function was independent of the choice of which representative of the class you chose. Thus $F(x) = 2x$ gives a good function on the quotient space since

215

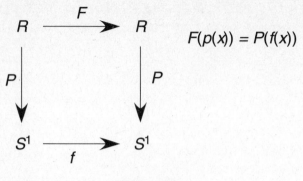

$$F(p(x)) = P(f(x))$$

Figure 5.3.

$y = x + n$ for some integer $n \Rightarrow F(y) = 2y = 2x + 2n = F(x) + 2n$.

On the other hand $F(x) = x^2$ would not give a function on the circle since, for example, $4/3$ and $1/3$ differ by an integer but $F(4/3) = 16/9$ and $F(1/3) = 1/9$ do not.

What is the point of these kinds of constructions? In the first place, we have a formalization of the notion of "sameness," here called equivalence. As usual, such a notion divides up the universe (here the set) into larger objects—it is a classification or naming scheme. Then mathematics makes up a new universe of discourse whose units are not the original elements but the classifications. Often the new universe inherits a structure from the old. The essence of what is going on involves multiple representations of mathematical objects—a process (equivalence) that becomes an object (equivalence class).

When the notion of equivalence is applied to whole categories of mathematical objects, groups, rings, topological spaces, and so on, we have the variety of sameness that is called *isomorphism*. Isomorphism is fundamental to any mathematical subject—in a sense it defines the subject. Each category of mathematical objects carries with it the appropriate notion of isomorphism, which is a formal way of saying that two objects are identical from the point of view of that particular subject. For example, two sets are isomorphic from the point of view of Cantor's theory if they have the same cardinality. If you are studying metric spaces, where the abstract notion of distance is defined, then the isomorphism will be called an *isometry*, a distance preserving mapping. If the study is topology, then the relevant isomorphism is called a *homeomorphism*. In fact, the general notion of "isomorphism" does not only refer to a relationship between two different mathematical objects—as we saw in Chapter 4, a mathematical object can be isomorphic to itself or to a part of itself in a complex way. Thus an infinite set has the same cardinality as a proper subset or a fractal is self-similar to a part of itself.

After you say what you mean by two mathematical objects being the same (or "isomorphic") in a certain context, you are faced with the problem of actually determining whether two specific objects are or are not isomorphic. This is the problem of classification and it is, in a way, the ultimate question in any area of mathematics. For example, one of the great successes of finite group theory involved the classification of all possible simple groups. The key question in topology might be "When are two topological spaces homeomorphic?"[19] that is, the classification of all topological spaces up to homeomorphism.

If the classification of objects up to isomorphism is the ultimate aim of a mathematical theory, in practice we usually settle for something less—determining when two objects are *not* isomorphic. This is done by means of *invariants*. An invariant is some number or other mathematical object that is the same for all elements of an isomorphism class. For example, two triangles that are congruent in Euclidean geometry have the same area. Thus area is an invariant of congruence.

The study of topology might proceed by producing invariants associated with the relationship of homeomorphism. The most elementary of these might be the number of connected components of the space, that is, the number of connected pieces that the space naturally is divided up into. Thus one circle is not homeomorphic to two circles, and the Roman numeral I is not homeomorphic to III. However this invariant can be used in a more subtle way. Consider a line segment, like the set of real numbers between 0 and 1, [0, 1]. If we take out any number except 0 and 1 the interval breaks down into two subintervals. Compare this to the circle: if we remove *any* point of the circle we still have only one piece. It follows that the circle is not homeomorphic to the interval. Similarly the figure X is not homeomorphic to the figure Y, nor is a figure eight homeomorphic to a circle.

Now the invariants associated with a given mathematical subject like a topological space may even be another mathematical object in its own right. While algebraic structures like groups, rings, and fields also have their own notion of isomorphism and thus classes of invariant objects associated with them, an entire algebraic structure may be considered as an invariant of a given topological space. This is the idea behind the field of algebraic topology, where the algebraic structure is seen as a measurement of some topological feature of the space such as the number of "holes" it possesses.

Geometry and physics are very much taken up with the question of determining the invariants, the things that do not change, under different groups of transformations. Transformations are functions that move the elements of the space around. Of course there is a dual relationship between invariants and transformations. A certain invariant determines the set of transformations that preserve that invariant. For example, area is preserved by translations and rotations of the plane. On the other hand a group of transformations implicitly determines a set of invariants associated with that group; for example translations and rotations of the plane preserve quantities like area, length, and the magnitude of angles. Each set of invariants determines a classification of objects; two objects are the same if the invariants have the same value for both objects.

The notion of sameness in its various guises of equality, identity, tautology, equivalence, and isomorphism is absolutely fundamental to mathematical thought. However, what is happening here is not merely the classification of preexisting data but that the various classification ideas themselves generate new mathematics. All of this is based on a fundamental metaphoric ambiguity—"A is B," which stands for "A is the same as B yet A is different from B."

EUCLIDEAN IDEAS

I return now to Eulcid's *Elements* and consider some of his more famous results in terms of the ideas they are built around. Proofs are not mere logical algorithms. Each proposition of Euclid contains an idea that could not have been predicted a priori from the statement of the result. The idea emerges in a discontinuous way—you either get it or you don't. The moment of insight arrives suddenly and is accompanied by a feeling of certainty—you know that the result is true and you know why it is true. In fact, the idea arrives as a result of the mathematician's need for understanding—her need to understand why things are the way they are. The ideas behind the truth of some of the famous results of Euclidean geometry are particularly easy to isolate because the nub of the idea is usually contained in a *geometric construction*. A number of famous example follow.

What do these examples have to do with my theme of ambiguity? Clearly the whole Euclidean enterprise is built upon two obvious frames of reference, namely, geometry and proof. A priori these two domains are disjoint—geometry is related to properties of the natural or geometric world whereas proof is a logical product of human thought. The miracle of Euclid, as indeed it is of much subsequent mathematics, is that these two domains interact in the most fruitful way—that human thought processes can say something intelligible about the world of experience. We usually forget how surprising this is. Geometric facts give us the impression of being right or wrong—either the sum of the angles of a triangle is equal to two right angles or it is not. What proof does is to embed this geometric fact in a larger context and relate it to other geometric and logical "facts" that we deem for the moment to be more elementary. There is no reason why these two worlds should come together, and yet the implicit conflict between the two is resolved in that synthesis of mind and experience that we call Euclidean geometry. But the two do not come together passively. The attempt to prove geometric results and fit them together into a deductive structure produces new results. It is not only that we understand geometry better by creating this axiomatic system but that what we know as geometry comes into being as a result of this process.

The angles of any triangle sum to two right angles

The Idea: *Draw a line CE parallel to the line AB.*

Referring to figure 5.4, $\angle A = \angle DCE$ and $\angle B = \angle ECB$ (because of the parallelism of the lines). Thus the sum of the three interior angles of the triangle is identical to the sum of the three angles

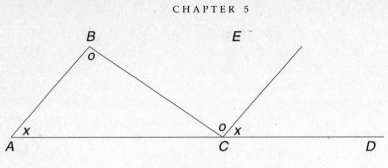

Figure 5.4.

that go to make up the angle at *C*, a straight line, that is, two right angles. Notice that what we are doing is creating an ambiguity—a second "frame of reference" for the sum of the interior angles, namely, the sum of the angles at the vertex *C*. The idea behind the result is clear, but it is only revealed through the idea of drawing the parallel line that reveals the idea. That is the inspiration. Make the proper construction, and the rest is the mere filling in of details. Of course these details depend on Euclid's painstaking construction of a deductive system. Why can we draw the parallel line with only straight edge and compass? Why is the angle at *A* equal to the angle *DCE*? These "facts" have been shown to be true in earlier propositions.

The Pythagorean theorem

This is another famous theorem whose proof depends on an ingenious construction. The Pythagorean theorem states that in the triangle *ABC*, if the angle at *B* is a right angle then we must have $AB^2 + BC^2 = AC^2$.

The Idea: *Draw a line BFG parallel to AE.*

Now observe first that the triangles *ADC* and *ABE* in figure 5.5 are identical (congruent) and so have equal area, but that the area of *ADC* is half the area of the square on the side *AB* and the area of *ABE* is half the area of the rectangle *AFGE*. Thus the construction has divided the square on *AC* into two rectangles, one of which is equal in area to the square on *AB*, the other to the square on *BC*. Again. the proof hinges on looking at the square on the line *AC* in two ways, the second of which is as the

220

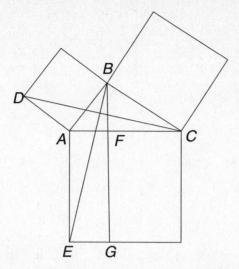

Figure 5.5. Proof of Pythagorean theorem

sum of two rectangles whose areas equals that of the two other squares; also a form of ambiguity.

Now there are many proofs of the Pythagorean theorem. I shall include a second one here to make the point that the question "Why is it true?" can have many answers. Here, again, there is a right-angled triangle with sides of length a, b, and c. We are trying to explain why $a^2 + b^2 = c^2$. Draw two squares with sides equal to $a + b$ (figure 5.6). The area of the first square is $a^2 + b^2 + 2ab$. The area of the second is $c^2 + 2ab$. The conclusion follows. The "idea" here is that the same square can be divided up in these two ways—that there is this ambiguous way of looking at the square on the line $a + b$. It is true that this second proof is more algebraic than geometric in tone and so not really in the spirit of Euclidean geometry.

THE FIVE PLATONIC SOLIDS

Euclid ended his compendium of all the known geometric results of his time with one of the most beautiful and remarkable results in geometry—that there exist precisely five regular convex polyhedra. A regular polyhedron is a three-dimensional figure each of whose faces is the same regular polygon (an equi-

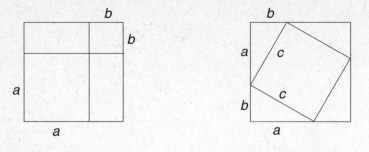

$(a+b)^2 = a^2 + b^2 + 2ab$ $(a+b)^2 = c^2 + 4(1/2ab) = c^2 + 2ab$

Figure 5.6. Alternate proof of Phythagorean theorem

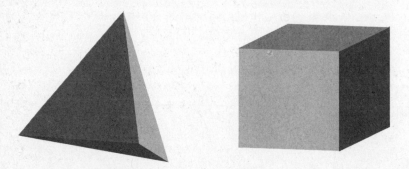

Figure 5.7. The tetrahedron and the cube

lateral triangle, a square, a pentagon, hexagon, and so on). A polyhedron is convex if the line connecting any two points inside the polyhedron is itself entirely interior to the polyhedron. For example, a *tetrahedron* has four faces all of which are equilateral triangles. A *cube* has six faces each of which is a square (figure 5.7).

Since there are an infinite number of regular two-dimensional polygons, one might imagine that there are also an infinite number of regular polyhedra. But that is not the case—there are only five! The idea is as follows.

Suppose that you have such a regular solid figure. Focus on one of the vertices and the polygons around that vertex. Then flatten out that portion of the solid onto a plane and note that by so doing you have created a new, two-dimensional frame of reference for the polyhedron. You will (possibly) increase the angles. The maximum angle would be 360 degrees. Thus we have:

The Idea: *The sum of the plane angles making up a solid angle is less than 360 degrees.*

The proof consists of enumerating all possible cases.

I. The faces are equilateral triangles. Since each angle is 60 degrees, you could have 3, 4, or 5 triangles at each vertex. (6 triangles would give a total angle of 360 degrees, which is not possible.)

I (a). 3 equilateral triangles at each vertex yield a *tetrahedron*.

I (b). 4 equilateral triangles at each vertex yield an *octahedron* (figure 5.8).

I (c). 5 equilateral triangles at each vertex yield an *icosahedron* (figure 5.9).

II. The faces are squares. The only possibility is three squares at each vertex, for four squares would give a total angle of 360 degrees.

III. The faces are pentagons. Again the only possibility is that there are three at each vertex (each interior angle of the pentagon is 108 degrees). This yields a *dodecahedron*. (figure 5.10)

These are the only possibilities and are the only regular convex polyhedrons. Now there is one question left. How do we know that there is only one example for each of the above five cases? For example, how do we know that the cube is the only polyhedron for which three squares meet at each vertex? The reason is connected to the Euler characteristic for a polyhedron that I shall talk about in connection to the work of Lakatos is the next chapter.

EUCLID'S PROOF THAT THERE ARE AN INFINITE NUMBER OF PRIMES

This is a surprising and wonderful result. If you had no experience with prime numbers and you were asked to guess whether there was a largest prime, then it might be difficult to know what to answer. As numbers get larger and larger, there are more and more numbers that might divide into them, so it is conceivable

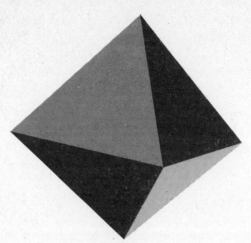

Figure 5.8. The octahedron

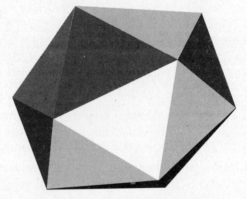

Figure 5.9. The icosahedron

that every sufficiently large number is divisible by some smaller number. You might guess that there are only a finite number of primes. Then, of course, you would have guessed wrong, as Euclid's argument will show.

The Idea: *Suppose we have a finite collection of prime numbers and we create a new number, N, by multiplying all the original numbers together and adding 1. Then either (i) N is prime or (ii) N is divisible by a prime number that is not in the original collection.*

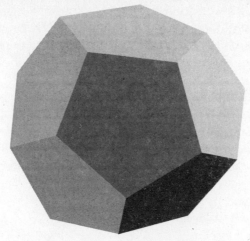

Figure 5.10. The dodecahedron

Suppose we have the primes: $p_1, p_2, \ldots, p_k$ (all primes are larger than 1) and $N = p_1 \times p_2 \times \cdots p_k + 1$. Since N is larger than any of the initial primes, if it is prime then we have found a new prime. If it is not prime, then it is divisible by some smaller prime, call it p. If p is equal to one of the original primes, p_i, say, then p divides evenly into N and $N - 1 = p_1 \times p_2 \times \cdots \times p_k$. Now the only positive integer that divides evenly into two successive integers is 1, so p and p_i must be equal to 1. But we began by assuming that each of the initial primes was greater than 1. Therefore p could not have been one of the original primes, $p_1, p_2, \ldots, p_k$. Again we have produced a new prime. Using this procedure over and over, we can keep producing new primes. For example if we start with $p = 2$ the procedure will produce 3; starting with 2 and 3 will give us 7; 2, 3, and 7 will give us 43; 2, 3, 7, and 43 will give us 1807, which is not prime but is divisible by 13, a new prime; and so on. Thus we have shown that there are a infinite number of primes.[20]

THE AMBIGUOUS IDEA OF EXPONENTIATION

Ambiguity involves the existence of multiple contexts, but as we have seen these contexts may not arise at the same time historically or in the learning process. Sometimes an idea involves the emergence of a second, more general context. We saw this in the case of the equation $x^2 + 1 = 0$, which seemingly evoked the complex numbers. In the fascinating case history that I shall now discuss, ambiguity of various sorts is the essence of what is going on.

The most elementary way of thinking about the exponential function, e^x, involves a number raised to a power—a generalization of more elementary power relationships, like $2^4 = 16$. Of course here e is the transcendental number 2.7182818... On the other hand e^x can be represented as a power series:

$$e^x = \sum_{n=0}^{\infty} \frac{x^n}{n!} = 1 + x + \frac{x^2}{2!} + \frac{x^3}{3!} + \cdots .$$

This latter formulation is extremely fruitful. Differentiating term by term we see that this function is its own derivative. Since $e^0 = 1$, e^x is the unique functional solution to the simplest possible differential equation, $dy/dx = y; y\,(0) = 1$. Thus we have two contexts— e^x as a power function and e^x as a power series. The first context does not explain what is so special about the number e or the function e^x as opposed to other power functions like 2^x, for example. The second singles out e^x as special in a certain way, but it does not immediately explain what the exponential function has to do with taking powers. The two ways of looking at exponentiation describe the same situation from two different perspectives. Both perspectives are valuable.

The power series representation is indeed very powerful. This can be seen by comparing the power series representation of the exponential function to that of the trigonometric functions $\sin(x)$ and $\cos(x)$. At first glance the trigonometric functions and the exponential function have nothing whatever to do with one another. The graphs of $\sin(x)$ and $\cos(x)$ are periodic and bounded between the values of -1 and $+1$, whereas the graph of $\exp(x)$ is nonperiodic and unbounded. The exponential function involves the mysterious transcendental number e. In the definition of sine and cosine e is nowhere to be found.

Yet there is a deep connection between these functions. So basic is this connection that it would make sense to say that the exponential function and the two trigonometric functions above are really the "same" function. By now this is the way we have

learned to look at an ambiguous situation: exponential and trigonometric functions are different—the exponential function is
the exponential function and the trig functions are just that—but
at a deeper level they are the same.

A hint of the deeper connection emerges when all of these
functions are written out as power series:

$$\sin(x) = x - \frac{x^3}{3!} + \frac{x^5}{5!} + \cdots,$$

$$\cos(x) = 1 - \frac{x^2}{2!} + \frac{x^4}{4!} + \cdots.$$

$$\exp(x) = 1 + x + \frac{x^2}{2!} + \frac{x^3}{3!} + \cdots.$$

Here the expansion for sin(x) has the odd powers of x, the
expansion for cos(x) has the even powers, and the expansion of
the exponential has *all* the powers. This is a hint that there is
indeed a connection to be made between these functions. Yet the
connection is not obvious because of the alternating positive and
negative terms in the first two series as opposed to the uniformly
positive terms of the latter one. Indeed, the connection does not
appear if x is restricted to "real" values. However, suppose we
expand our vision to include complex numbers and, in particular, introduce the imaginary number i, where $i^2 = -1$. The formula for the exponential function then becomes

$$\exp(ix) = x + ix - \frac{x^2}{2!} - \frac{ix^3}{3!} + \frac{x^4}{4!} + \frac{ix^5}{5!} - \frac{x^6}{6!} - \frac{ix^7}{7!} + \cdots$$

$$= \left(1 - \frac{x^2}{2!} + \frac{x^4}{4!} - \frac{x^6}{6!} + \cdots\right) + i\left(x - \frac{x^3}{3!} + \frac{x^5}{5!} - \frac{x^7}{7!} + \cdots\right).$$

That is, exp(ix) = cos(x) + i sin(x) the famous Euler equation.

It follows that the exponential function (for imaginary values)
is totally determined by the sine and the cosine. Conversely we
can solve for sin(x) and cos(x) if we know the values of the imaginary exponential. That is,

$$\cos(x) = \frac{1}{2}\left(e^{ix} + e^{-ix}\right),$$

$$\sin(x) = \frac{1}{2}\left(e^{ix} - e^{-ix}\right).$$

There exist in this situation many of the elements of the kind of ambiguity that has been described: an incompatibility at one level that is replaced by a unity at a higher level. The new point of view reveals things that would not even be suspected in the original framework. For example putting $x = \pi$ in the Euler equation gives us

$$\exp(i\pi) = \cos(\pi) + i\sin(\pi) = -1,$$

that is,

$$e^{i\pi} + 1 = 0.$$

This one equation reveals a completely unexpected relationship between five of the most important constants in mathematics.

This leads us to consider the complex exponential function $e^z = e^{x+iy} = e^x\, e^{iy} = e^x(\cos y + i\sin y)$. Now every complex number $z = x + iy$ can be considered to be a point in the plane with coordinates x and y. It is interesting for our discussion that points in the plane naturally have a dual representation. They can be represented by their x- and y-coordinates and also by their distance from the origin, $r = \sqrt{x^2 + y^2}$, together with the angle Θ that the line from 0 to z makes with the positive x-axis. The way we have written e^z above shows us that the angle in this case is y and the distance $r - e^x$. In particular, the Euler equation gives us a representation for the y in e^{iy} as an angle in the complex plane.

Now it is interesting that this unification of the trigonometric and exponential functions is not achieved at the level of the real numbers. As real-valued functions, the exponential and trigonometric functions are indeed "incompatible." However, this incompatibility is reconciled at the level of "complex" functions[21] and this higher level unity is made explicit by the Euler formula.

Since exponential functions have the nice property that $(e^{ix})^n = e^{inx}$, the Euler equation gives rise to the De Moivre formula: $(\cos\theta + i\sin\theta)^n = \cos n\theta + i\sin n\theta$. This formula helps us to resolve an anomaly of the real number systems that concerns the roots of numbers—for example, positive numbers like 4 have two square roots, namely ± 2, but negative numbers have no square roots at all. As for cube roots, in the system of real numbers all numbers have exactly one cube root, also one fifth root, and so on. The whole situation is irregular and vaguely dis-

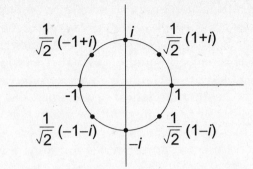

Figure 5.11. The eighth roots of unity

satisfying. The De Moivre formula tells us that if we are looking for the square roots of −1, for example, any such roots would have to satisfy

$$\cos(2\theta) + i\sin(2\theta) = -1.$$

Now, −1 has the angle π (or 180 degrees with the positive x-axis). The question becomes which angles, when doubled, give an angle of 180 degrees, giving us the two angles of 90 and 270 degrees. This gives rise to the two square roots of −1, the imaginary numbers $\pm i$. This method is completely general; for example the 8th roots of 1 are obtained by dividing the unit circle by the eight angles $\pi/4$ (45 degrees), $\pi/2$, $3\pi/4$, π, $5\pi/4$, $3\pi/2$, $7\pi/4$, 2π (or 0). These are the complex numbers ± 1, $\pm i$, $1/\sqrt{2}$ (± 1, $\pm i$) (figure 5.11). Every nonzero complex number has n distinct nth roots that are arrayed in a beautiful geometric pattern—evenly spaced around a circle of the appropriate radius. Our various multiple representations of complex numbers and exponential functions have given us a way of consolidating our understanding of questions that, although they can be formulated in the domain of real numbers, cannot really be resolved until we are pushed out of the reals into the complex realm. There they seem to be resolved in a natural way.

Analogy is a simple form of ambiguity, but it often produces important mathematical ideas—indeed, reasoning by analogy seems to be a very basic way of thinking. In his books on heuristics in mathematics, George Pólya[22] places a great deal of stress on thinking by analogy. An analogy is similar to a *metaphor* in that it involves a comparison between two situations. Like a metaphor, a fruitful mathematical analogy involves a leap—the realization that a certain kind of generalization may be fruitful.

In the power series representation for the exponential function one is tempted to ask (by analogy) what it would mean if the real variable, x, is replaced by a complex variable z. Making this substitution makes mathematical sense since the power series involves the ideas of addition, multiplication, and convergence, all of which makes sense for complex numbers as well as for real ones. If you use the power series to define a new complex function

$$e^z = \sum_{n=0}^{\infty} \frac{z^n}{n!} = 1 + z + \frac{z^2}{2!} + \frac{z^3}{3!} + \cdots,$$

it turns out that you now have another way to represent the complex exponential function that we earlier defined by $e^z = e^{x+iy} = e^x(\cos y + i \sin y)$. Almost all the interesting properties of the real exponential function carry over to the complex case. These include the various exponential laws like $e^x e^y = e^{x+y}$. The complex exponential function also has the vital property that its derivative is equal to the original function, that is, it is a solution to the differential equation $dw/dz = w; w(0) = 1$.

In fact there is only one natural way to extend the real valued function exp (x) to the complex function exp (z) = exp $(x + iy)$ = exp $(x)[\cos(y) + i\sin(y)]$. Thus one could say that the complex functions are implicit in the real functions and therefore that the Euler unification is also implicit even in the real number formulation.

Now a power series makes sense in many situations and so the analogy can be pushed further. For example, if we are given a square matrix, $\mathbf{A}$, we might define the *exponential of a matrix by setting*

$$e^A = I + A + \frac{A^2}{2!} + \frac{A^3}{3!} + \cdots.$$

As usual when one writes down an infinite series, one must ask whether the series converges. In this case one can show that the exponential of a matrix is indeed another matrix of the same dimensions as the one we began with. When we begin to investigate possible properties of the exponential of a matrix we start with the properties of the ordinary exponential function and attempt to prove analogous properties for the matrix exponential.

The real, complex, and matrix exponential functions are just the beginning of the story of the idea of the "exponential." This idea has applications in a disparate variety of mathematical situations from differential equations to differential geometry and Lie groups.

The Idea of "Continuity"

In the discussion of Euclidean geometry, I mentioned that there are hidden assumptions in Euclid that are never made explicit. One of these, I said, was the assumption that two arcs that appeared to intersect one another actually did intersect.[23] It is not surprising that this detail is missing in Euclid because it turns out to depend on the notion of "continuity." Continuity is one of the deepest and most interesting ideas in mathematics. Indeed, continuity is half of one of the most basic dualities in all of science—the discrete versus the continuous. Are space and time discrete or continuous? This was an important question for the Greeks, the basis for the paradoxes of Zeno.

Thinking about continuity has gone on for thousands of years and in mathematics has only reached a certain state of stability in the relatively recent past. A good place to begin is with a *geometric picture*.

The Idea: *A curve is continuous if it can be drawn without lifting the pen from the paper.*

As we shall see, there is a surprising amount of mathematics hidden in this geometric idea. Now, lifting the pen from the paper would leave a "gap," so the idea here is that continuity means "no gaps"—a continuous curve being one with no gaps. But what does it mean to say that a line has "no gaps"? How can we give this idea mathematical substance? It turns out that "no gaps" is quite a subtle idea that depends very heavily on the mathematical context in which it is considered. In the Greek world of rational numbers, all curves have "gaps" in the sense that curves that seem to intersect can, in fact, pass right through one another without touching. For example, the line given by the equation $y = 0$ (the x-axis) has no rational point in common with the parabola $y = x^2 - 2$, since the normal points of intersection have coordinates $(\pm \sqrt{2}, 0)$. These two curves pass through each other without meeting in the world of rational numbers. In other words, there is a hole or gap in these curves corresponding to every point with at least one irrational coordinate.

These kinds of gaps can be filled in by expanding our mathematical context from that of the rational numbers to that of the real numbers. Now, physically there is no difference between a "rational line" (a line that contains only points with rational coordinates) and a "real line" (a line with points with real numbers as coordinates). The difference between the two must therefore be a matter of deciding which is the more useful context for capturing the essential mathematical intuition of "continuity." The reals are "complete" whereas the rationals are not. Completeness is therefore a way of formalizing the idea of "no gaps." It was discussed on page 139 and means nothing more or less than that every sequence of numbers that should converge actually does converge to some specific number.

Does the completeness of the real numbers actually guarantee that the real line has "no gaps"? Most mathematicians would say that it does. Yet on closer examination we can see that it all depends on the context, that is, on what you mean by a gap. For example, in our discussion of infinitesimals (pp. 181–185), we say that this system allows "numbers" of the form $x + h$, where x is a real number and h is an infinitesimal to be placed between any real number x and any other real number. Thus, in this system, even the real numbers have "gaps"!

The idea that the graph of a continuous function must not contain a "gap," can be directly reformulated into the following:

Proposition 1: *Suppose a function $f(x)$ is continuous on the interval $[a, b]$ and has $f(a) < 0$ and $f(b) > 0$. Then $f(c) = 0$ for some c between a and b.* ∎

In geometric language, if the graph starts below the x-axis and ends above it, it must cross the axis. To miss zero you would have to raise your pen from the paper. It turns out that the existence of such an intermediate value, c, is guaranteed by the continuity of the function $f(x)$. Viewed in this way the proposition is really nothing but a restatement of the idea of continuity. Nev-

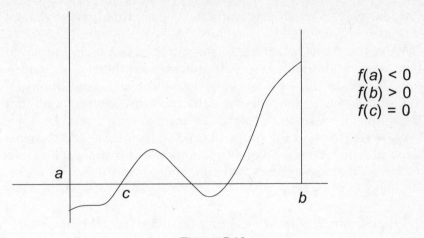

$f(a) < 0$
$f(b) > 0$
$f(c) = 0$

Figure 5.12.

ertheless for students in a modern course in mathematical analysis it is a fairly technical and difficult result.

This proposition could be restated in a more general form:

Intermediate value theorem: *If a continuous function defined on the interval [a, b] has f (a) < k and f (b) > k, then f (c) = k for some intermediate value c between a and b.*

Again, this is just a restatement of the idea of continuity. Or else we could derive it from the first proposition through the idea of setting the function $F(x) = f(x) - k$.

The intermediate value theorem may seem obvious, but it has interesting consequences. It is an example of an existence theorem. It tells you that a certain number exists (the solution to the equation $f(x) = k$) but not what it is. This is yet another example of the type of ambiguity that was discussed in Chapter 1, where one knows the number in one sense but not in another. Yet even if one does not know the solution explicitly just knowing that a number or a solution exists is important.

For one thing, the intermediate value theorem gives you a method to compute the number in question. Knowing $f(a) < k$ and $f(b) > k$ tells you that the solution lies between a and b. Then look at the midpoint between a and b. Call it a_1. If $f(a_1) < k$, then the solution, c lies between a_1 and b. If $f(a_1) > k$, then the solution lies between a and a_1. We can continue in this way, obtaining a sequence of approximations to c. This is a way (but not the most

233

efficient way) of obtaining numerical approximations to the solutions of equations.

We can use this reasoning to show that $\sqrt{2}$ is a real number. In fact, what we look for is a real number such that $a^2 = 2$ (that is, $a = \sqrt{2}$). We do this by observing that $f(x) = x^2$ is continuous, $1^2 < 2$, and $2^2 > 2$, so the intermediate value theorem tells us that there must be a solution to $a^2 = 2$ lying between 1 and 2.

Now why restrict the proposition discussed above to the situation in which the graph of a function crosses a straight line (the x-axis)? What if we have two functions, f and g, where $f(a) < g(a)$ and $f(b) > g(b)$?

Proposition 2: *If two functions, $f(x)$ and $g(x)$, are continuous on the interval $[a, b]$, with $f(a) < g(a)$ and $f(b) > g(b)$, then $f(c) = g(c)$ for some c between a and b.* ∎

The idea is merely to look at the function $f - g$, which is negative at a and positive at b, and apply the first of our propositions.

This idea gives us the most elementary example of something that is very important in mathematics, namely, fixed-point theorems.

Proposition 3: *If $f(x)$ is a continuous function that maps the interval $[0, 1]$ into itself, then f has a fixed point, that is, $f(x^*) = x^*$ for some x^* between 0 and 1.* ∎

The argument is immediate from figure 5.13. If $f(0) = 0$ or $f(1) = 1$, then we have a fixed point. If not, we apply Proposition 2 to the functions $f(x)$ and $g(x) = x$. Since $f(0) > 0$ and $f(1) < 1$, we must have a point where the two graphs intersect, that is, $f(x) = x$.

FUNDAMENTAL THEOREM OF CALCULUS

This important theorem was discussed in Chapter 1 as an example of ambiguity. Recall that it states that differentiation and integration are inverse processes. The argument contains the simple geometric idea that we have been discussing. The argument is interesting in itself, but it also allows us to distinguish be-

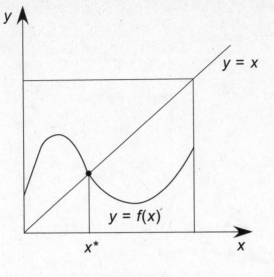

$$f(x^*) = x^*$$

Figure 5.13. Fixed point theorem

tween the result itself and the idea behind the proof, that is, to distinguish between what it says and why it is true.

In order to discuss why the theorem is true, we must first formulate the theorem in a way that allows the idea to be seen clearly. We shall look at integration followed by differentiation. Thus let

$$F(x) = \int_a^x f(t)\, dt.$$

The theorem states that the derivative of $F(x)$ equals $f(x)$, $F'(x) = f(x)$. Recall that the derivative of $F(x)$ is defined to be the unique number that is approximated by the difference quotient

$$\frac{F(x + h) - F(x)}{h}.$$

With reference to figure 5.14, this difference is equal to the area under the curve $f(x)$ between x and $x + h$.

The Idea: *The area under the curve is equal to $hf(\xi)$, where ξ lies between x and $x + h$.*

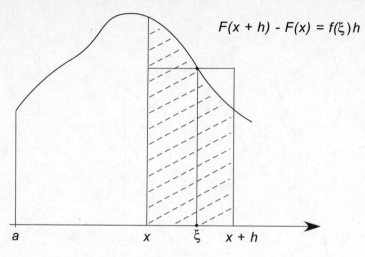

$$F(x + h) - F(x) = f(\xi)h$$

Figure 5.14.

Basically the idea is another variation on the theme of continuous variation. The area of the region in question $(F(x + h) - F(x))$ lies between mh and Mh, where m is the minimum and M is the maximum of the function on this interval. Thus it is equal to $f(\xi) h$, where ξ is some value between x and $x + h$. The quotient

$$\frac{F(x + h - F(x)}{h} = f(\xi),$$

and since ξ lies between x and $x + h$, as h tends to zero ξ is forced toward x and so the difference quotient tends to $f(x)$.

Pancake Theorems

There are innumerable applications of this idea that a continuous variable must go through all possible values within its range. One is called the "pancake theorem." Suppose an irregularly shaped pancake lies on a platter. Can one cut it in half with one slice of a knife? Consider positioning the cut initially underneath the pancake and gradually moving it up so that finally it lies completely above the pancake. Since the movement is continuous, there must be some intermediate position where the pancake is cut exactly in half.

Now what if there were two pancakes? Can they *both* be cut in half with just one cut of the knife? In mathematical language,

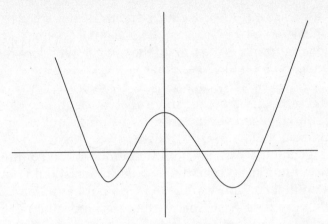

Figure 5.15. The graph of a generic quartic polynomial

Theorem: *If A and B are two bounded regions in the same plane, then there is a line in the plane which divides each region in half by area.*

For details of the proof, check the very nice little Mathematical Association of America book by Chinn and Steenrod. The main idea is precisely the intermediate value theorem.

THE FORMALIZATION OF CONTINUITY

Continuity begins as a simple geometric observation—certain curves can be drawn without lifting the pencil from the paper—and it can be carried quite far without heavy formalization. We can draw the graphs of the simple functions: linear, quadratic, in fact the graphs of all polynomials are easily seen to be continuous (figure 5.15).

However, the geometric definition of continuity runs into difficulties when one begins to consider functions that are defined implicitly, by power series or Fourier series. We have seen that the trigonometric and exponential functions have power series representations. But it is possible to show that for an arbitrary power series, that is, a series of the form

$$\sum_{n=0}^{\infty} a_n x^n,$$

there is a constant r, $0 \le r \le \infty$, such that the series sums to a finite value whenever $|x| < r$. Thus the series actually defines a function $f(x)$ whose value at x is the sum of the series. What does the graph of this function look like? We don't know. If we cannot draw the graph, then we certainly cannot use a geometric criterion to show that it is continuous. Another, more analytic definition of continuity is required. Whatever the correct idea turns out to be, it must satisfy a few criteria. First, the functions we "know" to be continuous by geometric means should continue to be continuous in this new definition. Second, the new definition should not distort our feeling that continuous functions do not have "jumps."

After a very long time, many attempts, and much discussion a consensus definition was arrived at. This turns out to be that $f(x)$ can be forced to approximate $f(a)$ by making x approximate a. Or, $f(x)$ is continuous at a if

For every number $\varepsilon > 0$ there exists $\delta > 0$ such that $|f(x) - f(a)| < \varepsilon$ whenever $|x - a| < \delta$.

This definition satisfies the two criteria stated above. It is a way of formalizing our "feel" for what a continuous function should be. However, when mathematics has (more or less) agreed on what the definition of a continuous function should be, a whole host of other questions are raised. Precisely which functions are continuous according to this definition? On the other hand, can we specify what discontinuities (points at which the function is *not* continuous) are possible? It turns out that writing down a formal definition of function and of continuity vastly enlarged the category of continuous function. Beyond the polynomials and the power series, all of which are relatively "nice," "most" continuous functions are much more complicated. This situation is similar to the situation that was discussed earlier of extending the rational number system to the real numbers. The advantage is that every possible decimal now gives you a real number. The disadvantage is that you now have all these new numbers (remember the transcendentals) that you don't really have any feeling for. Similarly the abstract definition of continuity gives rise to a huge family of continuous functions "most" of which are extremely complex and hard to describe.

The "idea" embodied in the above formal definition of continuity itself seemingly begs to be generalized—and it has been abstracted in many ways. To state just one possible generalization, the definition is based on the idea of distance: $f(x)$ gets "close" to $f(a)$ when x gets "close" to a. Hence we should be able to talk about continuous functions whenever we are in a situation where we can measure distance. It turns out that distance is a very general idea and can be defined in many situations. The most obvious is in higher-dimensional Euclidean spaces—dimension 2, 3, n— these all have various ways in which distance can be understood. Then there are curved spaces, so-called differentiable manifolds, like circles, spheres, toruses (doughnuts), and the like. These spaces are also metric spaces (distances can be defined). Thus, in all of these cases, it is possible to talk about continuous functions. Finally even function spaces, like the space of all continuous functions defined on the interval [0,1], can be given a distance. This enables us to talk about functions of functions and what it would mean for such functions to be continuous.[24]

This discussion only gives a hint of the complexity and ramifications of one deep mathematical idea, continuity. We began with a rather elementary, geometric picture of continuity. We then asked the question, "What is continuity, really? What is really going on?" Then the notion of continuity was formalized. After the notion was formalized there began a whole process of consolidating and generalizing the idea. In this case the formal definition was so complex that the question, "What is continuity? What does the definition really mean?" can now be asked again, this time with reference to the formal definition. This may lead to a new or deeper idea of what is going on, and these new ideas may themselves be formalized and the process will continue until such a time as it is felt that the ramifications of the idea have been sufficiently explored. Yet this is not the only possible way to formalize the idea of continuity—other definitions are possible. Thus the idea is not a single, well-defined object of thought but a whole process of successively deeper and deeper insights.

IDEA CARRIED BY PICTURES

Ivar Ekeland in his book *Mathematics and the Unexpected* discusses mathematical ideas that are carried by certain pictures. He says, "the power of certain pictures, of certain visual representations, in the historical development of science, will be the recurrent theme of this book. It is a power, in the early stages, to initiate progress, when the ideas it conveys are still creative and successful, and it becomes, later on, power to obstruct, when the momentum is gone and repetition of the old theories prevents the emergence of new ideas." Ekeland discusses, for example, the idea of uniform circular motion—a point moving along a circle with constant speed. This picture lies at the heart of the Ptolemaic system that Ekeland calls one of the major intellectual achievements of antiquity. Yet this fixed idea became an impediment to future progress since it made it difficult for astronomers up to the time of Johannes Kepler to see that the motion of the planets could be seen more simply as ellipses than as combinations of circles.

This is another reason why ideas need to be looked at from different points of view, that is, in an ambiguous way. This example makes the point that the "idea" is not rigid or absolute. It is valuable in a certain context but when it is forced onto a situation in an artificial way, when it is considered as true a priori, then it can just as well obstruct the development of the subject. Of course, in the latter case it is not a real idea any more. The true idea only comes alive in a certain situation. It has a dynamic aspect and appears on the scene in a sudden, unexpected way. When the idea is formalized, memorized, and applied by rote it loses its creative power and blocks the emergence of other ideas. Thus mathematics is a subject that is always alive and progressing with new ideas arising in response to changed interests and the ever-changing development of mathematics.

Metaphor

Many important mathematical ideas are metaphoric in nature. I discussed metaphor in Chapter 1. Here I just wish to emphasize the close relationship between metaphors and ideas. A metaphor, like an idea, arises out of an act of creativity. You have to "get" a metaphor just as you have to "get" an idea. For example, I might be considering an entire class of functions, say all the continuous functions, $f(x)$, defined for $0 \le x \le 1$. I might then metaphorically call each such function a "point" in the "space" of all such continuous functions. We have now metaphorically identified the analytic domain of a collection of functions with the geometric domain of "points" and "spaces." This is a powerful thing to have done. To give but one example, the integral can now be considered as a continuous function from the space of functions to the real numbers.

THE IDEA OF LINEARITY

Linearity is one of the most widespread and important ideas in mathematics. It begins, of course, with the straight line. Linear has the same linguistic root as line. So saying that certain data is linear would mean that the data points lie in a straight line. Now if you ask what functions have linear graphs, you obtain functions of the form $f(x) = mx + b$, where m and b are arbitrary constants. Setting $b = 0$ and so only considering functions whose graphs pass through the origin gives linear functions that satisfy the two properties

(i) $f(x + y) = f(x) + f(y)$,
(ii) $f(ax) = af(x)$.

By analogy with the one-dimensional situation, *any* function that satisfies properties (i) and (ii) is called a *linear* function. Thus a matrix can be thought of as a linear function (or transformation). For example the function that maps the pair of numbers (x, y) into the new pair $(x + 2y, 3x + 4y)$ through the process of matrix multiplication is linear because it satisfies (i) and (ii) above:

$$\begin{pmatrix} 1 & 2 \\ 3 & 4 \end{pmatrix} \begin{pmatrix} x \\ y \end{pmatrix} = \begin{pmatrix} x + 2y \\ 3x + 4y \end{pmatrix}$$

Notice that we now have two mathematical contexts in which to think of the idea of linearity. The first is the geometrical where lines are linear objects. So linear can have the meaning "non-curved," which allows us to think of a plane in 3-space as linear and this can be generalized to higher-dimensional linear spaces. The second context is linear functions that satisfy conditions (i) and (ii). The connection between the two is via the graph of the function—if the function is linear then the graph is linear; if the graph is linear and passes through the origin then the function is linear.

Because linear functions are the simplest mathematical functions, our knowledge about them is more extensive than it is for other (nonlinear) functions. A great deal of mathematics is involved in analyzing linear situations and problems.

However, many mathematical situations and applications are nonlinear. If we think about the power series representation for the exponential function that we discussed on page 226, then the first term, 1, is a constant; the second term, x, is the linear term; and all the subsequent terms, x^2, x^3, and so on are nonlinear. Nonlinear situations are much more complicated than linear ones. An important question is, therefore, whether nonlinear situations that we may not understand, can be approximated by linear ones that we do.

CALCULUS AS LINEAR APPROXIMATION

Now the calculus may be understood as a process of linear approximation. For example, the derivative of a function at a particular point may be understood as the tangent line to the graph of the function. Thus the derivative to the function $f(x) = x^3$ at $x = 1$ is equal to 3.[25] This amounts to saying that the tangent line to the graph of $f(x)$ through the point $(1,1)$ has slope 3. The derivative has a geometric interpretation as a straight line. It is the straight line that is the best approximation to the curve $y = x^3$ at the point $(1,1)$. Another way of saying this is that the linear function $L(x) = 3x$ is the linear function that best approximates $f(x)$ around the point $x = 1$ (actually the graph of the tangent line is $y = 3x - 2$, as we see in figure 5.16).

Calculus was not always understood in this way. It was when calculus came to be generalized from its one-dimensional roots to more complex situations such as higher-dimensional Euclidean spaces or differentiable manifolds that the idea of the derivative as linear approximation came to the fore. Consider again the function $f(x) = x^3$ at the point $x = 1$ whose linear approximation is the function $L(x) = 3x$. All the information about $L(x)$ is

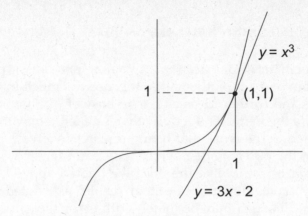

Figure 5.16. Linear approximation

contained in the one number 3. In higher dimensions, functions from 2-space to 2-space, for example, we would need four numbers to specify the appropriate linear function. If the nonlinear function was given as

$$F(x, y) = (x^2 + y^2, xy).$$

Then the linear approximation at the point $x = 1$, $y = 1$ would be given by

$$L(x, y) = (2x + 2y, x + y).$$

We can "linearize" a nonlinear problem by taking its derivative. We then have a linear problem which is much simpler and which we may well be able to solve. The question is, "What does the solution to the linear problem tell us about the original, nonlinear problem?" It turns out that, just as the tangent line tells us something about the graph but only when we are close to the point in question, so too the linear problem often gives us local information about the nonlinear problem. Thus, for example, to study the behavior of a system of differential equations in the vicinity of what is called an equilibrium point in the plane, one "linearizes" the system and obtains a linear system of differential equations for which the solutions are well known. We then infer (most of the time) that the solutions to the nonlinear system behave similarly to the associated solutions to the linear system as long as we stay close enough to the equilibrium point.

Some Concluding Remarks on Ideas in Mathematics

The mathematical idea acts as a dynamic organizing principle on a certain body of mathematics. It is used as much to generate questions and further ideas as it is to answer questions, that is, to generate theorems. The activity that is generated by the idea is open-ended in the sense that future research is always possible. However, the potential of the idea may be optimized in certain mathematical results like the ones we quoted above. A strong idea may have numerous optimal results associated with it. Moreover, this idea may be joined with other ideas to produce even stronger results. The whole matter is a dynamic flux of mathematical activity with occasional points of relative stability that we call theorems.

The usual sequence of definition, axiom, theorem, proof does not apply to the mathematical idea. The idea may not even precede the result. It is conceivable that the result comes first. In trying to understand the result (as in the case of Verrill's comments on "moonshine") we may become aware of what we have called the "idea." When we have isolated "what is really going on" we would then proceed to apply that principle as widely as our imaginations and knowledge will allow. I conclude that mathematics is not merely a body of facts arranged and justified by a stringent logical structure. The logical structure of mathematics gives theorems their stability, but when we remove logic as the focal point of mathematics and replace it by the "idea" we see that the seeming stability of mathematics is not absolute. The data can be structured in many different ways corresponding to what we are interested in and to the mathematical ideas that arise to do the structuring. These ideas arise in response to the question, "What is going on here?" That is, from the attempt to make sense of the phenomena in question.

This situation is alluded to in Thurston's fascinating article[26] in which he makes the following comments about the nature of mathematics:

1. Mathematics is that which mathematicians study.
2. Mathematicians are those humans who advance human understanding of mathematics.

He says, "What we (mathematicians) are doing is finding ways for people to think about mathematics." This perspective highlights the role of the mathematical idea. Thurston says, "When the idea is clear, the formal setup is usually unnecessary and redundant." He continues, "We mathematicians need to put far greater effort into communicating mathematical ideas." It is interesting that there is a circularity in these remarks, but that is inevitable when one steps out of a formal system. Our initial point was that mathematics is nontrivial and again, to talk about the nontrivial one cannot avoid things like circularities. But the main thing that I take from his comments are that the most basic phenomena in mathematics are the mathematical ideas. It is the mathematical idea that we must focus on if we wish to understand what is going on in mathematics.

APPENDIX: THE ORIGINS OF CHAOS

One of the revolutionary developments in science has been the discovery of the phenomenon of chaos. Two of the surprising things about chaotic phenomena are how complicated deterministic systems can be and how "normal" complexity is. These two properties can be illustrated by a simple theorem associated with the dynamics of one-dimensional maps. We are concerned here with the iteration of one-dimensional maps such as $f(x) = 2x(1-x)$. We begin with an initial point x_0 and consider its *orbit*:

$$x_1 = f(x_0), \; x_2 = f(x_1) = f(f(x_0)), \; x_3 = f(x_2), \text{ and so on.}$$

We obtain an orbit that consists of an infinite number of points unless there is a loop. For example if $f(x) = x$ we say that x is a *fixed point*. If $f^2(x) = x$ we say that x is a *periodic point* of period 2, if $f^3(x) = x$ the x is a point of period 3, and so on. Dynamical systems measure the way physical, biological, or mathematical systems change with time. They are concerned with the orbit structure associated with the generating function $f(x)$. Now the simplest orbits are the periodic orbits that we have defined above. To understand a dynamical system means in particular to have an understanding of its periodic orbits. The result that we shall describe below describes the periodic orbit structure for continuous functions defined on an interval of real numbers such as [0, 1].

I am interested both in the result and in the idea behind the result. I am interested in the result because it is surprising and fascinating and tells us a great deal about the behavior of these kinds of dynamical systems. I am interested in the "idea" because of what it tells us about how this piece of mathematics came to be and in general about how an idea can serve as an organizing principle for an area of mathematics.[27]

This discussion involves continuous functions $f(x)$ of one variable that map the points in some interval, $[a, b]$, of real numbers into themselves. For example the function might be $f(x) = x^2$. The interval in question might be the numbers x, $0 \le x \le 1$, since for such a number we also have $0 \le f(x) \le 1$. It is not too difficult to see that for this generating function 0 and 1 are fixed

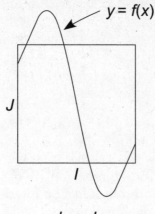

$I \to J$

Diagram 1

points, and that there are no other fixed points and no periodic points at all of any other period.

There is such a vast number of continuous functions that it might seem unrealistic to hope to say anything at all about the structure of their periodic points. Thus the following is quite unexpected.

Proposition 1: *If a continuous function has a point of period* 3 *it must have points of all periods* 1, 2, 3, 4, 5, ... ∎

Now the function $f(x) = 1 - x$ has a fixed point ($x = 1/2$) and points of period 2 ($x \neq 1/2$) but points of no other periods. Thus we might naturally wonder what is so special about 3? Why is a function with a point of period 3 so complex but a function with a point of period 2 less complex?

Unraveling the story of the periodic points depends on the following facts about continuous functions:

Lemma 1: *If $f(x)$ is a continuous function and I and J are closed intervals with $I \subset J$ and $f(I) \supset J$, then f has a fixed point in I.*

This follows from the intermediate value theorem just like Proposition 3 on page 234.

We use the notation $I \to J$ in the situation where $f(I) \supset J$, that is, the image of I contains J.

247

Thus in the situation of Lemma 1, we have

Lemma 2: *If there exists a loop of subintervals $J_1, J_2, J_3, \ldots, J_n$ such that $f(J_1) \supset J_2, f(J_2) \supset J_3, \ldots f(J_{n-1}) \supset J_n, f(J_n) \supset J_1$, (i.e., $J_1 \to J_2 \to \ldots \to J_n \to J_1$), then there must exist a point x in J_1 such that $f(x) \in J_2, f^2(x) \in J_3, \ldots$, and $f^n(x) = x$.*

The truth of this lemma follows from the observation that if $f(J_1) \supset J_2$ then there is a subinterval $K_1 \subset J_1$ such that $f(K_1) = J_2$. Thus $f^2(K_1) \supset J_3$, and so there is a subinterval $K_2 \subset K_1$ such that $f(K_2) = J_3$. Applying this reasoning repeatedly we get a subinterval K of J_1 whose points pass through all the intermediate subintervals and such that $f^n(K) = J_1$. This interval contains the required periodic point.

At this point, we have isolated the key idea.

Key Idea: *The existence of periodic points can be inferred from such loops in an ordered graph of subintervals.*

With this idea concrete results come easily.

Proposition: *If a continuous map of an interval into itself contains a point of period 3, then it contains periodic points of all periods.*

Proof: Suppose the orbit of period 3 consists of the three points $x_1 < x_2 < x_3$, where $f(x_1) = x_2, f(x_2) = x_3$, and $f(x_3) = x_1$. Then, letting I denote the interval between x_1 and x_2 and J the interval between x_2 and x_3, we have the graph

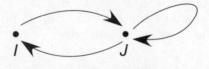

Diagram 2

This diagram obviously has loops of all possible lengths, namely, I followed by $(n-2)$, J followed by an I (the fixed point comes from the loop $J \to J$). ∎

248

Now the implication of this "idea" obviously extends beyond the case of period 3. It must say something about points of other periods. Thus the idea is already at work organizing this particular mathematical landscape. To begin with, consider a point of period 4 arranged in the following manner:

$$x_1 < x_2 < x_3 < x_4 ,$$

$$\begin{pmatrix} x_1 & x_2 & x_3 & x_4 \\ x_3 & x_4 & x_2 & x_1 \end{pmatrix}$$

This gives rise to the graph

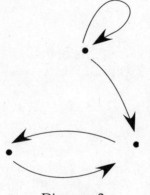

Diagram 3

This graph guarantees only points of period 1 and 2. Thus a map with a point of period 4 may have only periods of 1, 2, and 4 and no others. Maps with points of period 4 may be "simple" in the sense of having few points of different periods, whereas maps with points of period 3 are "complex." Which maps we may ask are complex? Which are simple?

With a little work our organizing principle will provide the answer. But before giving the complete answer let us consider another example. This one is a map with a point of period 5 arranged according to the following permutation:

$$\begin{pmatrix} x_1 & x_2 & x_3 & x_4 & x_5 \\ x_3 & x_5 & x_4 & x_2 & x_1 \end{pmatrix}.$$

This produces the following graph:

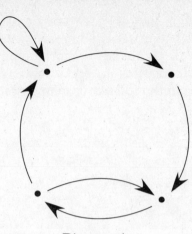

Diagram 4

This graph guarantees us points of period 1, 2, 4, 5, 6, . . ., in fact, points of all periods except 3. We can easily draw the graph of such a function:

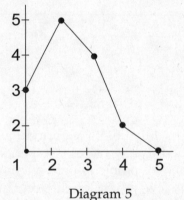

Diagram 5

We have shown that a function with a point of period 3 *must* have a point of period 5, but that a function with a point of period 5 need not have one with period 3. Thus we might say at this stage "something (systematic) is going on." What is in fact "going on" is described in a beautiful result due to A. N. Sarkovskii:

Theorem: *All the natural numbers can be ordered in the following way:*

250

$$3, 5, 7, \ldots, 2 \times 3, 2 \times 5, \ldots, 4 \times 3, 4 \times 5, \ldots, 2^n \times 3,$$
$$2^n \times 5, \ldots, 2^n, \ldots, 8, 4, 2, 1.$$

If a continuous function that generates a dynamical system on an interval contains a point with period n, then it must contain other points of periods for every integer to the right of n on the above list. Furthermore, for every integer n there is an example of a function with points of period n (and therefore all numbers to the right of n) but no periods to the left of n on the list.

So here we have an example of a mathematical idea operating as an organizing principle within a certain area of mathematical activity. The area is the dynamical systems generated by functions defined on one-dimensional intervals and, in particular, the kinds of periodic orbits that such maps can exhibit. The basic idea shows us that periodic points should not be viewed in isolation, but that the existence of one periodic point may force the function to have additional periodic behavior. The theorem that emerged above is not the definitive theorem, but it is a good theorem—it is surprising and it is elegant. However, it is conceivable that more information can be squeezed out of the basic idea. In this sense the above theorem points to further questions that could be posed about the structure of the periodic orbits of dynamical systems.

It is interesting that the unusual order that appears in the theorem appears in other situations involving the dynamics of one-dimensional maps. For example if one looks at the one-parameter family of maps

$$F_\alpha (x) = \alpha x (1 - x), \text{ where } \alpha \text{ varies from 1 to 4,}$$

one discovers that for small values of α the function has only fixed points that act as "attractors"[28] for all other orbits. As α increases one sees the appearance of points of period 2 and then 4, 8, and larger powers of 2. At the limit of all these powers of 2 is a precise value of α that marks the beginning of the "chaotic" regime. After this, with increasing values of α we traverse the theorem's list of integers from right to left, so to speak, culminating with functions that have points of period 3. This family of maps has lots of interesting mathematics going on beyond the structure of its periodic points, but it is interesting that the behavior we isolated in the theorem can be found in one specific family of maps. In fact, any family of maps that begins with simple maps with only fixed points and ends with maps with points of period 3 must pass through the ordering mentioned in the theorem.

To conclude, this mathematical idea or organizing principle can be seen as acting dynamically upon a certain body of mathematics. It is used to generate questions and further ideas as well as to answer questions, that is, to generate theorems. The activity that is generated by the idea is open-ended in the sense that future research is always possible.

Furthermore, the idea may not even precede the result. It is conceivable that the result comes first. In trying to understand the result we may become aware of what we have called the "idea." When we have isolated "what is really going on" we then proceed to apply that principle as widely as our imaginations and knowledge will allow. Thus mathematics is not a body of facts arranged and justified by a stringent logical structure. To give a dynamic metaphor that stems from the related field of fluid flow, it is like a turbulent river. The flow of the water is extremely complex, but every now and again you see the appearance of stable structures, eddies and whirlpools. These stable structures correspond to the propositions and theorems of mathematics. It is the logical structure of mathematics that gives these theorems their stability, but when we look at things in this way we see that the stability is not absolute. The data can be structured in many different ways corresponding to what we are interested in and in the mathematical ideas that arise to do the structuring. These ideas arise in response to the question, "What is going on here?"—that is, from the attempt to make sense of the phenomena in question.

CHAPTER 6

Ideas, Logic, and Paradox

INTRODUCTION

The previous chapter initiated a discussion of mathematical ideas. I considered a number of fairly straightforward examples of mathematical ideas from the areas of mathematics that I had developed in previous chapters. In this chapter I wish to extend this discussion in directions that are unexpected and perhaps even a bit shocking. This will demonstrate what differentiates the perspective I am taking from the usual one. Ideas arise from a context that may include ambiguity, contradiction, and paradox.

The proper domain of mathematics is this expanded region. This domain from which ideas may arise fits nicely with the expanded domain that I have been attributing to mathematics as a whole. In particular, the discussion of the role of the idea supports my repeated claims that mathematics is non-algorithmic.

To repeat, mathematical ideas are the heart and soul of mathematics. Mathematics comes into being through the medium of the mathematical idea, which organizes what would otherwise be an inchoate field of data into structures of extraordinary detail and subtlety. Building a discussion of mathematics on the foundation of mathematical ideas is a radical departure from the usual ways of looking at the subject. Formalism, for example, looks at mathematics as a static, finished product. Ideas are dynamic, not static. The field of ideas resembles a biological system. Ideas are continually being born, evolving, competing with one another, and disappearing when their period of usefulness is over.

Ideas are primordial. Without ideas the mathematical world would present itself to us as nothing but chaos. To paraphrase the Bible, we might say, "In the beginning was the idea." Of course the expression "in the beginning" is not to be understood temporally—what it means is that the idea is absolutely fundamental. It is through the medium of ideas that mathematics comes to be. So what I am doing here is starting with ideas and investigating what the subject of mathematics looks like when viewed from that perspective.

In trying to build up a view of mathematics around the centrality of mathematical ideas, I am moving in a direction opposite to the one of ordinary thought. Our normal thinking processes, especially in mathematics, move from assumptions to conclusions, from the simple to the complex. If we use the metaphor of the "stream of consciousness," then our normal movement is to go with the current, to go downstream. What I propose is to reverse this direction—to look upstream. That is, I ask where something comes from rather than where it is going. In considering the mathematical idea as the most basic object from which mathematics is born and around which it grows and develops, we are turning our attention upstream.

The centrality of ideas in mathematics is brought out by considering the difference between a page of actual mathematical text (a proof in a research paper, let us say) and a text in a formal language (the logically complete proof that might be generated by a computer program). Davis and Hersh say, "The steps that are included in such a [actual mathematical] text are those that are not purely mechanical—that involve some constructive idea, the introduction of some new element into the calculation. To read a mathematical text with understanding one must supply the new idea which justifies the steps that are written down." In comparison to the machine, for which "nothing must be left unstated. . . for the human reader, nothing should be included which is so obvious, so 'mechanical' that it would distract from the ideas being communicated."[1] Thus a research paper or a textbook should include the level of detail that is consistent with revealing the underlying ideas because it is the ideas that are important.

What then are the implications of moving the notion of the idea into the center of our discussion of mathematics? Taking ideas as primary allows us to focus on understanding, for example. Understanding is what we strive for when we learn mathematics, that is, when we attempt to master other people's ideas, but it is also what we are about when we are doing mathematics, when we try to understand some mathematical phenomenon. Recall David Blackwell's statement, "Basically, I'm not interested in doing research and I never have been. I'm interested in *understanding*, which is quite a different thing."[2] Or William Thurston, when he said "what we (mathematicians) are doing is finding ways for *people* to understand and think about mathematics."[3] This "thinking and understanding" is intimately connected with grasping the underlying ideas involved in a piece of mathematics.

Focusing on the idea as the entry point, so to speak, to mathematics forces one to regard mathematics as a creative activity for, as Weyl said, "mathematizing may well be a creative activity of man, like language or music, of primary originality, whose historical decisions defy complete objective rationalization."[4] It is only in this context that our discussion of mathematics can be based on mathematical practice, on what is really going on in mathematics, not on some a priori theory of what is going on. Working from the perspective of ideas will enable me to develop a novel approach to questions involving the relationship of logic to mathematics and the related questions of Platonism versus formalism and discovery versus invention.

Ideas are more fundamental than logic in the structure and practice of mathematics. Formal mathematical reasoning is blind to the dimension of the mathematical idea because the formalism of mathematics is downstream from the idea. Mathematical practice involves almost taking the formalism for granted. Thurston, for example, says, "When the idea is clear, the formal setup is usually unnecessary and redundant. I often feel that I could write it out myself more easily than figuring out what the authors actually wrote." When we think about the origins and nature of mathematical thought we must move upstream, past the formal proofs, past the axioms, into the realm of the mathematical idea. When we enter this realm we encounter a mathematics that looks very different from the mathematics we learned about in school. For one thing, it has lost the quality of absolute necessity though not the quality of truth.[5]

The term "mathematics" should most properly refer to what has sometimes been called "informal mathematics" as opposed to "formal mathematics." The latter would then be seen as merely one aspect of mathematics. Then it is not so surprising that mathematics, understood in this way, contains a large element of the contingent. A given body of mathematical data is capable of being organized in more than one way. Therefore theorems are not determined a priori by the given situation but emerge as a result of the ideas with which we organize the situation.

Ideas are organizing principles, but what are being organized are other ideas. Mathematics involves ideas organizing ideas in an iterative process that can attain incredible complexity. The ideas that are organized may be processes reified as mathematical objects. In fact if one goes back to the Pythagoreans, for example, one sees how complex and multidimensional were their ideas about what to us are the simplest mathematical objects—the integers one to ten.

Thus focusing on the idea rather than the formal structure will change our thinking about mathematics so that the subject we are describing is more in line with the mathematics that the mathematician experiences. Just as particular ideas organize areas of mathematics. so "the idea" (i.e., the idea of the idea) will organize our discussion of the nature of mathematics.

IDEAS ARE NEITHER TRUE NOR FALSE: CAN THERE BE A "GOOD" MISTAKE?

Mathematics is so commonly identified with its formal structure that it seems peculiar to assert that an idea is neither true nor false. What I mean by this is similar to what David Bohm means when he says, "theories are insights which are neither true nor false, but, rather, clear in certain domains, and unclear when extended beyond those domains."[6] Classifying ideas as true or false is just not the best way of thinking about them. Ideas may be fecund; they may be deep; they may be subtle; they may be trivial. These are the kinds of attributes we should ascribe to ideas. Prematurely characterizing an idea as true or false rigidifies the mathematical environment. Even a "false" idea can be valuable. For example, I earlier described the Shimura-Taniyama conjecture as the heart of the successful resolution of Fermat's Last Theorem. Goro Shimura once said of his late colleague Yutaka Taniyama, "He was gifted with the special capability of making many mistakes, mostly in the right direction. I envied him for this and tried in vain to imitate him, but found it quite difficult to make good mistakes."[7] A mistake is "good" precisely because it carries within it a legitimate mathematical idea.

Ironically, "formalism" itself is a mistake. The formalist conception of mathematics came about as a result of the efforts of mathematicians like David Hilbert to escape from the paradoxes that had arisen in the foundations of mathematics. He attempted to formalize *all* of mathematics, completing in this way what the Greeks like Euclid had begun. It turned out that this was not possible, as we shall see later on in this chapter. But the *idea that this could be done* was still a great idea!

Some of the greatest ideas are glorious "failures" in this way. In fact, in the next chapter I shall spend some time looking at these kinds of ideas in mathematics. The fact that so many of the key ideas of mathematics are of this type is not an accident but intimately connected with the nature of mathematics. In general, most sweeping conjectures turn out to be "wrong" in the sense that they need to be modified during the period in which they are being worked on. Nevertheless they may well be very valuable. The whole of mathematical research often proceeds in this way—the way of inspired mistakes.

Mathematical ideas are not right or wrong; they are organizers of mathematical situations. Ideas are not logical. In fact the inclusion should go the other way around—logic is not the absolute standard against which all ideas must be measured. In fact *logic itself is an idea*. In mathematics and in science in general, logic is the organizational principle par excellance. This is so true that the thought of putting logic in some sort of subsidiary role may evoke in certain people a sense of unease, even panic. This reaction may come from the fear of being thrown back into some sort of primeval chaos, a state that reason and logic appear to have delivered us from. Mathematics, when we learn it in school, has the attractive property of seeming to be black or white, right or wrong. This clarity, this precision, is very attractive to people who go on to become mathematicians and scientists. But to others this same precision may also be intimidating. It is the logical, black or white dimension of mathematics that gives rise to the "mathematical anxiety" that one observes in many, often highly intelligent, people.

The Use and Abuse of Logic

The above comments set the stage for a short discussion of the role of logic in mathematics. What does logical reasoning do for us? What can it not hope to do? When does it help and can the use of logic ever be harmful? These are important questions, and they can be clarified by thinking about logic as a particular kind of organizing principle.

Logic is essential to mathematics, and no one is proposing eliminating logic from mathematics. It is just that logic is not the defining property of mathematics. What *does* logic contribute to mathematics? Logic does a number of things that are essential for mathematics. In the first place, logic stabilizes mathematical ideas. Without logic mathematics would be in a continual state of flux, so it would be impossible to build up the intricate mathematical structures and arguments that characterize the subject. In the second place, logic organizes huge areas of mathematics. Without logical structures it is impossible to imagine the development of theoretical mathematics. Logic helps us to communicate mathematics by creating a common language in which mathematical ideas may be expressed. Finally, and perhaps most important, recall what was said in Chapter 1 about "controlled ambiguity." Logic provides the control without which theoretical mathematics would not exist. Thus in great mathematics one would expect to find a deep ambiguity balanced against a most intricate and subtle logical analysis.

Notwithstanding the essential role of logic in mathematics, the logical structure is not what is primary. This is why attempting to teach mathematics from a formal perspective creates such problems. Many of us can remember the pedagogical disaster of the so-called "New Math" that Morris Kline attacked so unmercifully.[8] He made a great deal of teachers telling students that "3 + 12" was equal to "12 + 3," let us say, because of the "commutative law." The "new math" was an attempt to start with the formal structure and deduce the properties of the operations of arithmetic from that formal, logical structure. In practice what happens is the reverse. The commutative law for addition (the fact that the order of addition is irrelevant) is an idea that structures a whole body of experience that children have with addition. They begin to "understand" this idea when they realize that you can add 12 to 3, by reversing the order, starting with the twelve and counting three more: 13, 14, 15. Usually one proceeds from the idea to the logical structure, not from the logical structure to the idea.

Logic itself, as I said above, is an idea, an extremely powerful organizing principle. It is connected to the notions of rationality and reason. In fact logic is the form in which reason appears in our scientific civilization. Thus logic is a prime example of the creativity of a profound idea and the ability of such an idea to literally change the world. However, we must distinguish between logic as an idea and *the use of logical argument to generate ideas*. Logical arguments do not generate ideas. As I said above, logic organizes, stabilizes and communicates ideas but the idea exists prior to the logical formulation.

Hannah Arendt had a great deal to say about the connection between ideas and logic. Arendt felt that logic replaced the "necessary insecurity" of thinking with the "negative coercion of deduction." She wrote that, "As soon as logic as a movement of thought—and not as a necessary control of thinking—is applied to an idea, this idea is transformed into a premise. . . . The purely negative coercion of logic, the prohibition of contradictions, became 'productive' so that a whole line of thought could be initiated, and forced upon the mind, by drawing conclusions in the manner of mere argumentation."[9] Here Arendt is engaged in her famous analysis of totalitarianism. Her point is that totalitarian regimes were so dangerous not because they were irrational but, on the contrary, because they were too rational, in a sense. They were ideological and ideology is the "logic of an idea." She said, "The danger in exchanging the necessary insecurity of philosophical thought for the total explanation of an ideology and its *Weltanschauung*, is not even so much the risk of falling for some usually vulgar, always uncritical assumption as of exchanging the freedom inherent in man's capacity to think for the straightjacket of logic by which man can force himself almost as violently as he is forced by some outside power."

The point here is certainly not to rail against logic, nor is it to glorify the irrational. Arendt's strong rhetorical tone is justified by the enormity of the disaster that she attempted to comprehend. What is relevant for our discussion of mathematics is the distinction between logic and ideas. Mathematical thought inevitably involves what Arendt calls "necessary insecurity," and this insecurity is a prerequisite to acts of understanding and mathematical creativity. Logic eliminates that insecurity. Therefore, introducing logic prematurely, much less making it the generative property of mathematical thought, is destructive.

Mathematicians recognize the poverty of logical thought without ideas by labeling such arguments or conclusions as "trivial" or "obvious." However, it must be said that not all mathematicians agree with this distinction between logic and ideas. As I mentioned in the introduction, there are those who dream of an algorithmic process that is so powerful that the importance of the human mathematician is reduced or eliminated. This dream of a machine-based intelligence is akin to the dream of a "final theory" that one finds in certain areas of physics. The dream is that we will find algorithmic ideas that are so powerful that they will be capable of generating all future mathematics *without the intervention of subsequent original ideas*. It should be clear by now that I consider this dream to be an illusion. In my view the existence of such an ultimate algorithm would mean the end of mathematics. However mathematics, from the perspective that I am taking, has no end. The goal of mathematics is not to "eliminate the need for intelligent thought." On the contrary mathematics *is* intelligence in action—without intelligence, and this means mathematical ideas, there is no mathematics.

Now one may read a proof and "get the idea" behind the proof. On the other hand, it seems that some mathematical writing seems to have the perverse purpose of hiding the ideas involved. Good mathematical writing brings out the ideas and suppresses the purely mechanical material that would obscure those ideas. Nevertheless, the question arises as to how to communicate mathematical ideas so that the reader "gets it." Certainly a perfect logical presentation does not ensure that the reader "gets it." In fact *no* presentation of an idea will ensure that the reader "gets it." "Getting the idea," which is akin to "seeing what is going on," is something that cannot be programmed. It is never certain. The very impossibility of communicating understanding perfectly, of communicating the idea, proves that the idea is more basic than logic, more basic, even, than any conscious formulation of that same idea. In exploring the idea we are plumbing the very depths of mathematical thought.

AMBIGUITY AND THE BIRTH OF THE IDEA

The point that "the idea" is more elementary than logic is borne out by the evidence of the chapters on ambiguity, contradiction, and paradox. Ambiguities, seeming contradictions, and even paradoxes can be the bases for powerful mathematical ideas. We saw this in the development of mathematical concepts like zero and infinity, and we shall describe below a number of instances where famous paradoxes have found their way into legitimate mathematical proofs. One such instance is the proof that the cardinality of the set of subsets of a given set is always larger than the cardinality of the set itself. But more generally mathematics is often generated out of a situation where two frameworks with a certain incompatibility rub up against each other. These situations of incompatibility, which in the extreme are contradictions, call out for reconciliation. This reconciliation, if it is possible, will arise out of a new mathematical idea. Thus ideas arise out of situations that we have called ambiguous. They do not arise out of logical deduction. The logical phase comes after the idea has been formulated and now must be verified, stabilized, refined, and communicated to others.

THE IDEA IS ALWAYS WRONG!

The given mathematical situation is never a perfect fit for the idea. A colleague of mine was describing this phenomenon when he said, "You are [the idea is] always wrong." Even after the idea emerges into consciousness a great deal of hard work remains to be done, one might say that it is at this stage that the real work begins! The idea now begins to work on all of the elements of the given mathematical situation. The idea may determine the appropriate definitions of terms, the appropriate hypotheses, and, of course, the conclusions of theorems. Everything in a piece of mathematics is in a state of flux; everything depends on everything else. Concepts can be defined or redefined with the object of bringing out the idea. Conclusions can be modified. Even the idea can be modified in the course of working on it. And this work can take days or centuries. It may involve one person or hundreds of mathematicians who are interested in the same problematic situation. Thus the notion that "the idea is always wrong" is consistent with the notion that the idea is an organizing principle but not with the notion that an idea is a well-defined, logically precise entity. This manner of looking at mathematics was developed in great historical detail in the following.

LAKATOS'S ACCOUNT OF THE DEVELOPMENT OF THE EULER FORMULA

Imre Lakatos was the author of a remarkable attempt to describe mathematics as it is actually experienced by mathematicians. Instead of a formal mathematical system built up from first principles, he presents us with human beings making arguments and being presented with counterarguments. "Instead of mathematics skeletalized and fossilized, he presents mathematics growing from a problem and a conjecture, with a theory taking shape before our eyes, in the heat of debate and disagreement, doubt giving way to certainty and then to renewed doubt."[10]

The context of Lakatos's work is a historical reconstruction of Euler's formula for polyhedra, three-dimensional figures, like the cube, each of whose faces is a polygon. This formula is

$$V - E + F = 2,$$

where V is the number of vertices, E the number of edges, and F the number of faces. For example, if the polyhedron is a tetrahedron then $V = 4$, $E = 6$, and $F = 4$, for a cube $V = 8$, $E = 12$, and $F = 6$ (see p. 222)

Is Euler's formula valid? Let us give just a taste of the beginning of Lakatos's brilliant historical reconstruction. The entire thesis consists of a hypothetical classroom discussion between a teacher and a number of students. The "teacher" begins by formulating a "proof-idea" that originated with Cauchy:

Teacher: In fact I have one [an argument for the validity of the formula]. It consists of the following thought-experiment.

Step 1: Let us imagine the polyhedron to be hollow, with a surface made of thin rubber. If we cut one of the faces, we can stretch the remaining surface flat on the blackboard, without tearing it. The faces and the edges will become deformed, the edges may become curved, but V and E will not alter, so that if and only if $V - E + F = 2$ for the original polyhedron, $V - E + F = 1$ for this flat network—remember we have removed one face. [Figure 6.1a shows the flat network for the case of the cube.]

Step 2: Now we triangulate our map—it does indeed look like a geographical map. We draw (possibly curvilinear) diagonals in those (possibly curvilinear) polygons which are not already (possibly curvilinear) triangles. By drawing each diagonal we increase both E and F by one, so that the total $V - E + F$ will not be altered. [Figure 6.1b]

Step 3: From the triangulated network we now remove the triangles one by one. To remove a triangle we either remove an edge—upon which one face and one edge disappear [Figure 6.1c], or we remove two edges and a vertex—upon which one face, two edges and one vertex disappear (Figure 6.1d). Thus if $V - E + F = 1$ before a triangle is removed, it remains so after the triangle is removed. At the end of this procedure we get a single triangle, For this $V - E + F = 1$ holds true. Thus we have proved our conjecture.[11]

Some of the "students" are convinced by the argument, some raise objections of various sorts. These objections range from questioning some particular step of the argument to questioning whether the "theorem" is true at all. In particular there is a discussion of "local counterexamples" versus "global counterexamples." A local counterexample is an example that refutes a lemma (a subsidiary result) without necessarily refuting the main conjecture. A global counterexample refutes the main conjecture itself. In particular, there ensues a discussion of global counterexamples such as (a) two pyramids that share a common vertex (figure 6.2) or a solid bounded by a pair of nested cubes, that is, a pair of cubes, one of which is inside but does not touch

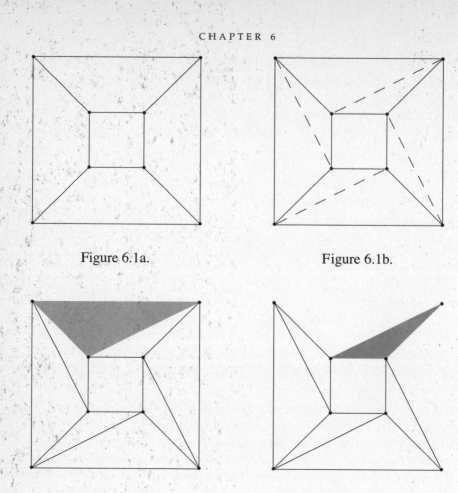

Figure 6.1a.   Figure 6.1b.

Figure 6.1c.   Figure 6.1d.

Figure 6.1. The Cauchy "Idea"

the other.[12] The former example, it turns out, is a true counterexample for $V - E + F = 4$. What does one do in the face of this example? Does one abandon the conjecture as false? Certainly not! The counterexample leads to a discussion of what the correct definition of a polyhedron should be. Lakatos calls this part of the discussion "monster barring." We have an idea, but what range of phenomena does the idea apply to? At one extreme we could *define* a polyhedron to be a geometrical object for which the formula $V - E + F = 2$ is true. This would make any idea unnecessary, and it would reveal nothing about the geometric structure of the objects we refer to as polyhedra. In fact the "monsters" bring to the fore the questions of what a good defi-

264

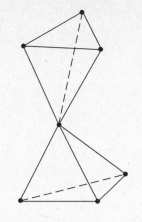

$$V = 7 \qquad E = 12 \qquad F = 8$$

Figure 6.2. Double pyramid monster

nition of a polyhedron should be. Thus the exploration of the "idea" forces upon us a deeper examination of the mathematical objects that we are studying.

At this stage one cannot say that the idea is either logical or illogical. It is certainly not illogical, but neither can we say that it is a formal argument. Nor would it be correct to say that the idea has no mathematical content. The idea has a great deal of implicit mathematical content. After the idea has been brought into existence there remains an immense amount of mathematical work to be done in teasing out and making explicit the mathematics that is contained in the idea. In this case the quantity $V - E + F$ generalizes to a number that is an invariant of the topological structure of a manifold, say. Thus another direction that our discussion may take us into is to allow a certain class of geometric figures to be distinguished by the invariant $V - E + F$. If a polyhedron has $V - E + F = 2$, how could we describe the geometric objects that have $V - E + F = 4$, say or 0?

IDEAS ON THE "EDGE OF PARADOX"

I promised earlier to look into the paradoxical as a source of ideas. In order to do this I shall look in some detail at a number of historically important results. In each of them a paradox *is* the

idea or at least forms a major component of the idea. I am bringing these examples up to refute the notion that the world is divided into two disjoint domains, the domain of logically demon-
strable mathematical truth and the domain that contains
ambiguity, contradiction, and paradox. In the conventional view,
mathematics is entirely contained in the first domain. Yet mathematical ideas do not fit easily into such a worldview. That point
is made more compelling by looking at some examples of how
mathematics uses paradox. It would appear that, far from
avoiding paradox and contradiction, powerful mathematics
goes right up to the edge. In complexity theory there is a saying
that interesting biological processes happen "at the edge of
chaos."[13] Now paradox is a form of chaos so it should not come
as too great a surprise to find that there is great mathematics on
the "edge of paradox."

The Cardinality of the Power Set

Once Cantor had made infinity into a well-defined mathematical
object, then, as we saw in Chapter 4, there was no reason why
there could not be more than one order of infinity. Cantor had
armed himself with a precise way of looking at cardinality, the
size of infinite sets, and it is to his credit that he made every
attempt to ferret out all he implications of his new way of looking at things. One of the extraordinary results he developed is
our next subject. It is of interest both for the result and for the
nature of the argument that is required. If the existence of two
orders of infinity is paradoxical, then how much more so is the
existence of *infinitely many* different infinities.

The key mathematical idea here is one that is very simple and
familiar. If we look at a set with three elements, say $\{a, b, c\}$, then
there are eight possible subsets, namely, $\{a\}$, $\{b\}$, $\{c\}$, $\{a, b\}$, $\{a, c\}$,
$\{b, c\}$, $\{a, b, c\}$, and the empty set. Of course, $8 = 2^3$. This is explained by noting that any subset, S, is formed by either including or not including any particular element, that is, given any
element, a, there are two possibilities: a is either an element of S
or not. The possibilities multiply and, since there are three possible elements, the total number of choices is 2^3. This reasoning
works for any finite set. In particular the number of subsets is

larger than the number of elements. It turns out that this latter result still holds true for infinite sets.

The proof is a beautiful example of the way contradiction and paradox are used to produce positive mathematical results. The paradox here has many formulations. Consider the following:

> There is a town with a single barber who cuts the hair of everyone who does not cut her own hair. Does the barber cut her own hair or not?

Of course if the barber cuts her hair then she does not. And if she does not then she does. A similar paradox arose in the foundations of mathematics at the beginning of the twentieth century. It is called Russell's paradox, and it involves the foundations of set theory, in particular, the question of what is a set. Normally we imagine that a set is specified by the rule that tells whether a possible element is or is not a member of the set. For example, the set of all even integers = {2, 4, 6, 8, . . .} and the set of all prime numbers = {2, 3, 5, 7, 11, . . .}. Now it is conceivable that a set has itself as an element, for example, the set of all sets. Call a set "normal" if it is *not* an element of itself. Now consider the set of all normal sets, N. Is it a normal set or not? Well, if it is normal then it is not an element of itself. But N is the collection of *all* normal sets, so N is not normal. Similarly if N is not normal, then N cannot be an element of itself. But then it *is* normal. Wow! There seems to be a major problem here. Somehow we get into trouble with the idea that anything can be a set. Maybe we should think some more about just what a set is or can be.

What is interesting for our purposes is the use Cantor makes of this paradox.

> **Theorem**: *Given any set A, then the cardinality of A (intuitively the size of A or the number of points in A) is strictly less than the cardinality of the set consisting of all possible subsets of A, 2^A.*

It is pretty clear that the cardinality of A is less than or equal to that of 2^A (you can match up every element a of A with the subset $\{a\}$). Thus the result will be true if we can show that the cardinality of A is not equal to that of 2^A. Now the cardinalities are equal (by definition) if we can match up all the elements of A with all the subsets of 2^A. We must argue that this is impossible, and the

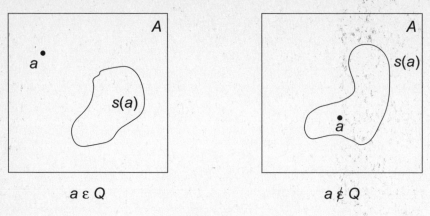

$a \, \varepsilon \, Q$ $a \, \notin \, Q$

Figure 6.3

way to do so is to argue by contradiction, that is, by assuming that there is such a matching and showing that a problem arises. We shall go through the argument in some detail because it's a fascinating argument and we want to examine the contradiction closely.

Suppose we have managed to obtain some way of matching the elements of A with the subsets of A. Let's indicate the matching in the following way.

$$a \leftrightarrow s(a),$$
$$b \leftrightarrow s(b),$$

and so on, where you can think of $s(a)$ as the subset that is matched up with the element a. Remember that every element of A is matched up with some subset (every a has a $s(a)$) and every subset, B, is matched up with some element of A ($B = s(b)$ for some b).

Now since $s(a)$ is a subset there are only two possibilities either a is an element of $s(a)$ or a is not an element of $s(a)$ (figure 6.3). Collect all the elements which are not elements of their matched subsets and name the collection Q. (That is $Q = \{a: a \notin s(a)\}$.) Now Q is a subset and so, since we are assuming that we have a perfect matching, it must be matched with some element. Let's call this element q. That is, $Q = s(q)$ or $q \leftrightarrow Q$.

Now is q an element of Q? We are now back in the situation of the barber who does or does not cut his own hair. If q is not an element of Q, which is, you remember, $s(q)$, then by the definition of Q, q *is* an element of Q. On the other hand if q is not an element of Q then q is an element of Q. Thus we have a contradiction.

What is the conclusion? The conclusion is that there could be no such perfect matching. The cardinalities are not equal. We conclude that the cardinality of the set of subsets, 2^A, is larger than the cardinality of A.

Case where A = N

In the case where the set in question is the natural numbers we have already proved this result using Cantor's diagonal argument. We shall compare that argument to this one. Let us begin by making the observation that every subset of the natural numbers can be considered as a sequence of 0's and 1's, where a 1 in the nth place signifies that n is a member of the subset and a 0 that n is not. Thus the subset $\{1, 2\}$ would be represented by the sequence 1, 1, 0, 0, 0, 0, . . . and the even numbers by the sequence 0, 1, 0, 1, 0, 1,. . . . The sequence 0, 0, 1, 0, 0, 1, 0, 0, 1, . . . would stand for the subset $\{3, 6, 9,. . .\}$.

Now the argument would begin by assuming that all subsets of the natural numbers could be put onto a list. Thus we would have a list

$$a_{11} \, a_{12} \, a_{13} \ldots ,$$
$$a_{21} \, a_{22} \, a_{23} \ldots ,$$
$$a_{31} \, a_{32} \, a_{33} \ldots ,$$
$$\ldots ,$$

where each element is either a 0 or a 1. The set Q would be formed by considering those elements n for which $a_{nn} = 0$. That is $n \in Q \Leftrightarrow a_{nn} = 0 \Leftrightarrow n \in S_n$. By assumption this set Q would correspond to some line above, suppose the kth line. Thus $k \in Q \Leftrightarrow a_{kk} = 0 \Leftrightarrow k \notin S_k$. However $Q = S_k$. And this completes the contradiction.

COMMENT ON THE CANTOR RESULT

This argument purports to produce an infinite sequence of distinct infinite cardinal numbers. The argument is by contradiction and contains as its central idea the Russell paradox. In a way it is saying that if we wish to avoid this paradox we must admit the result, the infinite hierarchy of infinities. The paradox is not so much avoided as deliberately provoked and, at the last moment, turned aside. We are left in awe at this masterstroke much as we might feel when a brilliant matador had allowed the bull's horn to graze his thigh before turning him and escaping unscathed.

GÖDEL AND THE CONTINUUM HYPOTHESIS

We have just generated a whole hierarchy of orders of infinity. It begins with the counting numbers, N, and proceeds by considering all the subsets of N, then all the subsets of the subsets of N, and so on. Expressed in this way the real numbers correspond to the set 2^N consisting of all the subsets of the natural numbers N. Remember that in the previous section I showed how any subset of N could be considered as an infinite sequence of 0's and 1's. This sequence of 0's and 1's in turn can be thought of as a unique real number by considering the sequence as the base 2 representation of the number. For example, the set of even numbers corresponds to the sequence 010101.... this sequence is converted into a real number in the following way:

$$.010101\ldots \text{(base 2)} = \frac{0}{2} + \frac{1}{4} + \frac{0}{8} + \frac{1}{16} + \frac{0}{32} + \frac{1}{64} + \cdots = \frac{1}{3}$$

The subset $\{1, 2, 3\}$ would correspond to the sequence 111000..., which gives us the number

$$\frac{1}{2} + \frac{1}{4} + \frac{1}{8} + \frac{0}{16} + \frac{0}{32} + \cdots = \frac{7}{8}$$

In this way every sequence of 0's and 1's gives rise to a unique real number. Conversely every real number has a representation as a base 2 infinite decimal which is a sequence of 0's and 1's and therefore can be considered to be a subset of the natural numbers. Thus the cardinality of the set of subsets of N is exactly that of the real numbers, R.

The question that arises is whether there are any infinite sets of real numbers that have cardinality strictly greater than that of N and strictly less than that of R. The *continuum hypothesis* asserts that there are none, and Cantor spent a great deal of time trying to prove or disprove this hypothesis (in fact he believed that it was true). It was one of the major intellectual defeats of his life that he was unable to resolve this question one way or the other.

Kurt Gödel[14] proved that the continuum hypothesis cannot be disproved using the standard Zermelo-Fraenkel axioms for set theory (including the axiom of choice). Twenty-three years later Paul Cohen[15] showed that the continuum hypotheses cannot be proved from these same axioms either. Thus the hypothesis is *independent* of these axioms—there could be a version of mathematics that assumes the continuum hypothesis is true (where there are no sets of real numbers that have cardinality between that of the rational numbers and that of the real numbers) and another for which the continuum hypothesis is false (for which sets of such intermediate cardinality actually exist).

Now the mathematical universe in which the continuum hypothesis is false is richer than the one in which it is true, since it would contain a more diverse variety of subsets of the real numbers. Perhaps for this reason Gödel himself believed the continuum hypothesis to be false. Since he was a Platonist he had no trouble asserting the (absolute) truth or falsity of a mathematical statement independent of its provability. It is interesting that we see here an important instance of a division between "truth" and "provability." In a sense the continuum hypothesis is an example of one of the paradoxes of infinity that we have been considering. Is it true or not? One has the feeling that one should be able to answer this question. It shouldn't just be a matter of opinion. We mathematicians feel a great sense of familiarity with the real number system, and the continuum hypothesis appears to be a very concrete property which that system does or does not possess. Does there or does there not exist a set of real numbers with the required intermediate cardinality? In speculating about this question one has the feeling that one is repeating another example of questioning the axiomatic status of a mathematical statement—the great crisis in the history of mathematics that was evoked by status of another questionable axiom—the parallel postulate of Euclid.

In fact, the paradox that is evoked by the continuum hypothesis really concerns the nature of mathematics itself. We have the sense that the properties of the real number system exist "out there" independent of axiom systems and human intervention. The "independence" proof of Gödel and Cohen might seem to imply that the truth of the continuum hypothesis is arbitrary—we can take it or leave it at our own discretion. But this is vaguely dissatisfying, as is the entire formalist approach to mathematics. Thus the continuum hypothesis brings us face-to-face with contemporary problems in the philosophy of mathematics. From a contemporary (or should we say post-modern) point of view it is perhaps not surprising that there should be certain statements that can neither be proved nor disproved within a given axiom system. But then what has happened to "Truth"?

Gödel and Incompleteness

Gödel's incompleteness theorem is one of the great intellectual accomplishments of the twentieth century. Its implications are so far reaching that is difficult to overestimate them. Gödel's result puts intrinsic limitations on the reach of deductive systems; that is, it shows that given any (sufficiently complex) deductive system, there are results that are beyond the reach of the system—results that are true but cannot be proved or disproved on the basis of the initial set of axioms. This new result might be proved by adding new axioms to the system (for example, the result itself) but the new strengthened system will itself have unprovable results. Gödel's argument demonstrates the existence of such results that "transcend" the limits of any deductive system, but, in the manner of other existence theorems, it does not make explicit their mathematical content. Subsequent work has shown specific mathematical results that fall into this category and one particular result is developed below.

Anyone who has studied the history of Western intellectual thought in general and mathematics in particular cannot fail to be shocked by this result. It appears to be a definitive end to that wonderful "dream of reason" that began with Euclid and continued down the centuries, in the work of Frege, Hilbert, and Russell and Whitehead. Yet even though Gödel's theorem is philosophically devastating it has had little or no effect on the work of the average mathematician. Nevertheless it has major implications for anyone concerned with the epistemology of mathematics.

Now Gödel's theorem may seem to be merely negative. However, what may be the end from one point of view can also be seen as the beginning from another. The collapse of the hope for a kind of ultimate formal theory can be seen as a kind of liberation. It is liberation from a purely formal or algorithmic view of mathematics and opens up the possibility of a view of mathematics that is more open and filled with creative possibilities.

One way of judging the value of any intellectual accomplishment is through the originality of the subsequent work it inspires. Here I must mention Douglas Hoftadter's Pulitzer Prize winning book, *Gödel, Escher, Bach: An Eternal Golden Braid*, written in 1979. Hofstadter sees the vast implications of Gödel and the relevance of his thought today. Gödel's theorem is a tour de force that operates on many levels. In the first place it is a technically brilliant theorem in mathematical logic. At a more universal level it has deep implications for the philosophy of science and mathematics and beyond that for the nature of human thought in general.

In a sense Gödel's result brilliantly illustrates and reinforces many of the themes of this book. The philosophical statements that I have been making about mathematics belong to the world of metamathematics. One of the extraordinary aspects of the Gödel result is that he manages to embed metamathematical statements within mathematics itself in a brilliant piece of self-reference. I broached the notion of self-reference and its relationship to the infinite in the introduction to Chapter 3. Thus not only does the *result* of Gödel's theorem overtly tell us a great deal about the nature of mathematics, but also the *method* of the argument tells us more. The result is pertinent at both levels.

The "Idea" Is a Paradox

The first important point to be made about the Gödel theorem is that the argument is built around a paradox. Let us consider what Gregory Chaitin, a brilliant successor of Gödel, has to say about Gödel's proof.

Gödel's incompleteness proof is very clever. It's very paradoxical. It almost looks crazy. Gödel starts in effect with the paradox of the liar: the statement "I'm false!" which is neither true nor false. Actually what Gödel does is to construct a statement that says of itself, "I'm unprovable!" Now if you can construct such a statement. . . in arithmetic. . . you're in trouble. Why? Because if that statement is provable, it is by necessity false, and you're proving false results. If it's unprovable, as it says of itself, then its true, and mathematics is incomplete.[16]

We have seen how the existence of certain paradoxical elements in mathematics led to the attempt to formalize mathematics as a way of doing away *in principle* with the possibility of paradox. A formal system would produce only true statements and, if it were complete, all possible true statements of mathematics. Around the turn of the twentieth century Frege's attempts to put arithmetic on a sound foundation was stymied by the discovery of Russell's paradox. We mentioned Russell's paradox above and noted that it is connected to the "barber paradox." As Chaitin points out, these paradoxes are related to the Epimenides paradox of the ancient Greeks, the paradox of the liar. The essence of this paradox is a statement that says of itself, "This statement is false." Is the statement indeed false? Well if it's false then it must be true. But if it's true then it's false. In other words, the statement is neither true nor false.

Gödel's brilliance is in part due to the way he managed to place this paradoxical idea at the heart of his argument. In this his work is reminiscent of the proof by Cantor described earlier in this chapter. But Gödel goes farther down this road than practically anyone would have believed possible. In the discussion of infinity I mentioned the possibility that the idea of infinity might be evoked by situations of self-reference. Gödel's proof is deeply self-referential. He manages to mirror statements *about* mathematics within mathematics itself. Thus Gödel uses the related notions of paradox and self-reference. The astonishing thing is that this use of paradox and self-reference is contained and controlled within an intricate and precise logical argument. The result is a result about the intrinsic limitations of formal, logical thought. It is all really quite extraordinary!

Formal Systems

The manner in which Gödel proceeds owes a great deal to previous attempts to formalize mathematics, attempts that one associates with the names of Hilbert, on one hand, and Russell and Whitehead, on the other. The defining characteristic of a formal system is that everything, absolutely everything, is pinned down. You specify what symbols (mathematical and logical) are acceptable (for example, "=," "0," or "⇒"), how these symbols may be combined to make statements, the precise rules under which certain statements may be said to follow from other statements, and so on. It turns out that all the symbols, statements, axioms, theorems, even arguments can be given a unique number. The simplest way to do this is to find a way to order all possible statements and lists of statements. Thus there is a first statement (which may be an individual symbol), a second, a third, and so on. It is possible to create such an ordering in many ways, some relatively straightforward and others more complex.

Nevertheless there is something strange about being able to list all mathematical expressions. This strangeness forms the subject of a paradox that was first enunciated by the French mathematician Jules Richard in 1905. Since each definition is associated with a unique integer, it may turn out in certain cases that an integer may possess the very property designated by the definition with which the integer is correlated. Suppose, for example, the defining expression "not divisible by any integer other than 1 and itself" is associated with the number 17; obviously 17 itself has the property correlated with that expression. On the other hand, suppose the defining expression "being the product of some integer with itself" were correlated with the number 15; 15 clearly does not have the property designated by the expression. We shall describe the state of affairs in the second property by saying that the number 15 has the property of being *Richardian*; and in the first example that 17 does not have the property of being *Richardian*.[17]

Now consider the property of being Richardian. It seems to designate a property of integers, so it must appear somewhere on our list. Suppose it appears at position n. Then we ask the question: *Is n Richardian?* Well we get the paradox that by now is familiar: n is Richardian if and only if n is not Richardian.

Now this particular contradiction can be gotten around. "We can outflank the Richard Paradox by distinguishing carefully between statements within arithmetic and statements about some system of notation in which arithmetic is codified."[18] Nevertheless the paradox leads to Gödel's idea that "it may be possible to 'map' or 'mirror' meta-mathematical statements about a sufficiently comprehensive formal system in the system itself."

Gödel invented a brilliant way of coding the elements of the formal system into numbers and arithmetical operations. How, one might ask, could all the intricate information contained in a mathematical argument be compressed into a single integer? Wouldn't we be losing a lot of information? Without going into the details, the key to the coding lies in the fact that even a single integer contains a lot of information if you look at the integer in terms of its decomposition into prime numbers. For example,

$$N = 2{,}337{,}185{,}664{,}000 = 2^{10} \times 3^8 \times 5^3 \times 11^2 \times 23.$$

The one integer N is built on five different prime numbers each of which is raised to a different power. Each of the primes and each of the powers may be thought of as different bits of information contained in the original integer N. This deeper structure of the integers is the key to coding (formal) mathematical statements and arguments into one (very large) integer. Conversely each such integer can be decoded back into the original piece of formal mathematics.

Not only did Gödel find a way of coding mathematical arguments into single integers with no loss of information. but he was then able to transform metamathematical statements, propositions about the relationship between different mathematical arguments into arithmetic. For example, "the meta-mathematical statement: 'The sequence of formulas with Gödel number x is a proof of the formula with Gödel number z.' This statement is represented (mirrored) by a definite formula in the arithmetical calculus which expresses a purely arithmetical relation between x and z."[19]

Gödel had created a situation of self-reference, or, in other words, a situation of controlled ambiguity, where there are two consistent contexts: on the one hand, the arithmetical consisting of integers and relations between integers, and, on the other, the formal and metamathematical consisting of propositions and proofs and statements about these proofs.

Let us assume that this numbering system has been created and let us follow Roger Penrose's[20] description of the central idea in the proof of Gödel's theorem. Consider propositional functions that depend on one variable that we denote by k.[21] All of these have a number, so let us denote the nth formula by

$$P_n(k).$$

Now each proof or potential proof of a mathematical proposition is a sequence of statements that can also be labeled by natural numbers. Let the nth proof be designated by

$$\Pi_n.$$

Now consider the following proposition:

$$\neg \, \exists \, n \, [\Pi_n \text{ proves } P_k(k)],$$

or in words, there is no proof for the proposition $P_k(k)$. Now because of the coding of all statements in arithmetical terms, this statement must itself be somewhere on the list of statements. Suppose it is in the mth position, that is,

$$\neg \, \exists \, n \, [\Pi_n \text{ proves } P_k(k)] = P_m(k).$$

We now look at what this says for $m = k$ (this is reminiscent of Cantor's diagonal argument):

$$\neg \, \exists \, n \, [\Pi_n \text{ proves } P_k(k)] = P_k(k).$$

Notice that this statement says of itself that it has no proof (within the system). Penrose discusses this self-referential, ambiguous formula as follows:

> The specific proposition $P_k(k)$ is a perfectly well-defined arithmetical statement. Does it have a proof within our formal system? . . . The answer to [this question] must be "no." We can see this by examining the *meaning* underlying the Gödel procedure. Although $P_k(k)$ is just an arithmetical proposition, we have constructed it so that it asserts that what has

been written on the left-hand side: "there is no proof, within the system, of the proposition $P_k(k)$." If we have been careful in laying down our axioms and rules of procedure, and assuming that we have done our numbering right, then there cannot be any proof of this $P_k(k)$ within the system. For if there were such a proof, then the "meaning" of the statement that $P_k(k)$ actually asserts, namely that there is *no* proof, would be false, so $P_k(k)$ would have to be false as an arithmetical proposition. Our formal system should not be so badly constructed that it actually allows false propositions to be proved! Thus, it must be the case that there is in fact *no* proof of $P_k(k)$. But this is precisely what $P_k(k)$ is trying to tell us. What $P_k(k)$ asserts must therefore be a true statement, so $P_k(k)$ must be *true* as an arithmetical proposition. We have found a *true* proposition which has *no proof within the system!*[22]

Now the negation of $P_k(k)$ cannot be proved either, since it is false and we are not supposed to be able to prove false propositions within a formal system. Thus neither $P_k(k)$ nor its negation can be proved within our formal system. This is what it means for a system to be incomplete.

Gödel showed that any formal system that is complicated enough to deal with arithmetic, that is, contains the counting numbers 0, 1, 2, 3, 4, . . . and their properties under addition and multiplication must either be inconsistent or incomplete. Thus Hilbert was wrong. There is no formal system for mathematics that produces all true results and only true results.

Goodstein's Theorem: An Example of an Unprovable Statement That Is True

Gödel's result tells us that any formal system has a valid result that is unprovable from the axioms of the system, but it does not give us the actual mathematical statement of a theorem that falls into that category. Nevertheless it is now possible to state such a theorem. First, following Penrose,[23] let us describe Goodstein's theorem.

The description of the theorem begins with the observation that it is always possible to write any given integer in terms of 1's and 2's alone (we shall call this its base 2 representation. For example,

$$45 = 32 + 8 + 4 + 1 = 2^5 + 2^3 + 2^2 + 1 = 2^{2^2+1} + 2^{2+1} + 2^2 + 1.$$

We could also represent this number to base 3, writing it with 1's, 2's, and 3's:

$$45 = 27 + 2 \times 9 = 3^3 + 2 \times 3^2;$$

or to base 4 (with 1's, 2's, 3's, and 4's):

$$45 = 2 \times 16 + 3 \times 4 + 1 = 2 \times 4^2 + 3 \times 4 + 1.$$

Now take any number and write it in its base 2 representation as above. Then

(a) Increase the base by 1.
(b) Subtract 1 from the number.

Alternate steps (a) and (b). What happens (in the first few steps) is illustrated below (in the case of 45, which could be called the "seed"):

$$45 = 2^{2^2+1} + 2^{2+1} + 2^2 + 1$$

Step 1: $3^{3^3+1} + 3^{3+1} + 3^3 + 1$
Step 2: $3^{3^3+1} + 3^{3+1} + 3^3$
Step 3: $4^{4^4+1} + 4^{4+1} + 4^4$
Step 4: $4^{4^4+1} + 4^{4+1} + 4^4 - 1 = 4^{4^4+1} + 4^{4+1} + 3 \times 4^3 + 3 \times 4^2 + 3 \times 4 + 3$
Step 5: $5^{5^5+1} + 5^{5+1} + 3 \times 5^3 + 3 \times 5^2 + 3 \times 5 + 3$
Step 6: $5^{5^5+1} + 5^{5+1} + 3 \times 5^3 + 3 \times 5^2 + 3 \times 5 + 2$
Step 7: $6^{6^6+1} + 6^{6+1} + 3 \times 6^3 + 3 \times 6^2 + 3 \times 6 + 2$

The question is what happens in the long run. It appears that the numbers are getting bigger and bigger, and it would be natural to speculate that the numbers grow without bound. Yet surprisingly Goodstein's theorem[24] proves that, no matter what "seed" number one begins with, this process eventually (and it is a very long eventually) ends at 0.

This is an extraordinary and initially counterintuitive result. The steps of type (b) (decreasing by 1) eventually balance off and

279

overcome the increase in the base. However, if one does some calculations one can get a feeling for why it might be true. We can see it already beginning to happen in step 4 above, where the fourth power becomes a third power and will eventually be eroded still further.

What is interesting about this result is that it is a concrete Gödel theorem for arithmetic plus induction. Mathematical induction was discussed in Chapter 3. Suppose we are working in a system that includes the ordinary operations of arithmetic and logic as well as the principle of induction. Then it has been demonstrated that the Gödel theorem for this system can be reexpressed in the form of Goodstein's theorem.[25] Thus it cannot be proved within the system, that is, using the principle of induction, yet it is true and can be proved using a more powerful assumption than ordinary induction. This example allows Penrose[26] to insist on the difference between our ability to access the truth of a mathematical result and what can be proved based on an a priori set of rules.

The Collatz $3n + 1$ Problem

There are many other problems that involve the kind of iterative procedure that one finds in Goodstein's theorem. One such problem was first posed by Lothar Collatz in 1937 and often goes under his name. I shall state it in the following way. Starting with an integer a_0, for $n = 1, 2, 3, \ldots$ define

$$a_n = \begin{cases} 3a_{n-1} + 1 & \text{if } a_{n-1} \text{ is odd}, \\ a_{n-1} / 2 & \text{if } a_{n-1} \text{ is even}. \end{cases}$$

Thus the "seed" a_0 gives rise to a sequence of values $a_1, a_2, a_3, \ldots$ For the smallest positive integer values of a_0 we get the following:

$$a_0 = 1 \; a_1 = 4 \; a_2 = 2 \; a_3 = 1$$
$$a_0 = 2 \; a_1 = 1$$
$$a_0 = 3 \; a_1 = 10 \; a_2 = 5 \; a_3 = 16 \; a_4 = 8 \; a_5 = 4 \; a_6 = 2 \; a_7 = 1$$

The list continues, where the first number in the list is the value of the seed:

4, 2, 1

5, 16, 8, 4, 2, 1

6, 3, 10, 5, 16, 8, 4, 2, 1

7, 22, 11, 34, 17, 52, 26, 13, 40, 20, 10, 5, 16, 8, 4, 2, 1, and so on.

The conjecture is that the sequence returns to 1 for *any* initial seed a_0, and it has been verified for very large values of the seed. Thus the conjecture "seems" true, but the question is how to prove it. It is the general consensus that the required proof will not be forthcoming any time soon. Paul Erdös is reported to have commented that "Mathematics is not yet ready for such problems." Jeffrey Lagarias concludes his survey paper with the following statement that highlights an ambiguity that, in the author's opinion, lies at the heart of attempts to resolve this question:

> *Is the 3x + 1 problem intractably hard?* The difficulty of set-tling the $3x + 1$ problem seems connected to the fact that it is a deterministic process that simulates "random" behav-ior.[27] We face this dilemma: on the one hand, to the extent that the problem has structure, we can analyze it—yet it is this structure that seems to prevent us from proving that it behaves "randomly." On the other hand, to the extent that the problem is structureless and "random," we have noth-ing to analyze and consequently cannot rigorously prove anything.[28]

Is the Collatz conjecture even provable? We don't know. It has been shown that a generalization of this problem is formally undecidable. Now there is a big difference between a conjecture that is not accessible to present techniques and a conjecture that is actually inaccessible in principle. Is this kind of problem the kind of inaccessible result that Gödel's theorem shows must exist? It might very well be. Could it even be that many mathematical results are of this type—well defined but unprov-able? The answer to this question has tremendous implications for our conception of mathematics. Could it be that the results that can actually be rigorously demonstrated are in some sense exceptional and that "most" mathematical results are of the Gödel variety?

COMMENTARY ON GÖDEL'S THEOREM

In summary I shall just collect and reiterate a few of the things that we can learn from Gödel's theorem. In the first place, the result is a meta-result, that is, it says something about the possibilities of a certain kind of knowledge; yet it is expressed within that body of knowledge itself. This is the mystery of Gödel's theorem—that within the context of logical thought one can deduce limitations on that very thought. Thus the theorem delves deeply into the dangerous but essential subject of self-reference.

Second, the result is built around a particular logical paradox but it is not itself paradoxical. In my view this use of paradox is not accidental but an essential characteristic of this kind of deep knowledge.

Again ambiguity is a crucial factor in the proof. The argument hinges on the identification of a surprising ambiguity. Integers are given a dual interpretation, first as arithmetical objects and second as mathematical propositions and arguments. As in so many of the examples in Chapter 1, Gödel creates a method of passing back and forth from one of these contexts to the other. Just as the equation $E = mc^2$ can be seen to mean that there is one "thing" that is matter when looked at in one context and energy when looked at in another, so statements in a formal system are ordinary arithmetical propositions when looked at in one context but metamathematical assertions when looked at in the other.

What has gotten so many people excited about this theorem are the implications they draw from it for science and for epistemological questions in general. It says something very specific about how formal, deductive systems fit into the ecology of human knowledge. One way of putting this would be to say that formal systems are local, not global. They may describe certain situations very well, but they are intrinsically limited. Thus the knowledge obtained from such systems is contingent, not absolute. We set up certain systems because we are interested in studying certain problems or investigating certain mathematical situations. The systems we set up may be vast, but they have limits and we inevitably rub up against these limiting situations as we pursue our investigations. The limiting situations are often characterized by paradoxes and other breakdowns and are precisely where we would expect "interesting" mathematical phenomenon to occur.

CONCLUSION

This chapter developed some of the implications of organizing a description of mathematics around the notion of the mathematical idea. In particular, it leads to a new relationship between mathematics and logic, for logic can be seen as another idea or organizing principle. Ideas that are "wrong" can still be valuable. Even paradoxes—those areas of extreme logical breakdown—are valid sources of mathematics. Brilliant mathematical ideas may arise from these very situations and be incorporated into the body of systematic thought.

Thus the relationship between mathematics and logic has been decisively altered. We are beginning to see paradox, not as something to be avoided and eliminated, but as a potentially rich source of ideas. Even the determination of what is correct versus what is incorrect, what is true versus what is false, turns out to be a matter of much subtlety.

In the next chapter I shall begin to consolidate a certain line of argument. Many of the ideas mentioned in previous chapters have a core that is ambiguous and sometimes contradictory or paradoxical. Looking back, I seem to have spent a good deal of time talking about certain ideas in which the "incompatibility" of the elements of the ambiguity was extreme. These are ideas that I sometimes said were "impossible" yet they exist. I shall give such ideas a name—"great ideas." In the next chapter, I have something more to say about "great ideas" in general and, more specifically, about the fascinating topic of randomness.

CHAPTER 7

Great Ideas

> All true good carries with it conditions which are
> contradictory and as a consequence impossible. He
> who keeps his attention really fixed on this impos-
> sibility and acts will do what is good. In the same
> way all truth contains a contradiction.
> —Simone Weil[1]

INTRODUCTION

Here we come to a key element in the view of mathematics that is being built up in this book—a point of view that will be made more explicit through a discussion of a certain variety of mathematical idea that will be called a "great idea." The discussion of "great ideas" was anticipated in the introduction and, in passing, in Chapters 1 and 5. How could anyone have the effrontery to attempt to define what is great in mathematics? This is clearly an impossible task from the point of view of the content of the mathematical idea. What will be done in what follows is to sketch out a generalized set of circumstances that accompany or set the stage for the emergence of certain fundamental mathematical ideas. These circumstances are so strange from our usual perspective that to take them seriously requires a radical shift of perspective—a great idea of our own.

In a sense the notion of a "great idea" emerges from a close consideration of the Simone Weil statement that opened this chapter, except that "great idea" replaces what she calls the "good." Thus the statement is shifted from the moral to the intellectual domain. The statement now reads, "a great idea carries with it conditions which are contradictory and as a consequence impossible." The meaning of this statement depends on what is made of the word "impossible." On the surface to say that something is "impossible" is to say that it cannot happen. It might then seem more accurate to replace "impossible" by "seems impossible." A great idea seems impossible when looked at from a certain point of view. In fact, a kind of impossibility is a necessary condition for the emergence of a great idea.

Why is it impossible? A "great idea" is "impossible" because embedded inside it we find a contradiction. It is impossible and yet it exists. What could this phrase possibly mean? And yet we have already observed this phenomenon on a number of occasions. Take the example of "zero" that was developed in Chapter 2. For the Greeks "zero" was impossible. The contradiction it contains is, of course, the existence of something that stands for nothing. "Zero" contains this contradiction yet for us today, there it is, a mathematical object we use every day without giving it a second thought. The fact that we now use "zero" so freely blinds us to the quality of "impossibility" that impeded its initial development.

"Impossibility" is the context out of which the "great idea" arises. The word "impossible" focuses attention on a property that one finds in certain situations of great originality. It is not to be taken on the level of pure logic, since, on that level, if it is impossible it does not happen—period. Perhaps instead of "impossible" one could say that the situations that are being described are "beyond what is possible" at least at a certain moment in time. They are impossible in the way that a new idea is impossible; it just hits you out of the blue—it comes as a surprise. "Oh, that's what it means, I get it. I never thought of it in that way before." It is perhaps inappropriate to use the word "miracle" in a book about mathematics, but a great idea is a miracle in the sense that it radically changes the intellectual landscape. There is something miraculous and therefore "impossible" about a great idea.

Recall the statement of Davis and Hersh in reference to infinity (pp. 118 and 121) "There is a tension. . .which is a source of power and paradox." Paradox refers to the quality I am calling "impossibility." On the other hand, "impossibility" cannot be separated from the power or creativity of the idea. In the discussion of the "great idea," impossibility and power are two sides of the same coin. From one perspective, the idea is impossible— it just cannot exist. From the other, there is the brilliantly original idea that transforms our understanding. This whole situation is reminiscent of a phenomenon that has been observed in various situations of ambiguity—what looks like an insurmountable block from one perspective becomes a situation of the openness and flexibility of multiple perspectives from another.

The juxtaposition of "impossibility" with creativity is essential to the "great idea," as it is to the entire point of view that is being developed in this book. A creative situation has negative and positive components. You could say that the old must be overcome in order to make room for the new. Every mathematical situation comes with a context, a frame of reference. The context that is adequate to express a problem, for example, may not be the frame of reference that is needed for its solution. The aspect of the situation that is being called "impossible" expresses how the situation looks from the initial frame of reference. It is this very "impossibility" that forces one to address the inadequacy of this frame of reference. Taking the "impossibility" seriously and yet not stopping there—continuing working on the situation in the face of the "impossible"—pushes the creative mathematician out of what you could call the "classical" point of view. It breaks down the initial frame of reference, and that is a necessary precursor to the emergence of the creative insight.

Remember our discussion of "infinity" and Gauss's protest against "the use of infinity as a completed [object]. . . The Infinite is only a manner of speaking." "Actual infinity" was as impossible for Gauss as it was for Aristotle. It remained impossible until the moment of Cantor's great success. In our earlier discussion of infinity there was an emphasis on how very strange this concept is, how difficult, how controversial it was at first. Cantor had such a hard time getting the mathematical world to take his ideas seriously! How could so many brilliant intellects have contemplated these ideas and backed away? Why did the great French mathematician Henri Poincaré call Cantorism a disease from which mathematics would have to recover?[2] Most mathematicians of the time, we must conclude, considered the idea of "actual infinity" to be impossible! In this way, it satisfies our criterion for being a great idea.

The ideas discussed in the book so far have been characterized by the property of ambiguity. Ambiguous ideas come with multiple contexts that are, or appear to be, in conflict. Great ideas take this conflict to an extreme. There is an irreconcilable conflict embedded in a great idea. This is why one wants to say that a great idea is impossible. And yet there they are! In discussing great ideas we get to the nub of the unorthodox approach to mathematics that I am proposing. Because great ideas have the dimension of impossibility, their acceptance by the mainstream is invariably slow and fraught with controversy. I mentioned in Chapter 1 how the normal reaction to a situation of ambiguity is to suppress one of the conflicting aspects and, in this way, restore the harmony that was threatened by the appearance of conflict. In the "great idea" the conflict is heightened, so the resulting tendency to ban one of the conflicting views is even more extreme. In these cases the idea is not only impossible and ridiculous but also threatening—threatening to people's mental health and perhaps even to the public order. The introduction of a great idea may even bring about a cultural crisis.

In the case of the ideas of "zero" and "infinity," the gap between the two contexts is vast. One of the contexts is actually denying the existence of the other. Not only denying, for example, that "zero" exists but also denying the very possibility that such an idea *could* exist. The gap between "zero the number" and "the zero that cannot be articulated" is really so huge that we can only gasp in awe at the human genius that managed this feat—that managed to convert this contradictory situation into an idea that could be articulated and worked with. It is difficult from the perspective of today, when such ideas have been so domesticated, to conceive of the threat they contained and therefore the courage and creativity it took to conceive of them. On the one hand, "actual infinity does not exist"; on the other, we have countable sets, uncountable sets, even a whole hierarchy of infinite cardinal numbers. How extraordinary!

In Chapter 1, I discussed the crisis that surrounded the proof of the irrationality of the square root of two. What was at the heart of that crisis? The problem may have been that the Greeks confounded the two senses of the word "rational." The Greeks were rational thinkers and rational thinking, by definition, is clear and sensible, unclouded by emotion or prejudice. Rational thought was not only the means through which one attained to true knowledge of the external world, but it reflected, for the Greeks, the hidden structure of that world—the world for them *was* rational. This rationality of the world was reflected in the assumption that the structure of the natural world could be captured by mathematics, by "number," and, as we saw earlier, by rational numbers. Mathematical "rationality" is quite a reasonable hypothesis, and the argument seemed to be clinched by the discovery that the musical harmonic series can be completely described using rational numbers.

The statement "all numbers are rational" was not so much a mathematical result as a statement of an entire worldview. This is the context in which we must consider the proof that the square root of two was not rational. We could say that a rational argument had shown that the world was not rational. How is that for a dilemma? From one point of view it was impossible for the root of two to be irrational. But, of course, there it was: the root of two existed as a geometric object via Pythagoras's theorem and there existed an airtight argument that "proved" that it was not a rational number. These are precisely the ingredients we expect to find when we talk about "great ideas." They are also the ingredients for a cultural crisis.

As another example, consider the crisis that was associated with the introduction of the calculus by Newton and Leibniz. On the one hand, there was the introduction of obscure ideas, infinitesimals and fluxions, which seemed to contravene all reason. Bishop Berkeley was right to take issue with these "results" (see Chapter 3). They were, he was saying, impossible. They made no sense. And yet they worked! This is all to say that the calculus is built around a "great idea." Now the calculus was obviously too valuable—the results were too good—to be overthrown by some narrow logical attack. Logic would just have to expand to account for this new mathematics. It took awhile, a century in fact, but that is exactly what happened. Again we have a "great idea" that is associated with a crisis that ultimately changes the world!

Those who feel that mathematics is merely rational and reasonable have difficulty accounting for great ideas and the cultural crises with which they are associated. Mathematics for them is an activity that could be performed by calculating machines or by human beings in a machine-like mode of using the mind. As was promised in the introduction, this book offers an alternative vision: a vision of ambiguity and paradox and therefore of creativity and great ideas. In my view you cannot have the one, the great creativity of mathematics, without the other, the ambiguity, the impossibility and therefore the crises. *It is in this place, the place of crisis, close to the impossible and the contradictory, that we will find the living heart of the mathematical enterprise.*

FORMALISM AS A GREAT IDEA

Great ideas are born of paradox and often engender a crisis. Return for a moment to that great crisis in mathematics that accompanied the discovery of non-Euclidean geometries. In Chapter 4 there was a discussion of what was at stake in this crisis, namely, the very nature of physical reality: was space Euclidean or was Euclidean geometry a mere model of space?

From today's point of view it may be hard to understand what the fuss was all about. Today we look at hyperbolic geometry, for example, as just another mathematical theory with its definitions, axioms, and theorems. Yet, at the time, Gauss was reluctant to publish his results for fear of the probable negative reaction from the scientific community. For most mathematicians at the time non-Euclidean geometry was "impossible." We find it difficult to appreciate this "impossibility" because we look at mathematics today from a point of view that came into being as a result of the subsequent crisis. It is always difficult to appreciate a cultural crisis in retrospect. One of the defining characteristics of such a crisis is that it retrospectively vanishes from view. This is the true measure of the "paradigm shift" that has occurred.

The original problem probably had something to do with the relationship between mathematics and truth. Is mathematics, Euclidean geometry in particular, giving us true information? Do the results of mathematics reflect what is real? Now one property of "truth" is that it has a claim to uniqueness. How can there be more than one truth? This is like asking how there can be more than one reality. For thousands of years Euclidean geometry "revealed" the detailed structure of physical space. From the point of view of "naive realism," space is out there and geometry describes what is out there—there is no separation between the two. Therefore there was an enormous shock associated with the idea that there could be multiple geometries—each of them with an equal claim to validity. What seemed to be at stake went beyond the nature of physical space. It put into question the relationship between mathematics and truth. In fact, "Truth" itself had been put into question, and this created an intolerable situation. Cultures exist, one might say, in order to give society a cohesive foundation, a common experience of reality. Schools exist in order to pass on these cultural assumptions to the young. Without such a common notion of reality the demons of instability and chaos threaten the fabric of social order. When fundamental cultural assumptions are questioned, as in this case, it is absolutely essential to reestablish a stable notion of truth.

The existence of non-Euclidean geometries had, for a brief period of time, revealed the underlying epistemological assumptions of mathematics and therefore opened them up to scrutiny. This is what a crisis accomplishes; it reveals a contingency that was always present but formerly went unrecognized. Thus, a crisis is interesting and valuable because of what it reveals about assumptions that were hitherto taken for granted. The response to the crisis is equally revealing. In the case in question, mathematicians were not slow to enter the breach and attempt to heal what had been ripped asunder. The problem was how to put mathematics on a completely secure footing—how to restore the "truth."

It was clear by the nineteenth century that not even Euclidean geometry was completely logically airtight—there were axioms that were missing, things that were assumed as "obvious" but never made explicit. One of these was discussed in Chapter 2— how can we be sure that two arcs of circles that appear to cross each other actually *do* have a point in common? Asserting the existence of such a point of intersection required, you will recall, an *axiom of continuity*. There are a number of other lacunae that need to be made explicit. David Hilbert accomplished that task. In his expanded version of Euclidean geometry everything had been made explicit—the set of axioms was complete and all results followed from a rigorous process of logical inference. The proofs were now really independent of the geometric diagrams and constructions that always are present in Euclid. These diagrams might remain as a guide to the intuition and to make the argument more transparent, but the logic of the argument did not rely upon them. Ironically, geometry had, in this way, been "liberated" from its geometric origins. It had been converted into a formal, deductive system of thought. Non-Euclidean geometry was merely *another* deductive system with an axiom system that was a minor modification of that of Euclidean geometry—the parallel postulate was different. In fact, any consistent set of axioms could in principle generate its own mathematical system. Thus *formalism* was born and, in the process, the whole notion of truth was radically transformed.

Formalism has all the characteristics of a "great idea." It was born out of the "impossibility" of non-Euclidean geometry. Then, because a "great idea" involves a profound creative insight, the domain of this great idea expanded beyond geometry into new territory. In fact Hilbert had the idea that what he had done for Euclidean geometry could be done for mathematics as a whole. This would involve finding a finite but complete set of axioms for all of mathematics. Were this to be successfully accomplished mathematicians would be certain that all of the theorems that were logically deduced from these axioms would be valid (at least as valid as the axioms themselves). It seemed reasonable at the time to expect that this formal system would be capable of deriving *all* of the valid theorems of mathematics.

Thus was born what we could call the "formalist dream" for mathematics—a deductive system that could produce all possible valid conclusions. What a powerful dream this was! In a way it is the old dream of Aristotle and Euclid, the Greek dream of reason moved forward by a couple of millennia. As I mentioned earlier, this dream is a kind of foundation myth for our culture. As a mathematician I was brought up on this myth, but all of us have this point of view so deeply ingrained in our consciousness that we have difficulty with the idea that it is "impossible." The dream is so compelling that even when someone like Gödel *proves* that it is impossible we refuse to give up the dream. Rather than relinquishing the dream we would rather put the impossibility on the side and continue as though nothing had changed.

Formalism would radically change the mathematical landscape. As a philosophy it would determine how each mathematical subject—from group theory to complex analysis to algebraic topology—was structured. It would determine how mathematics was presented to beginners and to the nonmathematical world but, more important, it determined how mathematicians described mathematics to themselves. For many people, formalism came to be seen as identical with mathematics; that is, the distinction between mathematics and the formalized version of mathematics was all but lost. Formalism is the aspect of contemporary mathematics that resides in conscious awareness. It provides an easy answer to questions about the nature of mathematics (deduce theorems from axioms) or what mathematicians do (prove theorems). However, like all easy answers it misses a great deal.

With every great idea it is important to bear in mind that something is lost, but, on the other hand, a great deal is gained. What is lost will be discussed later on, but for the most part it involves the tendency on the part of great ideas to over-extend themselves—to claim a kind of universality. The gains in the case of formalism include acquiring a freedom that mathematics never had before. The mathematician gained the freedom to create mathematical subjects that are derived from internal situations within mathematics itself, or are only loosely modeled on her experiences of the world, or are simply a product of her imagination. Mathematics could now be studied in its own right, for its intrinsic interest and beauty and not necessarily because it said something directly about some other science or the natural world. One might say that formalism gave birth to "pure mathematics."

Of course the tendency to produce arbitrary, abstract systems can go too far. It may come to be seen merely as a game and, as a result, fall into obscurity and irrelevance. Perhaps as a result of formalism, mathematics, which had been intimately connected to physics for centuries, went its own way for a time. Generations of mathematicians did not learn physics, for example, and physicists developed the mathematics that they needed in their own way and independent of the developments that were going on in mathematics. This was a problem, one that has been corrected in recent years when mathematics and physics have reunited with a vengeance.

Nevertheless the freedom that mathematics gained through formalism was valuable. It is one of the great mysteries of science that mathematics, developed for its own sake, for reasons internal to mathematics itself, turns out to have great value in other domains. One great example is Einstein's use of differential geometry to provide the mathematical foundations of the General Theory of Relativity. The mathematics was not produced for this purpose, the mathematics was already there, ready to be applied. This phenomenon is Wigner's "unreasonable effectiveness of mathematics in the natural sciences" that was referred to earlier and its presence is ubiquitous.

The whole progression from non-Euclidean geometry to formalism has many of the characteristics of a great idea. Chaitin summarized the situation as follows:

> Hilbert's proposal seemed fairly straightforward. . . .he wanted to go all the way and formalize *all* of mathematics. The big surprise is that it turned out that it could not be done. Hilbert was wrong—but *wrong in a tremendously fruitful way*, because he had asked a very good question. In fact by asking this question he created an entirely new discipline called metamathematics, an introspective field of math in which you study what mathematics can and cannot achieve. (italics added)[3]

Here we have a number of important characteristics associated with great ideas that will be commented on in the next section. In the first place there is the play between impossibility and actuality that has already been discussed. Then there is the tendency to "want to go all the way"—to go too far, in fact, because, finally there is the sense in which the great idea is "wrong." Finally there is the sense that the idea is wrong in a "fruitful," that is to say, creative manner. All these characteristics are present to varying degrees in the examples of "great ideas" that have been discussed: zero, infinity, irrational numbers, the derivative, and finally formalism itself. Certainly there is no claim here that these cases are identical. Nevertheless there are enough commonalities to justify the claim that there is something going on here that is worth paying attention to.

CHARACTERISTICS OF "GREAT IDEAS"

As was emphasized earlier a "great idea" begins with a gap—a gap that seems unbridgeable. In Chapter 3, I discussed the irreducible ambiguity that characterizes the concept of infinity by considering the concept of the "ineffable." By definition the "ineffable" is "that which cannot be put into words." Unfortunately (or is it fortunately?) "ineffable" *is* a word. Thus when we use the word "ineffable" we are articulating that which cannot, by definition, be articulated. The idea of the "ineffable" has this twist; it contains a gap that cannot be bridged. In mathematics the discussion of "zero" and "infinity" demonstrated the same gap. The gap exists in the idea and thus in people's response to the idea.

One way of thinking about this seemingly contradictory situation would be to say that "great ideas" can be approached from two sides. One side, which could be called the side of the "Absolute," says that the (absolute) infinite exists but cannot be articulated. This was the side that Aristotle and Gauss were coming from. It was the side that fueled Kronecker's bitter rejection of Cantor's ideas. The other side could be called the "relative side," the side of the explicit definition, and from this point of view there is no problem; moreover, it is difficult to see what the fuss is all about. From this point of view infinite cardinal numbers are just ordinary mathematical objects that reason manipulates like other mathematical objects, the integers for example.

The reader will not be surprised that I call this situation "ambiguous" and furthermore claim that both sides—the absolute and the relative—have their own legitimate claim to validity. For contemporary mathematicians, philosophers, and other intellectuals, the "absolute" is the more problematic dimension. The reason for this is that you cannot discern the absolute from within the formal aspects of mathematics and thus it is invisible to people for whom the formal is the only aspect of the mathematical situation that is real. I realize that the "absolute" may have an almost theological connotation for some, so they would exclude it from serious consideration. It is true that discussing an aspect of reality that can never be completely grasped in words or in conceptual terms is, in a sense, the ultimate ambiguity. It places a conundrum at the heart of the human drive to comprehend the world. Nevertheless it creates a context in which we can begin to understand certain developments in the history of mathematics.

However, there may be ways of thinking about the "absolute" point of view that will be helpful. From the absolute point of view certain ideas have a kind of unlimited depth. Thurston[4] lists ten different definitions of the derivative. Each of these adds something to our understanding of derivative, and yet the sum of these definitions does not necessarily complete our understanding. It is always possible that, in the future, someone may come up with a new way of thinking of the derivative and use it to do interesting mathematics. Many fundamental ideas in mathematics have this kind of depth, for example, real number and continuity, to name just two ideas that have been discussed earlier.

This situation of potential depth of ideas is a familiar one from a certain reading of quantum mechanics. The physicist Heisenberg[5] talks about the "in potentia" existence of the wave description of a particle before its collapse due to observation. The "wave" or probabilistic description would correspond to the deepest point of view, the observed particle to a specific formulation of that viewpoint. To give a more mundane example, there are certain aspects of our lives that cannot be put into words. This doesn't mean that one cannot try to put them into words— in fact, putting an experience into words may have the effect of enriching the experience just as one's success in coming up with a "good" definition of infinity has the effect of enriching mathematics. Nevertheless, the very attempt to articulate certain feelings, for example, changes them, makes explicit and precise that which does not live on that level. In just this way one can "explain" one's response to a favorite piece of music yet the explanation and the emotional response to the music lie on completely different planes.

The absolute cannot be realized definitively; it cannot be completely captured in a particular closed form. Cantor, for one, "deeply believed that infinity was God-given";[6] that is, he believed in something one could call "Absolute Infinity." Perhaps we should take our cue from the neo-Platonists and think of the absolute not as an object or a concept but as a process or a tendency that gives birth to concepts. In the discussion of the ineffable I said that not only could it not be expressed but also that there was an irresistible drive to express it. The infinite refers back to this inexpressible Absolute. In reading about Cantor's life one can see this overwhelming drive to give expression to his idea of infinity. Yet one can equally see that "infinity" is not any infinite cardinal or ordinal number nor is it contained in the definition of cardinality, for example.

For some the existence of the nonconceptual is a problem—a limitation on what can be "captured" by the human intellect. This could be summed up by saying (of the "Absolute" dimension of things), "Whatever you say it is, it is not." From this point of view, like Sisyphus, we are doomed to keep trying and keep failing. No matter how brilliant and subtle we are, the "Absolute" will never be reduced to a finite, symbolic form. However looking at the matter in only this way is a serious mistake. For one thing, it underestimates the accomplishments of Cantor and many others.

To understand the error I must anticipate the material of the following chapter and distinguish between a process and the results of the process. This is similar to the discussion of reification, whereby a process is transmuted into an object. Whereas the attempt to make a *definitive* object out of a "deep" concept cannot succeed, the drive to do so is itself an expression of the Absolute. Mathematics, we could say, is driven by the need to express the "Infinite" in finite terms, and this very drive is the "Infinite" in action. Great ideas, in particular, are situations in which we can most clearly see this phenomenon in operation. In a great idea there is a reaching out toward the unknown, toward what cannot be known but what our deepest nature needs to make known. As was said earlier about the "oneness" of Plotinus, the drive toward understanding the natural world and ourselves is an imperative, not a luxury. The fact that it can have no definitive resolution is not necessarily a problem. It implies that there is no end to creativity, no end, for example, to mathematics. Every great idea is a case of a huge creative leap. It is a moment when the world is definitively changed. It is a moment when the Absolute briefly makes its appearance in the human domain.

The whole discussion about the Absolute could be summed up by a couple of lines from a song by the poet and songwriter Leonard Cohen:

> There is a crack in everything.
> That's how the light gets in.[7]

The first line could be seen to refer to the fact that nothing is perfect, no theory or ideology can be identified with reality, that is, the Absolute cannot be captured definitively. The "light," of course, refers to the Absolute itself that, Cohen says, does not appear in the formal theory itself but in the places where the theory breaks down, the places we have been calling ambiguous. It is precisely because great ideas have a sort of unlimited depth that there is always room for the Absolute to make its appearance out of the "cracks" in the theory. This appearance of the absolute occurs in the act of creative insight.

It is important to reiterate that "Great Ideas" appear out of extreme ambiguity. They appear in the face of the impossible. Thus they require great courage on the part of their proponents. These people face the contempt of their peers; they face ostracism; they may even face violence, as in the case of the poor Pythagorean who was reputedly drowned for the crime of proving that the square root of two was not rational. Great ideas are the products of genius, but what is the essential quality of genius in these cases? It is the tenacity to face up to the impossible, to live for long periods of time with the tension of the ambiguous. It is, in short, a kind of frustration tolerance, the precise thing that we seek to eliminate by reducing human thought to the algorithmic.

As a result of an act of extraordinary creativity, the great idea is born. What it appears to have accomplished goes against the intuition of the culture out of which it arises. This accounts for the resistance that the idea initially encounters. Overcoming this resistance requires a conviction and strength of character, a feeling that "this is the way things are and they cannot be otherwise." On the part of the author of the breakthrough and his immediate followers, there may arise a sense of having discovered some absolute truth. Thus one of the characteristics of a great idea is the claims that that are made on its behalf. Take formalism, for example. First there is the notion of setting Euclidean geometry on a completely secure foundation. Then there is the idea that *all* of mathematics can be treated in a similar fashion. Finally there is the claim that formalism *is* mathematics. There is also the idea that all of science can be developed in this way. Also that the thinking process itself is in essence formalistic, that computers can think, and so on.

In this way, the introduction of a great idea is often followed by a period of inflation, a series of increasingly grandiose claims for the unlimited applicability of the idea. Of course, this phenomenon is not restricted to mathematics. Just think of the claims made on behalf of the theories of Karl Marx or Sigmund

Freud. This period of inflation is a natural and almost unavoidable part of the process that we are calling a "great idea." Because great ideas come out of an overwhelming need to grasp the absolute, to give truth a concrete form, so to say, the claim to have done so would appear to be inevitable.

It follows that the third stage in the life of a "great idea" will be the recognition that the idea does have limitations—that the claims made for it in the inflationary period were overstated. At this stage it may be asserted that the idea is false. However, it is not the idea so much as the *universality* of the idea that is false. Formalism is not wrong, but the claim that formalism could encompass all of mathematics was certainly wrong. Interestingly, in mathematics the "error" if we wish to call it that can be demonstrated in a completely rigorous manner. Of course, this was done in the work of Gödel that was discussed earlier.

Thus even in its limitations a great idea is ambiguous. To repeat Chaitin's remark about Hilbert and formalism, "Hilbert was wrong—but in a tremendously fruitful way because he had asked a good question." This is reminiscent of Shimura's comment about Taniyama's capacity to make "good mistakes." Great ideas are wrong but they are wrong in a brilliant and inspired way. Many of the seminal advances in mathematics and science are "wrong" in this sense.

A Door or a Window?

Great ideas are metaphorically closed doors that turn into open windows. The "closed door" phase refers to what we have called above the "impossibility" of a great idea. Then somehow the closed door is transformed into an open window that looks out on a novel and unexpected terrain. Mathematics is full of problems or questions that cannot be resolved within the domain within which they are initially formulated. The correct resolution arises from an expansion of the initial mathematical context to a new context that, in retrospect, was implicit in the initial problem.

Take, for example, the equation $x^2 + 1 = 0$. It makes perfect sense in the context of the real numbers, the rational numbers, or even the integers, but it has no solution within those domains.

The equation seemingly forces us to invent a solution, "$x = i$," but that single solution is not secure until the real numbers have been extended to the full set of complex numbers, that is, all numbers of the form $a + bi$, where a and b are real numbers. Unexpectedly it turns out that this new domain is algebraically closed, which means that if we now write down *any* algebraic equation, the "right number" of complex solutions exist. Thus for the equation $x^4 - ix^3 + (2 - i) x^2 + 3x + 2 = 0$, where the coefficients are all complex numbers, there are four solutions (counting multiple roots), all of which are also complex numbers. What is impossible in one context opens a window to a larger context, the complex numbers, which not only contains the solution to the original equation but is also a vast and wonderful new mathematical domain in itself.

Remaining with the real and complex numbers for a moment, take the function $f(x) = 1 + 1/(1 + x^2)$,[8] where x is a real number. This function can be written as an infinite series

$$1 + \frac{1}{1 + x^2} = 1 + \frac{1}{1 - (-x^2)} = 2 - x^2 + x^4 - x^6 + x^8 + \cdots .$$

Now this series converges for certain values of x and not for others; in fact it converges for all values of x (strictly) between -1 and $+1$. We say that its *radius of convergence* is 1 since it converges for all x such that $|x| < 1$. The radius of convergence is a very important property of the function $f(x)$. The mystery is how that number is to be discerned from properties of the original (real) function. Looking at the graph of the function, for example, where do we find the number 1?

However, when we consider the function as a function in the domain of complex numbers the secret is immediately revealed. When x is a real number, the denominator $1 + x^2$ is greater than 1 and thus never 0. However if we allow complex numbers then the denominator $1 + z^2 = 0$ for $z = \pm i$.[9] Therefore, the largest radius for which the complex function $f(z) = 1 + z^2$ is nonzero is

$$|z| < 1$$

The radius of convergence of the power series representation of a real-valued function, $f(x)$, can only be seen by considering the associated complex function, $f(z)$.

299

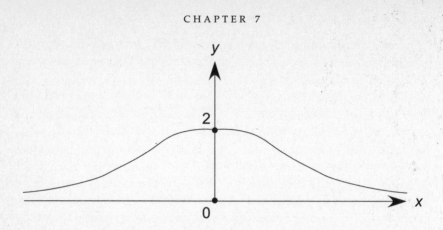

Figure 7.1. Graph of $f(x) = 1 + 1/(1 + x^2)$

This example reiterates the point that was made earlier about the relationship of the exponential function to the trigonometric sine and cosine functions. The "real" functions seem to be completely unrelated yet the associated complex functions are related by the Euler equation

$$e^{ix} = \cos x + i \sin x.$$

Here again a feature of the "real" function $f(x) = 1 + 1/(1 + x^2)$ is only explained by considering the associated "complex" function.

This phenomenon of a problem in mathematics that is stated in one domain yet requires a larger domain in order to be revealed properly is really quite general. For example, the most accessible proofs of the fundamental theorem of algebra[10] involve ideas from analysis or topology. Domains in mathematics like algebra and analysis cannot really be thought of as subjects that are independent of one another. Great progress in mathematics often comes at the interface between two mathematical subdomains. This, of course, is part of the phenomenon of ambiguity.

These kinds of situations also reveal the inadequacy of "formalism" as a description of mathematics. A development of a subject in terms of axioms and definitions, theorems and proofs leaves one with the feeling that the formal system "is" the subject and that the formal development somehow contains all the possibilities for the subject. It can never account for the realization that an idea from another part of mathematics entirely is just what is needed to resolve a question posed within the original system.

Of course, Gödel's theorem also makes the same point. Results may be "true" but not derivable from a given formal system. There was such an example given in Chapter 6. Mathematics is continually forcing mathematicians outside their initial frames of reference. It is the problematic aspects of mathematics that demand a creative resolution—one that is often completely unexpected and yet, in retrospect, can be seen to be inevitable.

These jumps to a higher viewpoint are not predictable. They cannot be programmed. Mathematicians cannot just sit around and speculate on where the next expansion is going to appear. The new viewpoint must be forced upon us by the situation, by a specific problem. The expansion from the real numbers to the complex numbers is a major breakthrough; it solves many problems, helps us to understand things that were formerly obscure, and suggests many new questions. The next expansion of that type from the complex numbers to the "quaternions"[11] is interesting but not as dramatic as the jump from the real to the complex numbers.

Randomness

In this section the previous discussion is illustrated by delving into one of the most fascinating ideas in modern scientific thought—the idea of randomness. To give a complete discussion of randomness would require a book in itself, maybe even a series of books. Thus the discussion here can have no claim to completeness. The aim is to give another, very nontrivial example of a great idea and its influence on scientific culture. Randomness is a topic that has great contemporary relevance. It has not been "resolved" in the sense that there is no clear consensus about the "correct" way to think about randomness like the consensus on the questions of analysis that were discussed earlier (even if it would be premature to say that the foundations of analysis have been resolved definitively and for all time). Randomness as a concept retains a certain rawness and, for that reason, it will be interesting to investigate it as a great idea. Remember that "great ideas" are characterized by a certain "impossibility" but simultaneously by an open flexibility. Our discussion will begin with the "impossibility" of randomness and move on to various applications of the idea to modern scientific thought.

What is randomness? At the level of our everyday life experience we call it "chance," something with which that we all feel familiar. It refers to something unexpected, something caused by luck or fortune, that is, without any apparent cause. Randomness is, in a sense, the opposite of determinism. It reflects the ordinary sense that some things are too complicated to admit of a simple explanation or even any explanation at all. Why do certain people with the healthiest of lifestyles die young, while others, who never take care of themselves, live to a ripe old age? Who has not been struck by the thought that such questions have no logical answer—life and death cannot be predicted, they involve an element of chance that seemingly cannot be avoided. A contingency seems to be built into our lives and our experience of the world. One might claim that human culture and religion have arisen as an attempt to control and mitigate somewhat this random factor in existence. In a way naming the contingent, calling it randomness, and developing mathematical and scientific theories based on this idea is among the most ambitious human cultural projects. If the contingent is the stuff of our nightmares, then conceptualizing the random is an attempt to strike directly at the heart of the beast.

The phenomenon of "randomness" is one of the primordial sources of mathematics. Ian Stewart, for example, lists "five distinct sources of mathematical ideas. They are number, shape, arrangement, movement, and chance."[12] He says that "chance" or "randomness" is the most recent of these sources. "Only for a couple of centuries has it been realized that chance has its own type of pattern and regularity; only in the last fifty years has it been possible to make this statement precise."[13] This is the normal way in which mathematical (and scientific) progress is understood—the "pattern" that is implicit in "chance" is made precise. This implies that chance always had a definite meaning, even though at a certain moment in time this meaning was not grasped explicitly. Then its essence was captured by an explicit mathematical structure.

In fact things are not so simple. The fact is that "randomness" has been made precise in a number of different ways, none of which are definitive. There can be no definitive definition of randomness. Each definition evokes another aspect of "randomness" and embeds it in a mathematical or scientific theory. Each of these tells us something about that aspect of reality evoked by the use of the term "randomness." The fact that randomness was conceptualized later than other sources of mathematics (like number, for example) is testament to some complexity that is associated with the random. In fact, as we shall see in what follows, there could be legitimate skepticism (as in the case of "infinity,") about whether chance is a legitimate object of mathematical inquiry.

Randomness has always been an important element of human culture. In the past it had the connotation of being "divinely inspired."[14] In many cultures (Greek, Hebrew, and Chinese, for example) there is a tradition of casting lots, coins, or yarrow sticks as a means of accessing a "higher" wisdom. "The purpose of randomizers such as lots or dice was to eliminate the possibility of human manipulation and to give the gods a clear channel through which to express their divine will."[15] It is interesting to speculate on the nature of "higher wisdom." What is the modern person to make of it? Perhaps the word "higher" is misleading and should be replaced with "more basic," in the sense that it was claimed that ambiguity and ideas are more basic than logic. What is randomness more basic than? It may well be that many cultures felt that "chance" evoked an aspect of reality that is not accessible to consciousness—much less to rational thought. To put this in another way, every human being is clearly locked into a bubble—their own private universe—consisting of their thoughts and sense perceptions. How can people discern the existence of a world outside themselves? One of the ways seems to be through "chance." No matter how complete our picture of the world may seem to be, reality always manages to surprise us. Something unsuspected always seems to turn up, something happenstance, something random. Thus the random is an escape from a solipsistic existence. These earlier cultures engaged in systematic random behavior—they put aside for a time their rational selves—and hoped that the divine would find a way of expressing itself to them or through them. The random was seen as a doorway to some "higher" or transcendent aspect of reality. Right away you can see the problem of conceptualizing randomness. How can we make a scientific concept out of something that, by definition, lies beyond the conceptual? Thus randomness, like infinity, is intrinsically problematic.

Let me try to further isolate the aspect of "randomness" that makes it problematic. My dictionary has a number of definitions of the word random but the first of these is "done, chosen, or occurring without a specific pattern, plan, or connection." Thus randomness refers to the condition of being devoid of pattern or regularity. Earlier in this book we discussed "pattern" and saw that in meaning it was very close to "idea." It was mentioned at that time that mathematics is sometimes defined as the study of pattern. But randomness refers to the *absence* of pattern. In other words, it would seem to be a subject that is outside the domain of mathematics. Yet randomness is not outside mathematics. As was mentioned above it is one of the key ideas in mathematics (and many scientific theories).

To repeat the conundrum: *Mathematics is about "pattern." Randomness is the absence of pattern. Mathematics studies randomness.* This same inner contradiction can also be thought of in the following way. Chapter 5 discussed how the mathematical idea revealed an order that was implicit in a mathematical situation. Now the definition of a situation of randomness is that there *is* no order. Yet randomness is an idea, so it must be organizing something, it must be revealing some order or regularity. What is it revealing?

Mathematics, far from being stymied by this situation, finds enormous value in it. The fecundity of "randomness" is astounding; it is an inexhaustible source of scientific riches. Could "randomness" be such a rich notion because of the inner contradiction that it contains, not despite it? The depth we sense in "randomness" comes from something that lies behind any specific mathematical definition. How extraordinary it is that mathematicians have discovered a pattern in situations characterized by the lack of pattern! What a subtle use of the intelligence is at play here! At first glance saying that randomness is the "pattern that is the absence of pattern" seems to be playing with words and does not seem to advance the discussion. Yet it does advance human understanding in a most important manner. It is indeed possible to give meaning to randomness in a number of distinct but profound ways. The fact that it is possible to do so should not blind us to the profound ambiguity that must be successfully navigated in order to conceptualize "randomness" in such a way that it can be used in mathematics and science. In the same way that "zero cannot exist" and "infinity cannot be conceptualized," so randomness contains an inner contradiction. It is "impossible" in this way. It is impossible and absolutely fundamental. These two dimensions make it into a great idea.

The quintessential random event involves flipping a coin. Saying that flipping a coin is a random event means that the two possible outcomes, heads (H) and tails (T), are equally likely but that the outcome of any single flip is not predictable. So even though we say that the outcome of a single flip is a random event, the only way to verify this randomness would be to perform the experiment of flipping the coin a great many times, thereby generating a sequence that might look like $HTTHHHTHTTTTTHHTT$. . . . If the sequence is truly random we should expect to see (at least) two things. The first is that there are no (finite) patterns that go on forever. Thus a sequence like $HTHTHTHT$. . . is *not* random. Second, in the long run there will be the same number of heads as tails. That is, if $H(n)$ is the number of heads in the first n flips and $T(n)$ is the number of tails, then the ration $H(t)/T(n)$ will be approximately 1 as n gets large. What is unpredictable on one level (the individual flip of the coin) is *completely* predictable at another (in the long run the ratio $H(n)/T(n)$ is 1).

There are a couple of comments to make about the above two criteria. The first criterion is phrased negatively. It says that randomness is the absence of order. In this sense randomness is like a whole series of concepts we have encountered: *irrational* number (not rational), *transcendental* number (not algebraic), *uncountable* set (not countable), *infinite* (not finite). The second criterion is phrased positively; it says that the 0's and 1's are uniformly distributed, and makes that precise in a definition that can be verified. Thus even in this simple example we are confronted with the ambiguity of randomness. Furthermore, this example shows that randomness of this sort involves that difficult concept, "infinity." Neither of these criteria can be inferred by studying very long but finite segments of the sequence of heads and tails, since a sequence of one million heads, for example, might still be the beginning of a random sequence.

Randomness is elusive because, though it corresponds to something real in ordinary experience, it seems to disappear on closer examination. Does randomness really exist? Does it correspond to something real? For example, are the tosses of a coin really random? If we really had complete data on the total situation of the coin flipping experiment, wouldn't it be possible to predict precisely the results of any flip using the laws of physics? In other words, is randomness an illusion, merely an indication of insufficient data? It is interesting in this regard that "young children do not accept the notion of randomness." Bennett notes that Piaget and Inhelder found that young children conceive of random results as displaying regulated but hidden rules.[16] In this they are in the same situation as any gambler who believes that she discerns some hidden pattern in a seemingly random event. The question "Does randomness exist?" is an important one that I return to later on. It has no easy answer but it is vital!

Paradoxes and Randomness

If the concept of randomness is inherently problematic, then it is not surprising that there are many paradoxes associated with randomness. Paradoxes are things that we have come to associate with great ideas. They capture the "impossibility" of the great idea. In her fascinating little book entitled simply *Randomness*, Deborah Bennett devotes her last chapter to "Paradoxes in Probability." She goes through a whole series of paradoxes.

For example, consider the following:

In a study at a prominent medical school, physicians, residents, and fourth-year medical students were asked the following question:
If a test to detect a disease whose prevalence is one in a thousand has a false positive rate of 5 percent, what is the chance that a person found to have a positive result actually has the disease?

Almost half of the respondents answered 95 percent. Only 18 percent of the group got the correct answer: about 2 percent.[17]

What is going on here is that one person in a thousand actually gets the disease but 50 in a thousand test positive for the disease. Thus even if a person tests positive the chances that he actually has the disease are still only one in fifty.

Why do intelligent people have so much difficulty with this problem? Bennett claims that "the key here seems to be the difficulty of understanding randomness. Probability is based on the concept of a random event, and statistical inference on the distribution of random samples. Often we assume that the concept of randomness is obvious, but in fact, even today, the experts hold distinctly different views of it."[18]

However, it follows from our discussion that there is a good reason for people's problems with probability and randomness. In terms of difficulty, probability and randomness should properly be compared to the ideas of "infinity" and irrational numbers, both of which are similarly difficult to get an intuitive feeling for. The problem here is not a question of bad teaching in the schools. It is that randomness is intrinsically paradoxical. It is open-ended—it cannot be understood definitively and then put away. Behind Bennett's comments lies the feeling that there exists some definitive way to understand randomness but it is precisely this that is being questioned.

THE RELATIONSHIP BETWEEN RANDOMNESS AND ORDER

The importance of the concept of randomness is illustrated by the fact that it now appears as a basic feature of a number of the most important scientific theories: quantum mechanics, the theory of evolution, and the theory of chaotic systems. Much earlier, randomness had found its way into mathematics through attempts to describe games of chance. These attempts led to the development of the modern theories of probability and statistics, the uses of which are ubiquitous. Each of these theories conceptualized "randomness" in a slightly different manner. In a way each can be seen as an investigation into randomness. It is bizarre that randomness, the absence of order, is used to create or discover the ordering of the natural world by means of these fundamental scientific theories. In a strange and paradoxical manner, randomness is inseparable from order. There is no more basic ambiguity than the duality that could be referred to as "randomness/order."

At first glance "randomness" and "order" would seem to refer to two complementary and opposite ideas. "Randomness" is nonorder and "order" refers to the nonrandom. Now do both of these notions refer to a phenomenon that is real or is one real and the other illusory? One could argue, for example, that the natural world has a built-in order, an order that science accesses and reflects in its theories. This is the conventional view in science. From this point of view, any perceived lack of order would merely reflect our ignorance, our inability, at a given time, to comprehend the basic order that is nonetheless present. This would be like the coin-flipping experiment—given complete knowledge of the situation the result of any individual flip should be predictable. From this point of view randomness does not exist and would be regarded as merely a strategy for dealing with complex situations. This point of view is close to what I have called the "dream of reason." Looking at matters in this way, it appears that the natural world has an implicit but objective order and that it is the job of science to make that order explicit.

There is another point of view. One could stand with the Old Testament and claim that God created an orderly world out of primeval chaos. That is, the most basic situation is that there is no order; the world is chaotic, a word which is, of course, a synonym for random. This is a common way of looking at things in most cultures other than our own. How, in this view, does the order that we perceive all around us arise? Well, one could take the position of certain psychologists and claim that the mind and the senses, that is, perception and cognition, create order. Thus, for example, if a person is shown a random array of black and white dots, then after a while patterns will seem to appear. In this way people see the constellations as pictures or see forms in the arrangements of clouds. The patterns here do have a physical basis, but the order that one perceives involves an interaction of the mind with the physical data. From this

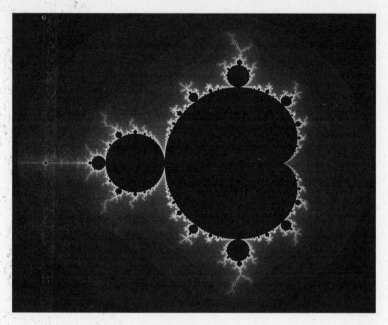

Figure 7.2

point of view, "ordering," the creation of order, reflects a natural tendency of the mind. It is a primary function of intelligence. If mathematics is indeed "the science of pattern,"[19] then one can see why mathematics is so important—it is one of the most important ways in which human intelligence interacts with, and orders, the natural world.

We have sketched two diametrically opposed points of view— one of which denies the existence of randomness, the other, of order. Perhaps the truth is more complex than either view, but it would seem that randomness and order are both inevitable parts of any description of reality. When we try to understand some particular phenomenon we are, in effect, banishing disorder. Before a piece of mathematics is understood it stands as a random collection of data. After it is understood, it is ordered, manageable.

A good geometric metaphor for the relationship between randomness and order is a "fractal," like the Mandelbrot set, for example (figure 7.2). Many people have a strong aesthetic response to the geometry of fractals—many books have been devoted to their mathematical and geometric properties. One such

309

book is even called *The Beauty of Fractals*.[20] What makes fractals so compelling? Fractals contain an extraordinary level of detail—so great as to be seemingly beyond the human ability to comprehend. Yet the detail is clearly ordered—a subtle order, it is true, one that requires the full mathematical description to fully appreciate. Yet the order and complexity are both present in the geometry. It is perhaps this ordered complexity that people find so attractive. So many facets of the natural world share this feature of structured complexity. Take the leaves of a tree, for example. They form a visual field that is extraordinarily complex, and yet a powerful order is present—not one leaf seems to be in the wrong place, the leaves all fit together in a manner that seems completely natural, perfect, in fact. This is how the world presents itself to our senses and to our minds—as a field of ordered randomness. Both properties—the randomness and the order—are present simultaneously. This is what should be called complexity. Complexity is ordered randomness.

What is extraordinary about fractals is that this extraordinary complexity is generated by relatively simple rules. In this respect a fractal is a wonderful metaphor for the world of science and for mathematics itself. Mathematics is a world of ordered complexity, in other words, of complex patterns. Of course these patterns are in the intellectual and not (necessarily) the visual domain. Nevertheless our aesthetic reaction is the same. Every mathematical result puts an order into an infinite collection of data. Our minds have discerned order within randomness and our reaction is one of wonder. We stand in awe of the ability that mathematics has—the ability that we all have—of appreciating the order within complexity.

A Mathematical Connection between Order and Randomness

More evidence of the intimate connection between order and randomness comes form an example due to the statistician M. S. Bartlett that is developed and commented on by Edward Beltrami in his book, *What Is Random?*[21] Bartlett introduces a simple procedure for generating a random sequence in the following way:

It begins with a "seed" number u_0 between zero and one and then generates a sequence of numbers u_n, for $n = 1, 2, \ldots$, by the rule that u_n is half the sum of the previous value u_{n-1} plus a random binary digit b_n. (The random digit b_n is either 0 or 1.)

If the random digits are 0,1,1,1,0, . . . and the "seed" is .5, then the random sequence would be .5, .25, .625, .8125, .90625, .453125, and so on.

Now I have been writing numbers in ordinary decimal notation (that is, in base 10). In base 2, that is, only using the digits 0 and 1,[22] our seed would be .1 and our terms would be .1, .01, .101, .1101, .11101, .011101, and so on. (Notice that the two sequences are identical: for example .1101 (base 2) = $1/2 + 1/4 + 1/16 = 13/16 = .8125$ (base 10).)

Converting from base 10 to base 2 is like moving from one language to another. Nothing mathematical is lost in the translation. What is gained is that the nature of the transformation is more transparent in the "base 2" language. If the seed number has binary (base 2) representation,

$$u_0 = .n_1 n_2 n_3 \ldots,$$

where each number n_k is either a 0 or a 1, then

$$u_1 = .b_1 n_1 n_2 n_3 \ldots,$$
$$u_2 = .b_2 b_1 n_1 n_2 n_3 \ldots,$$
$$u_3 = .b_3 b_2 b_1 n_1 n_2 n_3 \ldots,$$
$$\ldots$$

Thus at each stage we merely insert the random digit as the first decimal place and move all of the remaining digits one place to the right.

Now, let us consider what happens when we reverse this procedure. To go backward, say from u_3 to u_2, all that happens is that the digit b_3 is eliminated and therefore the new first digit (b_2 for u_2) is the old second digit (b_2 for u_3). In general, to move backward we eliminate the first digit after the decimal point and therefore move all the subsequent digits one place to the left. Thus .01110010. . . would become .1110010. . . which, in turn would become .110010. . .

Since we were writing our decimals in base 2 notation the reverse transformation could be defined by

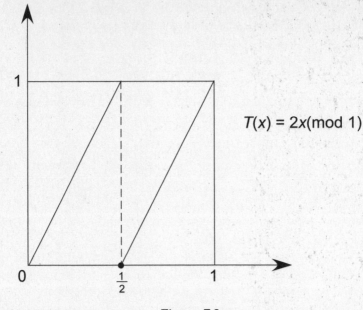

$$T(x) = 2x(\text{mod } 1)$$

Figure 7.3

$$T(x) = \begin{cases} 2x & \text{if } x = .0 \ldots, \\ 2x - 1 & \text{if } x = .1 \ldots \end{cases}$$

That is, $T(x) = 2x$ if x is less than $1/2$ and $= 2x - 1$ if x is greater than $1/2$. We usually write $T(x) = 2x$ (mod 1). (Remember that multiplying a decimal written in base 2 by 2 is accomplished by simply moving the decimal point one place to the right, just as multiplying an ordinary (base 10) decimal by 10 is accomplished by moving the decimal place to the right.) Thus $T(1/3) = 2/3$ and $T(2/3) = 2(2/3) - 1 = 1/3$.

Now $T(x)$ is a well-defined procedure. If you know x, then $T(x)$ can be calculated. The procedure $T(x)$ is deterministic—there is no randomness involved as there is in the definition of the original procedure. Strangely enough, the inverse of a random process is a deterministic process. Beltrami comments:

In effect, the future unfolds through Bartlett by waiting for each new zero or one event to happen at random, while the inverse forgets the present and retraces its steps. The uncertain future and a clear knowledge of the past are two faces of the same coin. This is why I call the Bartlett iterates a

Janus sequence, named after the Roman divinity who was depicted with two heads looking in opposite directions, at once peering into the future and scanning the past.[23]

The characteristic that is being referred to by using the name "Janus" is precisely the double nature that was characteristic of ambiguous situations. In fact Koestler also invokes the name "Janus" to refer to ambiguity. In this example, randomness and determinism are tied together—from one point of view, that of the original transformation, the process is random; from the other, it is deterministic. This seems counterintuitive. Yet the situation gets even more complex as we investigate things more closely.

We saw that the transformation, T, has a simple description when we write numbers in base 2 notation. If the "seed" is $v = .v_1 v_2 v_3 v_4 \ldots$ then $T(v) = .v_2 v_3 v_4 \ldots$, $T^2(v) = T(T(v)) = .v_3 v_4 \ldots$ and so on. In practice we can only determine our "seed" to a certain degree of accuracy and that accuracy is determined by the number of decimal places we include. Thus 1 decimal place will determine v to within $1/2$, 2 decimal places to within $1/4$, and n decimal places to within $1/2^n$. Suppose that we can accurately determine v to within n decimal places. This means that the digits after the nth place, v_{n+1}, $v_{n+2}, \ldots$, are essentially random. Now

$$T^n(v) = .v_{n+1} v_{n+2} \ldots .$$

After n applications of the transformation T all information about the initial value is lost. In this way a "deterministic" transformation generates data that are indistinguishable from the random.

There is another way to look at this situation. Suppose that two "seeds, " v and w, are initially very close together. That is, their decimal representations agree up to n decimal places (so the distance between them is at most $1/2^n$). Then $T^n(v)$ and $T^n(w)$ may be close together or far apart—it is impossible to determine the distance precisely. Thus a small error in determining v may result in a large error in the position of $T^n(v)$. This phenomenon is called "sensitive dependence on initial conditions," and it is one of the defining properties of a chaotic system.

Chaos and Determinism

"Chaos" is the term that is now used to indicate that simple functions when iterated (as above with T), produce complex, even random behavior. I remember the furor in the scientific community when it was demonstrated that the family of simple quadratic functions, $f(x) = 4\alpha x (1 - x)$, produced what we now call "chaotic" dynamics for a whole range of values of the parameter α. The scientific community at the time felt that it understood these types of functions perfectly and yet when they were iterated, that is, used to generate dynamical systems, they exhibited behavior that was surprisingly intricate.

The surprise was due in part to the introduction of a random or statistical element into what had previously been considered a classical deterministic domain. Compare this to the Newtonian metaphor of the natural world as a machine governed by systems of differential equations. Differential equations have the property that the present completely determines the future and the past. Thus, given any initial conditions for the system at time $t = 0$, the configuration of the system is then determined for all times, t, positive and negative. In contrast to this picture, chaos theory implies that each system has an "event horizon," a time beyond which it is impossible to predict the state of the system, in just the same way that our system $T(x)$ gave no information after the nth iteration. Chaos theory has reintroduced the random (or the statistical) as an unavoidable feature of macroscopic, that is, everyday, physical systems. This reverses the classical view that the statistical techniques of randomness are useful approximations to a complicated but deterministic reality. Here, on the other hand, one could claim that it is classical determinism that is an approximation to a reality in which randomness is an intrinsic feature.

On the Boundaries of Chaos

It is interesting that the theories that go under the name of "chaos" and "complexity" have developed their own unique approach to the duality between randomness and order. Classical systems are deterministic. If they are governed by systems of differential equations, then their long-term behavior is governed by a number of "attractors." An "attractor" is, as the name suggests, a set, A, with the property that every solution of the differential equation that starts in a certain vicinity of A gets closer and closer to A in positive time.[24] For classical systems of differential equations, these "attractors" were either individual "equilibrium points," where the solutions do not change with time, or "closed curves" that represent periodic solutions. Thus the long-term or "asymptotic" behavior of such classical systems consists of a finite set of states.

At the other extreme one has classically random behavior like the outcome of an ideal coin-flipping experiment. Here the long-term behavior is completely predictable: heads and tails are equally distributed—their ratio tends to 1. Ironically, randomness in the outcome of a single toss leads to predictability in the outcome of a long series of tosses. Thus both the classical and the random cases are relatively easy to understand. In one sense they stand in opposition to one another—the first is deterministic, the second is random—but in another they both reflect systems that are predictable albeit in different senses of the term.

In between these two extremes are the complex systems studied in the theory of chaos. These are often a blend of the deterministic and the random. Such systems may exhibit attractors that are vastly more complex than the ones that have been mentioned above, the so-called "strange attractors." For example, there is a mapping of the form $f(x) = \alpha x (1 - x)$ that gives rise to an attractor that is essentially the Cantor set that was mentioned in Chapter 4. The behavior of the system restricted to the "strange attractor" would often be random enough to be best described by some probability distribution. On the other hand, such a system might be "deterministic" in the sense that it has a finite number of "attractors" and that a solution starting from an arbitrary initial position would be attracted to one of these attractors.

Thus there developed the idea that "interesting" behavior for physical and biological systems was to be found "on the boundary of chaos," a region that was neither completely deterministic nor random.[25] This is the place where one might expect stable but complex structures to develop. In theories of complexity as applied, let us say, to simulations of biological systems, the "boundary of chaos" is the domain in which one might see the "emergence" of regularities that were not obviously predictable from a knowledge of the underlying transformation rules.

In the idea that interesting scientific phenomena emerge on the boundaries of chaos we see yet another example of the metaphorical power of mathematics. In particular, there is the uncanny ability of mathematics to be self-referential—to model itself, so to speak. In an earlier chapter there was a discussion of how significant mathematics also emerges on the "boundary" of the problematic, and the "chaotic" is certainly problematic. Without the existence of the phenomenon we are calling randomness, life would be boring; there would be no evolution, no innovation, no creativity. On the other hand, a world that consisted only of the random would be terrifying, unpredictable, chaotic with no regularities and therefore no life. Thus the "boundary of chaos" is where we live and where we need to stay if we are to be creative. Mathematics, too, lives on the boundary of chaos.

Algorithmic Randomness[26]

The normal mathematical way of approaching randomness is based on classical probability theory. Probability theory is the basis of statistics and has many applications throughout the sciences and engineering. However in recent years there has been another approach that is associated with the names of the great Russian mathematician A. N. Kolmogorov and Gregory Chaitin, who has been referred to earlier. According to their definition, *a series of numbers is random if the smallest algorithm capable of specifying it to a computer has about the same number of bits of information as the series itself.* For example, the series 0101010101010101 is not random, since it can be specified by saying "print 01 eight times"—if the sequence was longer then one would only have to alter the formulation by saying, for example, "print 01 a thousand or a million times." Thus one could say that the information in the sequence 0101. . .01 (a million times) is compressible. What if the information in the sequence is incompressible? What if the most efficient way to specify the digits in the sequence is to actually list them? Then we say that the sequence is random.

The "incompressibility" approach to randomness is really a very natural one. Chaitin refers to Ray Solomonoff's model of scientific induction, in which a scientific theory is seen as a compression of the data obtained through a scientist's observations. The "best" scientific theory would be the minimal program that could generate the data. A "random" theory in this sense would be no improvement on the actual raw data.

Using the algorithmic approach to randomness has led to the development of a tool for measuring randomness. This involves the concept of *complexity*. The complexity of a series of digits is the number of bits that must be put into a computer in order to obtain the original series as output, that is, it is the size (in bits) of the minimal program for the series. Thus a random series of digits is one whose complexity is approximately equal to its size.

Every approach to randomness teaches us something new about that essential but ungraspable condition of (primal) randomness. The algorithmic approach is fascinating and has a number of very deep implications. The first is that whereas one can show that most numbers are random we can never prove whether an individual number is or is not random. This last result is in the spirit of Gödel's incompleteness theorem, discussed in Chapter 6. The argument is also, like Gödel's, based on a paradox, in this case a paradox that goes by the name of the Berry paradox. Also, like Gödel's theorem, Chaitin's work has implications for the philosophy of mathematics. In particular, it puts restrictions on the information that can be derived from a formal, axiomatic system.

Algorithmic randomness, like so much of the mathematics I have discussed, tells us something about the nature of mathematics itself. "Gödel's work, Turing's work and my work are negative in a way, they're incompleteness results, but on the other hand they're positive, because in each case you introduce a new concept: incompleteness, uncomputability and algorithmic randomness. So in a sense they're examples that mathematics goes forward by introducing new concepts."[27] Chaitin's work brilliantly illustrates many of the themes that have been discussed in these pages. Ideas (concepts) are fundamental to mathematics, and thus mathematics is endlessly creative. There are

317

no theories of everything—all ideas (formal systems) have limitations. Mathematics does not avoid paradoxes but discovers them, plays with them, tames them, and uses them to produce new mathematics.

Randomness in Quantum Mechanics

Randomness is an unavoidable feature of the description of the world given in quantum mechanics. This is another example of how "randomness" captures some of modern science's deepest insights into the nature of reality. Yet, again, wherever the idea of randomness appears, it is accompanied by questions of meaning. Here these questions involve the nature of quantum reality. How can we understand the nature of the reality that is so successfully described by the mathematical formulations of quantum mechanics?

At the turn of the twentieth century there was a famous debate between Einstein and Bohr in which Einstein made the famous comment that "God does not play dice." At issue here were our by now familiar positions on the nature of the random: does it refer to some intrinsic, irreducible aspect of reality, or is it merely a measure of our present ignorance which will inevitably be dispelled when a more complete theoretical construct is developed?

Take, for example, the nature of a subatomic particle, sometimes called a quon, an electron, say. Is the electron a single particle or is it intrinsically a wave—a probability distribution. Does it have particular properties like position and momentum with precise values? To make the analogy with flipping a coin, is the electron analogous to a single flip or should it be identified with a whole series of flips? The nature of the electron is fundamentally ambiguous and appears to depend on the act of measurement.

> It looks like a particle whenever we look. In between it acts like a wave. Because measured electron is radically different from unmeasured electron, it appears that we cannot describe this quon (or any other) as it is without referring to the act of observation.

If we ignore observations for the moment, we might be tempted to say that an electron is all wave, since this is how it behaves when it's not looked at. However this description ignores the massive fact that every observation shows nothing but little particles—only their patterns are wavelike. If we say, on the other hand that between measurements the electron is really a particle, we can't explain the quantum facts. How does each electron on its own know how to find its place in the (wavelike) Airy pattern? What does a single electron "interfere with" to produce Airy's dark rings?[28]

If we say that "the electron is all wave," then we are saying that the probabilistic description is the more basic—in other words, that randomness is intrinsic. If we hold to the particle description then we have the hope that some further description, some "hidden variable," will reveal the true nature of things. Then the probabilistic description reveals our present ignorance. However, there exists in quantum mechanics a fundamental principle called the "Heisenberg uncertainty principle," which puts intrinsic limitations on experimentalists' ability to obtain precise information about the complete set of observable quantities associated with a given quon. This is all to repeat what was said in Chapter 1—the nature of the natural world as revealed by quantum mechanics is ambiguous. The heart of this ambiguity is directly connected to the ambiguous nature of randomness—further evidence that randomness is indeed a great idea.

RANDOMNESS IN THE THEORY OF EVOLUTION

The idea of randomness plays a crucial role in another of the great scientific theories—the theory of evolution. Evolutionary theory has provided a framework for approaching the most fundamental and perplexing questions one can ask about the nature of human beings—why are we the way we are and how did we come to be like this? It is interesting to note the manner in which randomness is a central feature of this theory. A simplistic description of the theory of evolution is that genetic mutations occur in a random fashion. "Survival of the fittest" refers to the mechanism whereby the mutations that survive are those that augment the species' chances for survival. Thus the equation seems to be randomness plus competition equals evolution. Of course things are not at all that simple, but randomness remains the motor that drives evolution.

Now the theory of evolution is a scientific theory, and therefore it follows that its basic terms should be clearly defined. However, though most people have a common sense feeling for what "chance" refers to, on closer examination it appears that randomness is not so clearly defined. What definition of randomness do we have in mind when we say that a mutation is random? Is it random in the sense that flipping a coin is random? If this were so, then any of the "letters" of the DNA alphabet would be equally likely to mutate into any other, and in the long run, all possible mutations would occur with equal frequency. Or are some mutations more likely to occur than others? It also appears that complex organisms have some process that stabilizes the DNA by eliminating "errors."[29] Why then do some mutations survive and others disappear?

In fact one could ask the same questions with respect to "random mutations" as about the randomness that arises in other situations. When we speak of a random mutation, are we talking about our ignorance or about some intrinsic limitation on our ability to know? Is the biological situation deterministic? Could we predict the results of a mutation given complete knowledge of the situation? In fact could we predict which mutations would occur in that hypothetical situation of total information? Or is the factor of randomness that enters into evolutionary theory an intrinsic limitation on our ability to predict?

Though most educated people take the theory of evolution as established scientific fact, there remains a rear-guard movement that argues in favor of an older point of view that is sometimes called "intelligent design." This doctrine holds that there is some intelligence (divine or natural) that drives the evolutionary process. This is still an active debate in certain quarters, a debate that in my opinion revolves around the right way to think about randomness. On the one side is the notion that randomness is the most basic level; on the other is the notion that the randomness of evolution only masks our ignorance, that on a deeper level (which may or may not be accessible to human intelligence) what is going on is not random but subject to this "intelligent design."

To put it another way, in the traditional theory of evolution, mutations are random and the consequences follow in an almost mechanical way—the fit survive and the less fit disappear. In the alternate viewpoint the process involves a kind of creative intelligence. Now it is true that this intelligence sometimes involves an omnipotent "God," but of course it is conceivable that this feature of intelligence could arise from within the natural world itself, without recourse to any supernatural force from outside the natural world. Is it reasonable to say that natural processes are intelligent? This would require seeing the evolutionary event as analogous to an act of creativity. The resolution of a problem that arises in research can sometimes be brought about by some chance occurrence, but this random event will have no effect if the scientist has not been totally immersed in the problem for some considerable time. Thus the random event has no effect unless it occurs within a mental field of considerable tension.

In the same way an evolutionary event may be the successful resolution of some tension or problem that is brought on by a seemingly random event that enters into the preexisting tension. The resolution of this problem would itself be an act of intelligence. Of course this is not the usual way of thinking about intelligence. To be consistent with the approach that I have advocated throughout this book, intelligence would be defined to be the sudden emergence of a new order within the field of opposing forces that is characteristic of a situation of ambiguity. In mathematics we have seen that this new order arises through the emergence of an idea. In evolution the creative act is the evolutionary adaptation. Thus the two points of view—classical evolutionary theory and evolution through intelligence—are not necessarily in opposition; in fact they complement and complete each other. A great deal depends on one's take on randomness.

CONCLUSION

This chapter has covered a good many complicated ideas, so it may be worthwhile to make a few last, summarizing comments. Order, disorder, and creativity are the basic constituents of the mathematical universe. Order arises out of acts of creativity. Though this order may be algorithmic in nature, the manner in which it comes into being is not algorithmic. It arises out of ambiguity, one dimension of which is the problematic or, at its most extreme, what we called the "impossible." Intelligence itself could be defined as a tendency toward order. Mathematics, at its most basic, involves such acts of intelligence—the discernment of pattern.

The order that mathematics discusses is a very subtle affair. It does not reside in the natural world in its finished form waiting around to be discovered. Mathematics is capable of making order out of the very lack of order. Beyond the question of whether the natural world is basically ordered or chaotic there is the indisputable fact that science has made randomness into a most fruitful scientific concept. One could argue that it has done this, not in one single way, but in a whole host of different ways, all of which have a certain commonality. And at the heart of these diverse approaches to randomness lies a black hole of power and paradox. This is the way it is with great and audacious ideas.

An interesting aspect of this discussion is the extent to which it is self-referential. Thus the great scientific theories that incorporate randomness as an essential feature—evolution, quantum physics, chaos—incorporate in the theory the very processes out of which they arise. Mathematics contains an ordered investigation of disorder. Chaos theory contains within itself implications for the creation of a theory of chaos. It is not surprising, therefore, that all the above theories have been used as metaphors for situations that go well beyond their natural scientific domains. In particular, the complementary pair "randomness/order" has something to tell us about the nature of mathematics—what it is and the manner in which it comes into being. Mathematics is intertwined with metamathematics. The content of mathematics cannot be definitively separated from how mathematics is created and understood.

SECTION III

THE LIGHT AND THE EYE OF THE BEHOLDER

In this section I start to draw out some of the implications of what has been developed in the earlier chapters. The "light" that is associated with creativity in mathematics is not to be found in any formal rendition of mathematics. It is to be found in the human creativity of the mathematician. If mathematics is a human activity, then mathematics is not objective in the normal sense of the word. And yet mathematics is surely not merely subjective. This points me toward an unusual take on mathematical truth—one that includes the mathematical mind. René Thom once said that something in mathematics is true if the best five mathematicians in the world say it is true. What could he possibly have meant by that comment?

This will lead me to take another look at some of the traditional philosophies of mathematics—formalism, Platonism, constructivism (new and old). Each of them points to some important aspect of mathematics. Of course each of them also omits something important as well. There is an ambiguity at the core of mathematics. For me the role of the philosophy of mathematics is to look into this ambiguity as it is manifested in all sorts of mathematical activity.

Then I turn my attention back to a comparison of the algorithmic and the creative—the trivial and the deep. The algorithmic needs no light, but the creative is always connected with "turning on the light" to some extent or other. This has clear implications for differentiating the activities of mathematicians from those of the computer, the human mind from the machine mind.

CHAPTER 8

The Truth of Mathematics

THE LIGHT OF "IMMEDIATE CERTAINTY"

Mathematics is about truth: discovering the truth, knowing the truth, and communicating the truth to others. It would be a great mistake to discuss mathematics without talking about its relation to the truth, for truth is the essence of mathematics. In its search for the purity of truth, mathematics has developed its own language and methodologies—its own way of paring down reality to an inner essence and capturing that essence in subtle patterns of thought. Mathematics is a way of using the mind with the goal of knowing the truth, that is, of obtaining certainty.

Unfortunately the very notion of truth has become problematic today. We hear that "truth is relative" or that "truth is constructed." Even in mathematics some have put forth the "heretical" idea that "mathematics is fallible" because it is a human activity. There is a good deal to be said for this point of view. It is certainly consistent with much that has been said in this book, especially in the discussion of mathematical ideas. Nevertheless to completely discard the notion of truth is to abandon the vital source of mathematics. Thus the question that will be faced in this chapter is how to have a meaningful discussion of the role and nature of truth in mathematics in the light of current mathematical practice.

The discussion of truth in mathematics is a story of the relationship between the natural world and the mind that seeks to understand that world. This is one of the oldest and greatest of mysteries. Mathematics is a window on that mystery. One of the oldest puzzles about mathematical truth is whether it is discovered or invented. Where does the truth of mathematics reside? Is it in the external world (or some other domain that is outside the human mind) and so must be discovered, or does it lie in the mind and is therefore invented? In other words is the truth of mathematics objective or subjective? What is truth anyhow? This is a profound and wonderful question. Thinking about mathematics will give an unusual perspective on the nature of truth and its relation to human intelligence.

In Chapter 2, Bertrand Russell was quoted as saying, "At the age of eleven, I began Euclid. . . . This was one of the great events of my life, as dazzling as first love." Euclidean geometry has had an extraordinary effect on generations of thinkers. What is there about this subject that led so many intelligent people to a kind of intellectual awakening? More generally, what is it that draws people to the study of mathematics? Russell says elsewhere, "I wanted certainty in the kind of way in which people want religious faith. I thought that certainty is more likely to be found in mathematics than elsewhere."[1] This is it! This is an important part of what draws people to mathematics—the sense that the results of mathematical activities are definitive; that it is possible to arrive at certainty in mathematics. The certainty of mathematics is different from the certainty one finds in other fields; it is somehow purer and therefore more powerful. To be certain is to *know*. What an extraordinary feeling it is to know and to know that you know! What strength there is in that position! What confidence one has when one sees into the truth of some mathematical theorem! The angles of a triangle add up to two right angles. It is not a matter of "probably," "almost always," "in our experience," or "as far as we know." None of the usual caveats apply—there is no quibbling. If you are not completely certain, if you have the slightest doubt, then you just don't get it. Mathematical truth has this certainty, this quality of inexorability. This is its essence.

Take the following famous quote from the great French mathematician Henri Poincaré. Poincaré had been working on proving the existence of a class of functions that he later named Fuchsian:

> Just at this time I left Caen, where I was then living, to go on a geologic excursion under the auspices of the school of mines. The changes of travel made me forget my mathematical work. Having reached Coutances, we entered an omnibus to go to some place or other. At the moment when I put my foot on the step the idea came to me, without anything in my former thoughts seeming to have paved the way for it, that the transformations that I had used to define the

Fuchsian functions were identical with those of non-Euclidean geometry. I did not verify the idea; I should not have had time, as, upon taking my seat in the omnibus, I went on with a conversation already commenced, but I felt a *perfect certainty*. On my return to Caen, for conscience's sake I verified the result at my leisure.

Poincaré goes on to say:

Then I turned my attention to the study of some arithmetical questions apparently without much success and without a suspicion of any connection with my preceding researches. Disgusted with my failure, I went to spend a few days at the seaside, and thought of something else. One morning, walking on the bluff, the idea came to me, with just the same characteristics of brevity, suddenness, and *immediate certainty*, that the arithmetic transformations of indeterminate ternary quadratic forms were identical with those of non-Euclidean geometry.[2]

Poincaré's "immediate certainty" is an essential but often neglected component of mathematical truth. Truth in mathematics and the certainty that arises when that truth is made manifest are not two separate phenomena; they are inseparable from one another—different aspects of the same underlying phenomenon. Now the certainty that is being discussed here is not the certainty that results from a correct calculation or chain of reasoning although both of these may be involved in preliminary or subsequent work. "Immediate certainty" is part of what is often called the "aha" or "eureka" experience—the high point of the creative process in mathematics and elsewhere. Often, when this phenomenon is discussed, people emphasize the suddenness with which the solution reveals itself. Immediacy is certainly present—insight often reveals itself in a flash—but so is certainty. One is absolutely certain that the solution that has just sprung into one's mind is the correct one. The problematic situation that one has been working on for a long time is suddenly resolved. It is as though all the disconnected data that one has been looking at have now finally jelled into one coherent picture. For all these reasons, Wiles, in the statement quoted in the introduction, uses the metaphor of "turning on the light." This metaphorical light is identical to "immediate certainty." Using the metaphor of "light" emphasizes that some underlying structure is revealed. It all suddenly makes sense. The certainty, the light, and the coherent picture are not different phenomena; they all refer to the same situation—a situation that is revealed in both the objective, formal domain as well as in the cognitive, subjective one.

When the word "certainty" is used in this chapter, it will usually refer to the phenomenon of "immediate certainty" that Poincaré refers to—a phenomenon that could also be called "creative certainty." Usually in descriptions of creative insights, the focus is on the object of the insight, the solution that is discovered, the problem that is solved, or the theorem that is established or enunciated. Of course, in mathematics when we are certain, we are certain about something; when we "know," we know something. In other words, "knowing" is usually considered in its transitive sense. What is proposed here is to consider its intransitive dimension. Focusing on "immediate certainty," as something that is worth paying attention to in its own right, will enable us to look at mathematics in a new way and resolve various questions about the nature of mathematics that are otherwise opaque.

Certainty and Truth

Truth is normally seen as knowledge that is certain, stable, and therefore reliable. To be "true" is, by dictionary definition, to be in conformity with reality or fact. In this definition "reality or fact" is the independent variable and "that which is in conformity" depends on the "reality or fact" for its existence. Thus when "truth" is used in the conventional manner it postulates a duality. In science, "that which is (or is not) in conformity" might be a law or theory such as, for example, Newton's law of gravity. This would then be shown experimentally to agree or disagree with empirical evidence. In mathematics, where the subject matter consists of concepts such as numbers and functions and the structures that contain them, the question of what is in correspondence with what is more difficult. Nevertheless it is interesting that there remains a duality of sorts, not so much in the mathematics itself as in the attitude of mathematicians toward the mathematics they do. Most mathematicians will readily admit that they feel as though the truth or falsity of a mathematical proposition exists outside their own minds, in some objective realm. This belief in an objective, Platonic, realm that contains the truths of mathematics shall be discussed later on in this chapter.

Truth, understood as certain knowledge, has always been a principal human objective. All disciplines in their own way pursue the elusive phenomenon of certainty. When most people talk about mathematical truth, the truth they are referring to is an objective truth. It is the same truth for all people. Many even have the sense that mathematical truth is universal—that any being, human or extraterrestrial, endowed with a sufficiently developed intelligence would recognize the truth of certain mathematical theorems. It would seem at first glance that the essence of truth lies in its objectivity.

These days it is quite normal to claim that absolute, objective knowledge is unattainable. In this view absolute truth may never be grasped but can only be approximated. Therefore science is seen to create models of reality. Even though these two points of view, objective truth and approximate truth, are very different, they both tend to agree that there is an objective domain—in science it is the natural world—wherein the truth resides.[3] As I said above, mathematicians proceed as though the situation in mathematics was analogous to that of the sciences, that is, that mathematical truth resides in some objective domain.

Now compare the inner "immediate certainty" that Poincaré speaks of to the "absolute, objective knowledge" that is the goal of science and mathematics. What is the connection between these two? At first glance they are almost opposites: one refers to something that goes on "in here," that is, in the subjective domain of the mind, whereas the other is "out there" in the objective world—although in mathematics it is not exactly clear where that objective world is located. Yet the two domains are connected—the inner certainty must reflect something that goes beyond the merely personal. In anecdotal accounts, like that of Poincaré, the drama is precisely the fact that the two domains can be linked so intimately. In a typical aphorism Einstein remarked paradoxically, "The most incomprehensible thing about the universe is that it is comprehensible."[4] This astounding connection between mind and matter—the "knowability" of the natural world—is precisely the content of Einstein's religiosity.[5] Why does mathematics describe the natural world as well as it does? In the famous question of the physicist Eugene Wigner, what accounts for "the unreasonable effectiveness of mathematics in the natural sciences"? There appears to be a profound connection between the structures of the natural world and the mathematical intelligence that investigates these structures and brings them to our conscious awareness. It is this connection that is the subject of our investigation.

Consider the question, "What comes first, 'immediate certainty' or truth?" The conventional point of view is that the truth comes first and that certainty arises from coming into contact with this truth. I shall stand this relationship on its head for a moment and investigate the consequences of claiming that the sense of certainty comes first and gives birth to what we call the truth. The most obvious objection to this formulation is that that the certainty that one feels, or thinks one feels, may turn out to be imperfect, even incorrect. Of course saying that it is incorrect already presupposes the existence of some objective truth against which correctness or incorrectness can be judged. The existence of this domain of "absolute truth" is precisely what is in question. When "good mistakes" were discussed in Chapter 5 or in Chaitin's comment (Chapter 7) that Hilbert was "wrong in a tremendously fruitful way," there is implicitly another way to look at the phenomenon of insight and creative certainty. Is Euler's formula, $V - E + F = 2$, for polyhedra correct or incorrect? It all depends—on what you call a polyhedron, for example. Recall the argument that a theorem was a kind of optimization of a mathematical idea. Thus it is at least arguable that what we are doing in mathematics is creating truth, and therefore that truth does in some way follow from certainty. This does not imply that truth is arbitrary, but it leads one to question the naive belief in the existence of an "objective truth."

"But," some will argue, "isn't this inner certainty that you are talking about a merely subjective phenomenon whereas the truth is surely objective?" On one level this is definitely the case. Certainty arises in the mind, and one definition of "subjective" is "existing only in the mind and not independent of it." Subjectivity is normally held to be something that is based on someone's feelings or opinions and not on facts or evidence. Certainty, a state of the mind, is subjective in this sense.

However, the "immediate certainty" Poincaré refers to is also objective. It does not change from situation to situation. Poincaré, in the above excerpt, talks about two different instances of this sensation, and it is clear that it is the same certainty that he feels on both occasions. It may well be that "immediate certainty" that one experiences in such situations is the same for all people at all times. Certainly when one discusses such experiences with one's colleagues and friends it seems that everyone has had such an experience, and that these experiences have a commonality that is independent of the particular situation in which they arise, the particular problem that is being studied, the personality of the researcher, and so on. It would then appear that "certainty" refers to something that is real.

Even animals may experience such certainty. Take, for example, the experiment with a talented chimpanzee named Sultan by the psychologist Wolfgang Köhler.

> Beyond the bars, out of arm's reach, lies an objective [a banana]; on this side, in the background of the experiment room, is placed a sawn off castor-oil bush, whose branches can be easily broken off. It is impossible to squeeze the tree through the railings, on account of its awkward shape; besides, only one of the bigger apes could drag it as far as the bars. Sultan is let in, does not immediately see the objective, and, looking about him indifferently, sucks one of the branches of the tree. But, his attention having been drawn to the objective, he approaches the bars, glances outside, the next moment turns around, goes straight to the tree, seizes a thin slender branch, breaks it off with a sharp jerk, runs back to the bars, and attains the objective. From the turning round upon the tree up to the grasping of the fruit with the broken-off branch, is one single quick chain of action, without the least "hiatus," and without the slightest movement that does not, objectively considered, fit into the solution described.[6]

Sultan not only used a stick as a tool, he *created* the tool by breaking off the branch. His actions, after he had arrived at what can only be called an insight, were characterized by an economy of action and a sense of purpose that are reminiscent of what I have been calling "immediate certainty." Problem solving, creativity, and certainty are most likely attributes of all life forms. Thus the "immediate certainty" that accompanies flashes of mathematical insight is something that is at a more primitive level than the particular form of the insight. The conclusion must be that this kind of "creative certainty" is an objective phenomenon that arises in the mind. The objection that "creative certainty" is subjective does not necessarily disqualify it as an object of study.

The certainty that arises in mathematical activity is the subjective correlate of truth. It is the sense that the mathematical truth one has seen into is beyond all doubt. It is that glorious feeling that comes with any successful mathematical activity—that we have entered a timeless world. Certainty is the central irreducible aspect of mathematics. People who merely use mathematics may find it helpful—even indispensable—to their work. They may even be amazed at the "unreasonable effectiveness" of mathematics. However, the utility of mathematics is a secondary quality that somehow follows from what is more elementary—truth and certainty. These factors are inextricably joined together: the cognitive with its properties of "certainty" and "knowing," on the one hand, and truth, on the other. Truth is found in mathematics with a purity and clarity that is unique to the subject. Mathematics exposes the Truth, but the Truth that mathematics reveals is a mysterious, subtle affair.

What Does a Proof Prove?

One of the main ways in which certainty arises in mathematics is through proof. Is mathematical truth contained in the proof itself or does the truth exist independently of the proof? Normally one would claim that the proof of a mathematical proposition establishes its truth. However, because there may be a number of independent proofs for the same mathematical proposition the proof cannot be thought of as identical to the truth of that proposition.

In fact, does the existence of a proof for a mathematical proposition make it true in any absolute sense? Take the proposition that the sum of the angles of a plane triangle is two right angles. Is this an absolute or is it relative truth? The modern answer to this question is that it is a relative truth. It is stated within the context of Euclidean geometry and, as we saw earlier, this proposition does not hold for non-Euclidean geometries. In other words, the truth of this theorem depends on the system of thought within which it is embedded—in this case Euclidean geometry. This is the formalist position: there is no truth; there are only logical inferences.

On the other hand, there is solidity to mathematics; this is why we love it; it stands against the contingency of the world of experience. In his autobiography the neurologist Oliver Sacks tells about his childhood when he was sent away from home during the Second World War to a Dickensian boarding school where he was beaten, starved, and otherwise tormented. He needed a way to escape from the pain of his life. "For me, the refuge at first was in numbers. . . . I liked numbers because they were solid, invariant; they stood unmoved in a chaotic world. There was in numbers and their relation something absolute, certain, not to be questioned beyond doubt." This certainty is what many have sought and found in mathematics.

The Greeks had the idea that this theorem, that the sum of the angles of a triangle is equal to two right angles, is true, period. They felt that it is an absolute, objective truth—a property of the natural world. This is not just an old-fashioned idea. It is related to "Platonism," which will be discussed later in this chapter. Mathematicians have this sense they are not merely playing logical games but that what they are dealing with is real. "Real" is just another way of saying "true." The interior angles of a planar triangle do add up to two right angles. There is truth in mathematical results. To a mathematician it is not so surprising that the applications of mathematics work as well as they do. Some have claimed this is because the world is basically mathematical. But you don't have to go that far to accept that there is something real that is going on in mathematics. Mathematics is not arbitrary!

When one "gets" the idea of a proof—in the case of the theorem about the sum of the angles, when one draws the proper parallel line construction (see Chapter 5)—one "sees" with an immediate certainty that the proposition must hold. The actual

335

proof is an afterthought. This is what Poincaré means by the comment, "for consciences sake I verified the result at my leisure." It is interesting that the word "verified" is used in this context. Verification is mechanical and does not require understanding. Thus though most proofs are built around some mathematical idea, unfortunately this idea may not be explicit in the proof. In fact certain proofs manage to conceal the idea on which they are based. Such a proof might begin with the words "suppose $f(x)$ is defined by. . ." and go on to verify that "$f(x)$" has just the properties that are needed to establish the result in question. How the author came up with that particular choice of "$f(x)$" is often a mystery to the person reading the proof. One is forced to accept the validity of the argument and therefore that the theorem is true, while nevertheless remaining in the dark as to *why* it is true. Verifying a proof is one thing and understanding it is quite another. Verification does not require any moment of "immediate certainty" but "understanding" does. Thus the proof is one thing and the idea another. One can accept a proof and yet never experience "immediate certainty."

For most mathematicians, the idea is the deepest level. It is "what is really going on." In Chapter 5 mathematical ideas were discussed in their role as "organizing principles." They organize mathematical situations by revealing relationships that would otherwise be hidden. A mathematical idea is an insight into a relationship between mathematical objects and procedures, a pattern within a mathematical domain.[7] The certainty that we feel when we become conscious of a mathematical idea arises at the same time as the idea. The formal proof then becomes a way of objectifying and communicating this certainty. Perhaps the "truth" of the mathematical situation is accessed by means of the mathematical idea.

It is therefore conceivable that inner certainty arises out of the birth of an idea that structures the domain in question. Then the "truth" would be a consequence of the existence of some mathematical idea. Certainty would accompany the emergence of the idea into full consciousness. In previous chapters there was a discussion of various stages in the development of the idea. Some of these stages are unconscious or at best semi-conscious. However when the idea bursts forth into full consciousness in that glorious moment described by Wiles as "turning on the light," it is accompanied by the characteristics described by Poincaré—suddenness and immediate certainty.

Nevertheless it is necessary to add a qualifier to the previous paragraphs insofar as they may appear to posit an absolute dichotomy with "proof " on one side and "ideas" on the other. In some places I may even seem to be saying that proofs are bad but ideas are good. That is not my intention. I stressed the limitations of proofs because logical argument has the ability to freeze up the natural dynamic tendency of thought in the manner that Arendt warned about. When people first come into contact with the power of systematic thought, they may well be transformed by its possibilities. The world, for them, will never be the same again. Later on this very revelation can become a subtle prison, as, for example, when one tries to get an understanding of an unfamiliar mathematical situation by merely rearranging the elements that one previously grasped into logically acceptable patterns. You will never go from understanding addition to understanding multiplication in this safe way—multiplication must eventually be understood in its own terms. It involves taking a cognitive leap. Eventually one has to let go of the old way of thinking and take such a leap. To the extent that logic may impede our ability to take that leap, my first task is to deconstruct, if you will, our tendency to be transfixed by logic.

However, as I have repeatedly stressed, it is not my intention to denigrate proof but to put proof in a more realistic perspective. Thus not all proofs are created equal. The best proofs, including many of the ones I have referred to in this book, are built around a key mathematical idea. This is why a good proof is so satisfying. If mathematics is completely objective and formal then there is no important difference between one proof and another—what matters is whether or not a proof exists, period. However, in practice most mathematicians recognize the difference in quality between different proofs. A "good" proof, one that brings out clearly the reason why the result is valid, can often lead to a whole chain of subsequent mathematical exploration and generalization.

In practice there is a subtle and dynamic relationship between proofs and ideas. One couldn't get started on a proof if one had no idea if or why the theorem in question is true. Nevertheless, as Joseph Auslander pointed out to me, mathematical ideas frequently arise when one is in the process of trying to write up some mathematical result. One begins with an idea and then attempts to give that idea a logical structure, that is, to prove a result. Often this attempt leads to a problem, an ambiguity or contradiction, and one is forced to grapple with this problem. Grappling with the ambiguity leads, if one is lucky, to a new idea which one then attempts to write up and the whole process is iterated. Hopefully this series of events converges in a finite number of steps to a solution of the original problem.

When one approaches a mathematical conjecture or unsolved problem not only does one ask oneself, "Is it true?" but also, "How would I go about proving it?" As I have discussed earlier, the truth of a proposition is not always equivalent to its accessibility. In fact, according to Gödel, a statement may be true but completely inaccessible to proof. Nevertheless, the very attempt to write down a proof often seems to evoke one or a series of mathematical ideas. Thus it is in practice impossible to draw a line and say that ideas are on one side and proofs are on the other. In fact, proofs structure ideas and ideas contain and are evoked by proofs. The whole process is an interactive one. "Doing" mathematics is a process characterized by the complementary poles of proof and idea, of ambiguity and logic.

Objective or Subjective?

Introducing ideas and bursts of certainty into a discussion of truth moves the discussion beyond the realm of the completely objective. Strangely enough, in order to understand mathematics and the subtle nature of mathematical truth it may be necessary to give up our attachment to the idea of "absolute objectivity." The existence of such an objective domain is an assumption that we all unthinkingly make, especially in science. But a moment of thought will reveal that it can only be an article of faith—it can never be proved. The only contact human beings have with reality is through the impressions that are received by the senses and the mind. The "objective" world is not, *as far as human beings will ever be able to tell*, completely objective. We know it through acts of perception and cognition. Postulating an absolutely objective realm is just that—an assumption—not something that can possibly be empirically validated.

Giving up on "absolute objectivity" will seem to many people to be a radical and perhaps dangerous thing to do. We are all at home in the paradigm of this Cartesian duality of subjective versus objective, so much so that even trying to imagine another possibility is difficult and sometimes irritating. Yet this tension that we all feel when we contemplate such an idea is the familiar one that precedes a shift in paradigms, that precedes a great idea. Nevertheless, such a shift is the whole trajectory of thought from the time of the non-Euclidean revolution until today. One finds echoes of such a shift in certain cognitive scientists, for example. In their book *Metaphors We Live By*, George Lakoff and Mark Johnson propose an alternative to what they call "the myths of objectivism and subjectivism."[8] They propose an alternative view that they call, the "experientialist alternative," based on their investigation of metaphor. They deny that there is an absolute truth, but they claim that the existence of such absolute truth is not necessary to deal with "the concern with knowledge that allows us to function successfully and the concern with fairness and impartiality." There is clearly a deep connection between their "metaphoric" approach to truth and my discussion of ambiguity, for metaphors are clearly ambiguous in the sense of the term that I am using in this book.

The existence of a realm of "absolute objectivity" is something that physics has had to confront in trying to access the deeper meaning of quantum mechanics and general relativity. In quantum mechanics it is the problem of the collapse of the wave function. In relativity it is the dependence of certain concepts like "simultaneity" on the state of motion of the observer. Both of these seemingly force a reassessment of the notion that there is a preferred point of view that is absolute and objective. Perhaps it is the time for mathematics to face up to the same situation and give up on the natural desire for absolute objectivity and absolute truth.

There is a danger here, and the danger is that we may throw out the baby of truth with the bathwater of absolute objectivity. This forces us to contemplate a "truth" that exists but is not completely objective. The existence of this primordial truth must be inferred and cannot be objectively demonstrated. It can be inferred through the shadows that it casts into the domains that we normally call "objective" and "subjective" and think of as "external" and "internal" (to ourselves). The internal shadow is what we have been calling certainty; the external shadow includes systematic thought and, in particular, proof. The existence of such a primordial truth should be taken as a hypothesis in the same way that the absolute objectivity of scientific realism is a hypothesis. What must be done is to add up the pluses and minuses of these two perspectives. Every conceptual framework inevitably filters out some aspects of things and highlights others. What kind of framework does this view of mathematical truth provide for past, present, and future mathematics?

Remember that giving up on absolute objectivity is not equivalent to giving up on objectivity itself, since, as was discussed above, ideas and sudden certainty are objective phenomena. However, in order to develop a way of looking at mathematics that includes the usual objective definitions, theorems and proofs as well as other form of mathematical activity such as computational mathematics not to speak of the creative insights of the mathematician it has been necessary to look more closely at what is meant by "objectivity" and "subjectivity." The supposed "objectivity" of mathematics includes elements that most people would call subjective but are necessary to a complete picture of what is going on. "Sudden or creative certainty" resides in a domain of "objective subjectivity," and mathematical truth includes both traditional objectivity as well as this "objective subjectivity."

The switch to including the element of "sudden certainty" in a description of truth was signaled when I used the expression "knowing" as almost synonymous with the term "certainty." However, these two terms have different connotations. For one thing, "certainty" may give an impression of something that is static, whereas "knowing" is dynamic. We expect certainty to be fixed, even timeless. This is a quality associated with the object of our certainty, "what we know" as opposed to "that we know." But the phenomenon that is being indicated by using the word "knowing" is not static. I can remember going back to read an old paper of mine, and not being able to make heads or tails of it. At the time of writing the paper I was totally immersed in the subject, so every statement in the paper evoked a whole host of associations. Later, going back, I had lost many of these associations. In order to really understand what I had written earlier, I would have had to go back and rebuild my understanding of the subject. The point is that mathematical statements make sense within a given context. What is clear to you at some point in time is not at all clear at another. Thus "knowing" or "immediate certainty" is not a static phenomenon. It comes and goes. It is knowledge, the content of the knowing, that gives the impression of being absolute and unchanging.

"Knowing," as certainty, does not have to be justified in terms of something more basic; it is self-validating. Knowledge needs to be validated, but the process through which that knowledge is acquired is sufficient unto itself. Why do mathematicians work so hard to produce original mathematical results? Is it merely for fame and fortune? No, people do mathematics because they love it; they love the agony and the ecstasy. The ecstasy comes from accessing this realm of knowing, of certainty. Once you taste it, you can't help but want more. Why? Because the creative experience is the most intense, most real experience that human beings are capable of.

Why, one might ask, were the Greeks so taken with geometry? Why did Plato's academy state, "Let no one ignorant of geometry enter here?" It probably was not because of the practical applications of geometry. Euclid's *Elements* is an exercise in pure mathematics—enjoyed for its own sake. Every time one successfully sees into a theorem or deduction of Euclidean geometry one momentarily enters into this magical realm of "knowing." Very few of the results in Euclid are purely algorithmic—most require a construction or other idea.

Now it is true that the quality of mathematics that I am attempting to bring out is not popular with everyone precisely because it puts the emphasis on the creative as opposed to the algorithmic dimension of mathematics. It was mentioned earlier how some people emphasize the benefits that accrue to humanity by the introduction of powerful algorithms into human thought. It is interesting that Descartes was such a person.

> Throughout his life, Descartes was very critical of the works of the Greeks in general, but their geometry especially irked him. It could get awkward and appear needlessly difficult.

He seemed to resent the fact that, the way Greek geometry was formulated; he had to work harder than necessary. In his analysis of a problem posed by the ancient Greek Pappus, Descartes wrote that, "it already wearies me to write so much about it." He criticized their system of proofs because each new proof seemed to provide a unique challenge, which could be overcome, as Descartes wrote, "only on the condition of greatly fatiguing the imagination.[9]

Here you have the tension between the creative and the algorithmic approach. You could say that Descartes won this particular battle (he succeeded in arithmetizing geometry) but lost the war in the sense that, though specific parts of mathematics can be "algorithmetized," all of mathematics certainly cannot be. What Descartes misses is that this use of what he calls the imagination and I call the creative intelligence is a value in itself. Yes it is fatiguing. Yes it is hard. But it is rewarding in a way that mechanically solving a problem will never be. But saying that creative mathematics is rewarding is not putting things strongly enough. The emergence of the truth is inevitably accompanied by the subjective aspect that I am calling knowing or certainty. This is as true for a small child learning some new bit of mathematics as it is for Poincaré. The truth that appears with the characteristics Poincaré describes is a totally different phenomenon from the truth of some algorithmic process of verification. The truth I am talking about—"creative truth"—is at a higher level than what could be called "algorithmic truth." The two are usually identified, and this leads to confusion. This confusion arises from considering the truth as lying entirely in the objective domain without realizing that truth has both objective and subjective dimensions.

So there are a number of reasons to consider the phenomenon of "creative certainty" as something that is worth looking at in its own right and not merely because it reflects the light of some definitive truth. It may have seemed that in this discussion the "subjective" component was given priority over the "objective." This is because this element of "objective subjectivity" is not generally acknowledged as being a valid factor in a discussion of mathematics. However, the actual situation is a little more complex. Remember the expression "to know the truth." This expression captures the dichotomy that was mentioned above: the subjective "knowing" and the objective "truth." Yet this duality is merely a feature of language not of the reality that language is attempting to describe. "Knowing the truth" is a single unity—both an object and an event, objective and subjective. Knowing and truth are not two; they are different perspectives on the same reality. There is no truth without knowing and no knowing without truth. In other words, the truth is not the truth unless it is known. Nevertheless "truth" and "knowing" are not identical. They form an ambiguous pair that could be written as "knowing/truth"—one reality with two frames of reference.

THE RAINBOW THEORY OF MATHEMATICAL TRUTH

Recall Simone Weil's statement, "all truth contains a contradiction," with which the previous chapter began. In that chapter it was pointed out that a "great idea," far from avoiding contradiction, might actually "contain" a contradiction in one way or another. In this chapter Weil's statement will be considered again, this time applied to "mathematical truth." The idea that "mathematical truth" might not be the antithesis of contradiction is an unusual one. In order to get away from one's instinctive rejection of any such connection the word "ambiguity" has often been used in place of Weil's "contradiction." Nevertheless discussing "truth" in close proximity to "contradiction" forces a reevaluation of what is meant by truth. It makes absolutely no sense if the truth is thought of as absolute, objective, and timeless. But if the truth, like a mathematical idea, comes into being, if it can evolve, then it is possible for truth to have an intimate connection with contradiction.

Discussions of the nature of mathematics are often characterized by the desire to produce a description of mathematics that is completely objective—objective in the sense of being independent of human beings, independent of thought, independent of mind, independent, certainly, of creativity. But how can mathematics, much of which is clearly a construct, be totally independent of thought? It is this attempt to "square the circle," so to speak, that has made the attempts to produce a viable philosophy of mathematics so difficult to accomplish.

Instead of an "objective and absolute mathematical truth," what is being proposed is an "ambiguous" theory of mathematics. Now ambiguity is not just an objective element of a mathematical situation; as should be clear by this stage of the book that *ambiguity is a way of using the mind*. As such it is not static but dynamic. As usual an ambiguous situation comes with two "frames of reference," here designated as "objectivity" and "objective subjectivity." Relegating traditional objectivity to the role of merely one point of view is, I realize, difficult to understand and even more difficult to accept. It seems to put into question the view that sees mathematics as the essence of absolute, objective truth. Yet it is a way of looking at things that finds echoes in other fields. I shall quote from a discussion of the nature of physical reality as described by Nick Herbert for the Copenhagen interpretation of quantum mechanics.

> An obvious feature of the ordinary world is that it seems to be made of objects. An object is an entity that produces different images from different points of view... but all these images can be thought of as being produced by one central cause.... Its division into objects is a most important aspect of the everyday world. But the situation is different in the quantum world. . . .
>
> The separate worlds that we form of the quantum world (wave, particle, for example) from different experimental viewpoints do not combine into one comprehensive whole. There is no single image that corresponds to an electron. The quantum world is not made up of objects. As Heisenberg put it, "Atoms are not things."
>
> This does not mean that the quantum world is subjective. The quantum world is as objective as our own: different people taking the same viewpoint see the same thing, but the quantum world is not made of objects (different viewpoints do not add up). The quantum world is objective but objectless.
>
> An example of a phenomenon which is objective but not an object is the rainbow. A rainbow has no end because the rainbow is not a "thing." A rainbow appears in a different place for each observer—in fact each of your eyes sees a slightly different rainbow. Yet the rainbow is an objective phenomenon; it can be photographed.[10]

Like the rainbow and the situations described by quantum mechanics, mathematical truths are objective. Yet, in Herbert's terms, they are not objects. This means that there is no absolute and immutable point of view. To use a geometrical metaphor, there is no absolute Euclidean context within which all mathematics can be definitively located. Thus when one encounters some mathematical entity it is never devoid of a point of view. This point of view needs to be taken into account when we talk about mathematics. When we talk about continuity, let us say, the sense that we are talking about something real and objective is not an illusion. "Continuity" corresponds to an objective property of the real world. However, when we attempt to pin it down, to define it, to prove theorems about it and place it in some larger system of deductive thought, then we must inevitably assume a specific point of view toward "continuity." Other mathematicians may assume a different point of view that is also valid and productive of new insights into the nature and consequences of "continuity."

This "rainbow" or ambiguous perspective already arose in the discussion of the nature of "pattern" in Chapter 5. There I asked where a given mathematical pattern was located—was it in the mind or was it in some objective domain? It would seem that to come down on either side—the objective or the subjective—would be incomplete, would not describe the situation as mathematicians experience it. Yes there is an element of "mind" in a pattern. One "sees" a pattern in the same sudden way that one "sees" an idea. However, the patterns that mathematics studies clearly have an objective dimension. Thus patterns live in that "rainbow world" that is neither completely objective nor completely subjective yet contains both objective and subjective perspectives.

Again, I realize that normally any hint of "subjectivity" is incompatible with the idea of (objective) truth. But, of course, this is the way it is with ambiguous situations, especially "great ideas"—from one point of view the incompatibility that is a necessary component of the ambiguous situation appears to be a barrier. Here the barrier is precisely the idea that the defining characteristic of truth is its absolute objectivity. The second perspective is the "immediate certainty" that has been discussed, a form of "objective subjectivity" that emerges as a basic feature of mathematics. Taking this position does not force us to give

345

up the objectivity of formal mathematics by imposing an idio-syncratic mathematics where everybody chooses his or her own truth. What it does is to render traditional views of mathematics incomplete. Mathematics is more than what is contained in any formal theory, as Roger Penrose asserts so passionately when he says, "it seems to me that it is a clear consequence of the Gödel argument that the concept of mathematical truth cannot be en-capsulated in any formalistic scheme. Mathematical truth is something that goes beyond mere formalism."[11] Penrose never manages to locate this "something" that goes beyond formalism because he is still committed to a traditional view of "objective truth." The formulation proposed above does indeed locate this additional factor, but in so doing it turns many traditional ideas upside down. The advantage of doing so is that it then becomes natural to consider learning, understanding, and creating as a legitimate part of mathematics.

Looking at mathematics in this way forces a reevaluation of what is meant by "truth." Truth is ambiguous in the same way that mathematics is ambiguous. To be consistent with what has been said in the previous paragraph, it will be necessary to deny the existence of some absolute and objective truth. The trick is to give up on absolute truth without giving up on truth, without descending into a realm of pure arbitrariness. Thus there is a truth and human beings are capable of accessing it, although it resists being completely captured by any formalism. The truth cannot be completely objectified; it is not completely "out there." Again the notion of "objective subjectivity" or maybe "subjec-tive objectivity" comes in here. A postmodernist might say that truth does not exist, but this is not what is being said here—it is only one part of what is being said, the part that says the truth cannot be "captured." The other part is that the truth is accessi-ble in moments of "creative certainty." Or, to put it negatively, without the flash of insight there is no truth just as there is no understanding, which is, after all, just another word for this quality of certainty that we are discussing. These moments of clarity herald the birth of the mathematical idea that is the entry into the truth. And yet, what is going on here is very subtle, for the content that is accessed is not the (complete) truth.

What is unorthodox in my approach is the claim that the truth contains this irreducible element of subjectivity that I have been calling "immediate certainty." Of course, this subjective element is not the whole story—there is also an objective element. This is the reason that ideas need to be verified—to see whether they are consistent with the larger context of which they form a part. Nevertheless, the creative moment must be part of any story that we tell about truth. It is a necessary component that is not subsidiary to the normal objective component. In fact, the creative person knows this at a gut level and that is why she spares no pains to return to this state of inner truth. It is not a means to an end—it is an end in itself.

Is mathematical truth permanent or does it come and go? This is just a reformulation of the question about objectivity and subjectivity. A totally objective mathematics is permanent. Mathematics characterized by "objective subjectivity" does not claim continuity. There is an aspect of it that is discrete. For example, understanding mathematics arises in a discontinuous manner— as a series of small or large breakthroughs, or insights. This happens in the same way as the creative breakthroughs of research activity. Now there can be research without insights—the so-called "turning of the crank" research. This kind of activity may well be continuous, but the other kind, truly creative activity, is discontinuous. Self-referentially, the duality of the discrete and the continuous is one of the great ambiguities of mathematics. It arises as the algebraic versus the geometric or the analytic. It is interesting that these two points of view arise not only within the content of mathematics but also in the context of a discussion about mathematics itself.

REVISITING SOME TRADITIONAL PHILOSOPHIES OF MATHEMATICS

Mathematics is one single thing. The Platonist, formalist and constructivist views of it are believed because each corresponds to a certain view of it, a view from a certain angle, or an examination with a particular instrument of observation.[12]

In the above quote Davis and Hersh have it exactly right! Mathematics is one unified subject, but the philosophies of mathematics are like the proverbial blind men examining the elephant—they each describe only one part of the beast. In the language I have been using, mathematics is ambiguous. Each of the traditional approaches to the philosophy of mathematics is a self-consistent framework or point of view. It is true that these approaches conflict with one another, but we have learned that this is an inevitable stage in the development of a more unified view. Each of the three philosophies that will be considered in what follows—Platonism, formalism, and constructivism—represents a legitimate insight into the nature of mathematics. Each one is revealing something important and valid. Yet each one is incomplete on its own. So what follows is a brief review of the strengths and weaknesses of each of these approaches with an eye to unraveling the central truth that each one reveals. In a sense Platonism, formalism, and constructivism each constitute a way of organizing and therefore understanding mathematics.

The nature of mathematics is a fundamental question on a par with the other mathematical questions I have considered. It is considered here because I propose to approach it in the same way as the previous mathematical questions were approached. For example, the concept of "zero" arose out of a struggle with certain ambiguous situations. It arose out of the need to articulate a certain fundamental human experience. Yet that experience was not logically consistent and clear. The ground out of which the concept "zero" arose was messy and complex, containing many contradictory elements. The case I have been making all along is that the concept "zero" does not arise merely from clarifying a situation that is latently present all the time. That approach would liken the intellectual task to pruning an unruly garden: you just cut back the weeds and then the plants that you are interested in stand out. The "zero" concept was not there to begin with—it arose out of the struggle with the incompatibilities of the human experience that could be called the experience of "nothingness" or "emptiness." Yet in retrospect it seems not at all arbitrary—it seems inevitable. What would an "inevitable" philosophy of mathematics look like?

A new, unified view of mathematics will arise out of a struggle with the appropriate ambiguous situation. Mathematics as a whole results from a need to articulate basic features of the landscape of that primal world of human experience—a world that is neither completely objective nor completely subjective. The features of the world that mathematics concerns itself with have been enumerated elsewhere and include quantity, pattern, and chance. These fundamental concepts of mathematics are each basic components of the human experience. Mathematics is itself an articulation of these basic themes. However, together with the development of mathematics there arose incompatibilities and crises at both the mathematical and the meta-mathematical level. One of the most basic of these concerns the questions of truth and meaning.

Take the question of whether mathematics is discovered or invented, a question I will take up later on in this chapter. It revolves around the most fundamental of mysteries faced by any thinker: does the human mind give us accurate information about reality? In other words what is the connection between our subjectivity and the objective world. How do we bridge the fundamental duality of human consciousness? If there were ever a fundamental ambiguity, it is this. It is easy to discern the poles of this ambiguity. They can be expressed as "discovered/invented," "objective/subjective," or in many other ways. The resolution of this ambiguity—the "higher viewpoint" that would unify these two seemingly irreconcilable points of view—is the true goal of the philosophy of mathematics. Each traditional "philosophy of mathematics" can be read as an approach to this ambiguous situation. The incompatibility between these various approaches stems most of all from their attempt to resolve the ambiguity by suppressing one of its poles.

Every philosophy of mathematics arises out of the sense that mathematics touches something that is profound yet difficult to make explicit. For example, you could say that mathematics is about "quantity" or about "pattern." But what is "quantity" or "pattern"? There is a mystery here,[13] and anyone who loves mathematics has felt a certain sense of awe in the presence of this mystery. As the Pythagoreans understood, even the simplest mathematical concepts—one, two, three, four, straight lines, circles—are nontrivial; in fact, they are deep and mysterious. A philosophy of mathematics is an attempt to plumb these depths, to penetrate the mystery of mathematics. We know that mathematics is one of the pillars of our civilization, but what is it and why does it work? Each legitimate approach to the philosophy of mathematics provides an answer that must be approached as we would approach an important idea; that is, each contains a valid insight into the nature of mathematics. Thus each of the approaches to mathematics that I now enumerate can be thought of as emphasizing one aspect of a larger "ambiguous" or "rainbow" perspective and therefore denying other, conflicting, aspects of mathematics.

PLATONISM

Platonism holds that the objects of mathematics are real, that they exist in some objective realm independent of our knowledge of them. The truths of mathematics all exist in this Platonic realm whether or not the mathematician is aware of them. In this view, the job of the mathematician is to discover these preexisting truths. In this sense a mathematician is like a scientist. In fact there is an analogy between Platonism in mathematics and naive realism in science. Just as most working scientists are realists—they believe that the entities that they work with, atoms, electrons, mass, energy, for example, are real, so working mathematicians are Platonists in their approach to their working lives. They believe in the reality of the objects they work with—continuous, nondifferentiable functions, infinite sets, space-filling curves, and so on.

Most mathematicians do not go public with their Platonism—it seems too fantastic to claim that extraordinarily complex mathematical objects exist in the same way that sticks and stones exist. Yet there have been notable exceptions: One was Kurt Gödel, who said,

> the objects of transfinite set theory. . .clearly do not belong to the physical world and even their indirect connection with physical experience is very loose. . . . But, despite their remoteness from sense experience, we do have a perception also of the objects of set theory, as is seen from the fact that the axioms force themselves upon us as being true. I don't see any reason why we should have less confidence in this kind of perception, i.e., in mathematical intuition, than in sense perception, which induces us to build up physical the-

ories and to expect that future sense perceptions will agree with them. . . . They too may represent an aspect of objective reality.[14]

With reference to geometric and not set theoretic objects, the great French mathematician René Thom said,

Mathematicians should have the courage of their most profound convictions and thus affirm that mathematical forms indeed have an existence that is independent of the mind considering them.[15]

Finally there is the position of another eminent mathematician and mathematical physicist, Roger Penrose:

I shall have something to say about another world, the Platonic world of absolutes, in its particular role as the world of mathematical truth. . . . Some people find it hard to conceive of this world as existing on its own. They may prefer to think of mathematical concepts merely as idealizations of our physical world. . . . Now this is not how I think of mathematics, nor, I believe, is it how most mathematicians or mathematical physicists think about the world. They think about it in a rather different way, as a structure precisely governed according to timeless mathematical laws. Thus, they prefer to think of the physical world, more appropriately, as emerging out of the ("timeless") world of mathematics.[16]

Penrose succinctly reveals the point of view of many mathematicians. The Platonic world is the world of mathematical truth and the physical (objective) world emerges out of the Platonic world, not the other way around. What are thinking people to make of such a seemingly counterintuitive point of view? For example, what is the nature of Penrose's Platonic world of mathematical truth? Why does he, together with so many of his colleagues, have the deepest conviction that this Platonic world is real, when to the skeptic the Platonic world is clearly an illusion—a construct and therefore unreal? Platonism in mathematics is the conviction, based on the experience of doing mathematics, that mathematical objects are real and that mathematical truth has a certain stability, in short, that the truths of mathematics reside in an objective domain. Notice that I say "conviction, based on experience." The mathematician has a subjective feeling that mathematics is objectively so. In my opinion the Platonism of the mathematician is a testament to the fact that the truth of mathematics exists but that it is not objective in a classical way.

The mathematician has great faith in her intuition, and for good reason. This faith in the Platonic reality of mathematics is what sustains her efforts, her voyages into the unknown realms of mathematical research. It would be wrong to disregard this "inner certainty" that mathematics is real, for it is telling us something important about the nature of mathematics. Platonism seems to be trivially false when we take the position that the objects of mathematics are constructs and therefore not as real as the objects in the physical world. There are two rejoinders to this. The first is that the objects of physics, let us say, are also constructs. The second is that Platonism is not affirming the existence of some abstract heavenly world. It is telling us something about the very world that we live in. It is saying that the nature of this world is that of an "objective subjectivity" with its dimensions of "certainty" and objectivity. As pure and absolute objectivity, the Platonic world is a vague and mystical intuition. As a world with cognitive as well as objective dimensions, it is the concrete world of mathematical activity. Thus the view of mathematical truth that I am advocating is supported by this revised view of Platonism. Conversely it allows one to retain Platonism as a valid view of the nature of mathematics.

Platonism in mathematics can be viewed as a response to the ambiguity of subject and object. To escape from the "subjectivity" of mathematical truth, an objective, ideal domain is postulated. This ideal domain is, of course, a world of ideas, ideas that are real. Since ideas are, naively speaking, subjective; this ideal realm must be a domain of "objective subjectivity" whose objective elements are projected onto some external, ideal but objective domain. Ironically, Platonism denies the subjective and pretends to be objective yet it is clearly grounded in an aspect of the subjective. Referring back to my earlier discussion of different aspects of subjectivity, Platonism is correctly denying that "subjective subjectivity" is relevant to mathematical truth. However, Platonism can be construed to be affirming the existence of this "objective subjectivity." To repeat, Platonism, in my opinion, has the intuition that the truth of mathematics exists. What it does not see is that this truth has both objective and subjective dimensions. In this way it manages to avoid dealing with the full ambiguity of the activity that we call mathematics.

FORMALISM

Formalism makes "proof" into the defining characteristic of mathematics. Formalism has been mentioned at various places in the book, in particular in Chapter 7, where its strengths and weaknesses were discussed. The strengths involved the acquisition of a new freedom that mathematics had not possessed before, a freedom that ultimately led to a view of mathematics as the science of abstract patterns. Ironically enough, this view leads directly to a mathematics whose nature goes beyond formalism, since formalism is merely the science of *logical* pattern. However, for my purposes here it is more important to stress that formalism arose out of a crisis and succeeded in reestablishing a stable notion of mathematical truth. More accurately, it transformed the notion of truth into the notion of valid logical inference. Thus logical considerations came to be seen as the essence of mathematics—the content of mathematics, mathematical ideas and creativity, became (formally) irrelevant.

In formalism truth *is* logical rigor. A kind of certainty remains—the certainty of an idealized machine. Indeed, computers can in principle verify rigorous proofs, for the ideal formal argument contains no gaps or omissions. Such proofs are complete, but that completeness is purchased at the price of meaning, for a rigorous proof is strictly speaking not about anything—it is just a succession of valid inferences.

Formalism is an ambitious attempt to remove "subjectivity" from mathematics because, of course, subjectivity is held to be the antithesis of truth. The dimension of "immediate certainty" is seen as a psychological phenomenon that is completely distinct from "real," that is, formal mathematics. Certainty does not disappear, but what I have called "creative certainty" is replaced by the logical certainty that is to be found in a proof. This kind of certainty seems to provide a secure underpinning not only for mathematics but also, in conjunction with logical positivism, for scientific theory in general. Thus from the standpoint of my earlier discussion of mathematical truth, formalism is an important advance and a crucial influence. It is an attempt to arrive at an ideal state of "objective knowledge" through an emphasis on logical criteria above all others. No wonder people like Frege or Russell and Whitehead attempted to demonstrate that mathematics could be totally derived from logic; the feeling that this was possible for all of mathematics is consistent with the worldview of formalism.

The world of mathematics is not so simple; it is infinitely richer and more interesting than the picture that arises out of formalism. The attempt to access objective truth and certainty by means of logic is fundamentally flawed, and not only because of the implications of Gödel-like theorems. Logic does not provide an escape from subjectivity. After all, what is logic, in what domain does it reside? Surely logic represents a certain way of using the human mind. Logic is not embedded in the natural world; it is essentially a subjective phenomenon. Logic is an idea—an organizing principle, *the* organizing principle if you are looking for an algorithmic account of thought. Logic introduces an order into mathematics, but not even formalism can make mathematics secure in the sense of totally banishing all aspects of the subjective. It does more or less succeed in banishing what we might call "subjective subjectivity," namely, individual arbitrary opinions, but it does not and cannot banish what I called "objective subjectivity" since one could only do this at the price of destroying the essence of mathematics itself.

In fact a moment of thought demonstrates that formalism misses most of mathematics. Where do the axioms come from that form the foundations of a formal system? Can there be "good" axiomatic schemes versus "bad" or "trivial" sets of axioms? What about mathematical definitions? How do they arise? What makes one definition better than another? Within a formal system there can be no such differentiation. Why has so much effort been expended in mathematics over the proper definition of "continuity," for example, or what it means for a function to be differentiable? Where, as I have said repeatedly, do the ideas of mathematics come from? Sure, it is not difficult to verify that a proof is correct, but where does the idea for the proof come from? Certainly none of these things—axioms, definitions, and proofs—come from the formal dimension of mathematics. It is like doing Euclidean geometry without the geometric pictures. It can be done, but when you omit the diagrams you lose the essence of the subject—you are not doing Euclidean geometry any more but some other (dry and uninteresting) subject. So it is with mathematics in general. You can take out the ideas, but what you are left with is not mathematics but some other subject entirely.

Formalism is based on the assumption that there exist some areas of mathematics that are objectively certain. It then proposes to build up these areas through logical deduction to ultimately arrive at a situation where all of mathematics is certain. Unfortunately, it is questionable whether *any* mathematics, even the most finite and elementary, is objectively certain. The simplest of concepts would be the small integers—0, 1, 2, 3—and I have stated repeatedly that these are not so simple. Formalism is committed to the proposition that knowledge is only secure if it can be built up from first principles. It is interesting that in practice mathematical research begins with certain results that are accepted by the community of experts and works from there. Though, in principle, one could go back to the axioms, in practice one never does. One works in some intermediate domain. Perhaps all of mathematics occurs in such an intermediate domain. There are no "atomic" truths in mathematics, no axioms that are "most" elementary in any absolute sense. All nontrivial mathematical objects contain some inner structure and thus can be said to be complex. For the purposes of some specific discussion one may "assume" that some concept or result is elementary and use it to build up more complex structures. The axioms and definitions in mathematics come about as a result of a wealth of mathematical experience. They contain all of that history, all of the situations out of which they have been abstracted. It is always conceivable that all of this mathematical experience could be thought of from a new and different point of view. Then we would have new axioms and new definitions.

There have been so many attacks on formalism as *the* philosophy of mathematics in recent years, and it *is* so obviously incomplete, that I cannot help but end this section with a statement about its value and importance. Formalism, from my point of view, springs from the conviction that certainty and truth are at the center of mathematics. Davis and Hersh refer to Hilbert's "conviction that mathematics can and must provide truth and certainty or 'where else are we to find it?'"[17] In point of fact, Hilbert's point of view was much more nuanced than that taken by more extreme formalists, sometimes called "logicists," who, as the name suggests, believe in proof without meaning. Hilbert did not deny the role in mathematics of intuition that is derived from the natural world. He said that "while the creative power of pure reason is at work, the outer world again comes into play, forcing upon us new questions from actual experience." Although he insists on "rigor in proof as a requirement for a perfect solution to a problem," he follows this with "to new concepts correspond, necessarily, new signs. These we choose in such a way that they remind us of the phenomena which were the occasion for the formation of new concepts." Hilbert is really making an argument for the power and value of the axiomatic method; he is not suggesting that the axiomatic system is all there is to mathematics. In Hilbert's own mathematical work, one sees clearly the drive for truth and certainty that forms the core impetus for mathematical activity. Formalism highlights this fundamental drive. In a sense, it pushes matters to an extreme—cutting away all (seemingly) extraneous matters in its search for an objective and, therefore, stable truth.

The ultimate lesson of formalism lies in its implications for the future. Formalism arose from a need to establish a new, stable notion of truth in the face of the challenge of various crises to the previous idea of truth. It seems to me that mathematics is again facing just such a challenge. This challenge comes from various directions, beginning with the implications of the work of Gödel. An explicit challenge to formalism was laid down by the work of Lakatos and his stress on the similarity between informal mathematics and science, in particular what has been called the "dubitability" of mathematics. But formalism also has to contend with the rise of new, experimental mathematics that does not necessarily prove results but uses the computer to obtain conclusions inductively in the way of science rather than deductively. All these factors have destabilized the formalist vision of mathematics and have led to various demands for a new way of describing mathematics that is more in line with how mathematics is experienced by the contemporary practitioner.

At this moment in time we are faced with a new crisis in mathematics, and this once again calls for a new and more stable idea of mathematical truth.

Constructivism

Constructivism traditionally referred to a point of view that originated with the Dutch mathematician L. E. J. Brouwer, who felt that all mathematics should be based constructively on the natural numbers. Constructivists opposed those "proofs by contradiction" that were used to establish the existence of infinite sets of irrational and transcendental numbers. They were even able to show that the elementary and intuitively "obvious" "law of trichotomy," according to which every real number is either positive, negative, or zero, does not hold from a constructivist point of view. This reveals that the real numbers are actually quite complex, after all the decimal representation of any real number carries an infinite amount of information. Since the reals are so complex, it seems strange, not to say paradoxical, to be able to show that "most" real numbers are transcendental (using a proof by contradicition) without being able to give more than a few concrete examples of actual transcendental numbers. Thus the constructivists were certainly on to something. Their point of view is significant in a number of ways. First they pointed out the vast implications for mathematics of these "proofs by contradiction," that is, of using the negative principle of banning contradiction as a positive principle guaranteeing the existence of complex mathematical objects. Next they forced a reevaluation of the work of Cantor, work that had demonstrated that real numbers were very complex objects in themselves and that infinite sets of real numbers were even more complex and, from many points of view, counterintuitive.

Nevertheless, the constructivists eventually lost the battle to the formalists. Just as one cannot imagine mathematics giving up on the number zero, so mathematicians refused to give up on the properties of the real numbers that Cantor had demonstrated. What one is comfortable with, namely, the natural numbers, seems simple and reasonable. What is new often seems complex and unintuitive. As I said above, maybe even the natural numbers are complex mathematical objects. At any rate we

357

owe a certain gratitude to these early constructivists for bringing up the question of what it means to say that a mathematical object exists. Do they all exist in the same way, or are some mathematical objects more "real" than others? Finally the constructivist position is of interest because of the advent of the computer, which is a kind of constructivist machine. One can input natural numbers (but not all natural numbers) or finite decimals, but not irrational numbers or infinite decimals. It may well be that the constructivists were talking about another, equally valid type of mathematics.

Radical Constructivism

There is a more modern movement that also goes by the name of constructivism. I shall discuss the part of this movement that is called "radical constructivism," but only insofar as it has something to say about the nature of mathematics. This movement takes as its point of departure the simple but controversial observation that "an observer has no operational basis to make any statements or claim about objects, entities or relations as if they existed independently of what he or she does."[18] I earlier paraphrased this statement by saying that every human being lives in a bubble. This bubble contains all their perceptions and cognitions. What exists outside the bubble is not knowable. Radical constructivists "do not make claims about what exists 'in itself,' that is, without an observer or experiencer."[19] This is a point that I also made earlier when I claimed that there exists no mathematical knowledge that is completely objective. Mathematical knowledge and truth must be considered as a package with both objective and subjective aspects. The belief in "objective mathematical knowledge," that is, knowledge that is independent of the beings who know it, is itself a belief and therefore nonobjective. There is no knowledge that is independent of knowing. There is no absolute, objective truth.

If it is not possible to prove the existence of any objective reality that is independent of the observer, then it follows that people construct their own realities—their own understandings, knowledge, and meaning. In particular, radical constructivism has implications for the teaching and learning of mathematics.

Is teaching merely the transfer of information and techniques from one who knows to one who does not? Do students' mistakes have any significance, or is everything in mathematics either right or wrong? What is going on in a student's head when he is struggling to learn some new piece of mathematics? What is the best way to teach a student the nature of some mathematical concept? One's response to these questions and to the whole enterprise of learning and teaching mathematics changes if one takes the position that a student constructs meaning, constructs his own understanding.

Now the word "construct" has a mechanical ring to it. I want to emphasize that the construction of meaning in mathematics is not mechanical. It arrives in discontinuous leaps of what can only be called creativity.

From this point of view we each have our own understanding of mathematics. Great mathematicians may well have a way of thinking about some part of mathematics that is substantially different from that of their peers.[20] But before we go too far down this road it is important to stress that understandings may differ but they are not arbitrary. If one stresses the construction of knowledge, there is always the danger of straying into a form of relativism where you have your truth and I have mine. It is to save us from that possibility, which is inherent in much of the postmodern point of view, that I have stressed that the subjectivity I am referring to is an "objective subjectivity"—a way of thinking about mathematics as containing both subjective and objective elements.

The importance of radical constructivism to this discussion is that it brings to the fore the cognitive aspect, the "knowing" that was discussed earlier. This is why it is so controversial: it is a critique of absolute objectivity. This is also the reason that it appeals so much to educators who are forced to confront the human dimension of learning.

Is Mathematics Discovered or Invented?

I promised earlier that I would come back to the fascinating but difficult question of whether mathematics is discovered or invented. I shall approach this question from the "ambiguous" point of view that I have been developing. In particular, the question of discovery or invention is connected to my discussion of "mathematical truth." Now the question of whether mathematics is discovered or invented should not be reduced to a simple choice, "mathematics is either discovered or invented." This is the usual approach—if mathematics is discovered it is not invented and vice versa. It also should not be reduced to a complementarity. This would imply that mathematics has two distinct domains—the "discovery" domain and the "invention" domain—and that mathematics is somehow the sum or union of these two domains. Both approaches would miss the conflict that is inherent in "discovered versus invented" and, as a result, would forego the possibility of a creative resolution to that conflict.

I propose to think of "discovery" and "invention" as evoking two different perspectives on the nature of mathematics. "Discovery" is one way to look at mathematics—the Platonic way. "Discovery" reflects the sense that something is going on in mathematics that is "out there," independent of individual personalities and idiosyncrasies. As I have discussed earlier, it reflects the sense that mathematics is not arbitrary but "universal," that it is connected to the truth. "Invention" is another frame of reference. It also reflects something real about mathematics, something that is suggested by formalism, that mathematics is a construct embedded in human culture. These two frames of reference are inconsistent. They are reminiscent of the "two cultures" division that permeates our entire civilization. On the one hand there is the "naive realism" of science that is analogous to the "discovery" take on mathematics; on the other hand the postmodernism of the humanities that is similar to "invention." In this sense mathematics contains within itself a reflection of this larger cultural battle. "Discovery" and "invention" evoke equally valid, consistent frames of reference that are clearly in conflict with one another. That is, the question "Is mathematics discovered or invented?" is ambiguous in precisely the sense that we have been using the term "ambiguous" throughout the book.

The either/or approach represents the easiest and most common approach to any ambiguity. It is the side of things from which the conflict that is present in the ambiguity appears to be a barrier that can only be crossed by means of a choice. But making that choice means giving up on one valid way of looking at mathematics. The alternative is to look for a creative resolution. Where is this resolution, this unified perspective, to be found? What is the higher viewpoint that will unify these seemingly opposing tendencies? This is the key question and a major goal of this book. In individual mathematical events, we saw how a higher point of view arises through an act of creativity. Now we are seeking to apply this paradigm to mathematics as a whole. Since the question we are dealing with is so basic and has such an affinity with other questions in the philosophy of science, it is possible that such a resolution would have implications beyond mathematics.

Now I maintain that for someone who has been reading this book attentively and therefore grappling with the subtle and seemingly intractable notion of ambiguity, the resolution is staring us in the face. The resolution involves refusing the choice of "either discovered or invented" but not slipping into strict complementarity. Mathematics is that one unified activity that looks like discovery when you think of it from one point of view and appears to be invention when regarded from another. We are back again to the "rainbow" approach to the nature of mathematics.

How are we to think about this unified conception of mathematics? Mathematics is a field of creative activity, one that is not static but continually coming into being. This complex field of activity can be broken down along many different axes: proof and idea, process and object, subjective and objective, to name just a few that have been considered in these pages. Each of these dualities tells us something about the nature of mathematics. Here we are considering the axis of discovery/invention. What is revealed about mathematics by looking at it through this particular lens?

"Invention" represents the freedom that exists in mathematical activity, the freedom to go beyond any previous point of view, any earlier theory. There is a sense of free invention in mathematics, a sense of the unconstrained play of the mind. It is like a beautiful, intricate, all-absorbing game. It is unpredictable. Developments are often totally unexpected; there is no saying where the next breakthrough will come from. "Invention" stresses that mathematics is created; it comes into being;

it evolves; it is dynamic. It evokes the process dimension of mathematics.

Yet, on the other hand, once the breakthrough has occurred, when we understand the new definition, theorem, or point of view, it seems perfectly natural, obvious, and inevitable. There is the sense that it was there all the time, waiting for us. Mathematics, as I said earlier, has a certainty and solidity; it has a truth that we would be making a big mistake to ignore. Thus mathematics is also discovered. "Discovery" evokes a sense that there is a static, unchanging aspect of mathematics. It emphasizes the objective dimension of mathematics.

Do you remember the Gestalt picture of the young/old woman in Chapter 1? I said then that both of these, the young woman and the old woman, were characterizations of the entire situation. So it is with our double perspective of discovery and invention. Mathematics in its entirety can be consistently characterized under the rubric of "invention" but also under the rubric of "discovery." This is why we feel that we *must* make a choice. However, in a situation of ambiguity we resist making this choice. Thus "discovery/invention" is a true ambiguity. Mathematics *is* (metaphorically speaking) "discovery/invention."

The ambiguity "discovery/invention" opens out onto a perspective of a more complete and unified conception of mathematics. As we have seen, each of the traditional philosophies of mathematics highlights one perspective on the question of discovery and invention. Yet it would be wrong to think that these different perspectives are completely disjoint from one another. For example, at first glance "invention" looks subjective whereas discovery looks objective. However, as we have seen in the earlier discussion, invention or creation involves elements that are certainly objective whereas discovery, involving as it does the agent who makes the discovery, involves an inevitable subjectivity. Each point of view gives a description of the whole of mathematics, yet each is incomplete without the other and each contains within it elements of the other perspective. Together they describe mathematics as a vast dynamic field of creative activity that is intimately related to "truth," that is ever changing and ungraspable in any fixed and absolute network. Mathematics is always reinventing itself. It is the oldest and the newest of disciplines.

OBSTACLES TO LEARNING MATHEMATICS

The inadequacies of traditional philosophies of mathematics come to the fore when we are confronted with the problems of teaching and learning mathematics. The essential factor involves the question that this chapter has been grappling with—is mathematics completely "out there," completely objective? Many of us were brought up mathematically in an atmosphere dominated by formalism. The consequences of this point of view for the classroom can be caricatured by the following fictional situation, elements of which are unfortunately all too common in university classrooms. The professor goes through some intricate proof of a mathematical theorem. The class is silent. If the professor were to turn around and look at the students, their blank faces would demonstrate that, in the vast majority of cases, they do not understand what he has been saying. The students say nothing, probably because they are afraid to reveal their lack of understanding but possibly because they have learned that it can be dangerous to ask questions. Finally some brave soul picks on some particular aspect of the argument and asks the professor to explain it. The professor is amazed that anyone could not understand something that is so obvious to him. Being a good sport he proceeds to "explain" by going over the section in question. But what does he do? He merely repeats the same words, the same argument, maybe talking more slowly and possibly even filling in a detail that he had omitted. "There," he says, "you can see that it's trivial!" The student remains in the same state of ignorance she was in before she asked the question and will think twice about asking another such question in the future.

This whole fiasco arises because teaching theoretical mathematics is often identified with communicating its formal structure. Understanding lies behind the formal structure but is not captured by that structure. The formal structure is blind to the ideas in mathematics and, as a result, the teacher may feel that it is not his job to communicate the ideas. In truth the above anecdote is overdrawn because it does not include the problems or exercises the student is asked to work on for homework or exam preparation. What actually happens is that the teacher develops the formal, theoretical structure in her lectures and the student is expected to develop an understanding of the "ideas" through working on problems. Nevertheless, "understanding what is going on" is rarely a conscious educational objective for teachers of mathematics. This leads to the huge problems one finds today in teaching and learning mathematics at all levels but especially the university level. These problems stem directly from the implicit philosophy of mathematics that is held by most mathematics teachers.

In Chapter 1, I made the point that ambiguities and paradoxes are not "resolved" in the normal sense in which that term is used. I meant that in two ways. In the first place, "resolving" an ambiguity usually is taken to mean eliminating it, often by embedding it in some formal, logical system. But from the point of view of learning and understanding, this is not what happens at all. What happens is that "ambiguity as barrier" gets replaced through understanding by "ambiguity as flexibility and openness." Thus the goal of teaching cannot be to eliminate ambiguity. The goal is to master the ambiguity, and this is not at all the same thing. To master the ambiguity is to be comfortable with it and learn to use it in a constructive way.

In Chapter 1, I also made a comment that the paradoxes from the history of mathematics, such as those that have been discussed throughout the book, become the "epistemological obstacles" of the present. What is an epistemological obstacle? Here is what Anna Sierpinska had to say:

> This is where Bachelard's[21] concept of epistemological obstacle turned out to be very useful. Students' thinking appeared to suffer from certain "epistemological obstacles" that had to be overcome if a new concept was to be developed. These "epistemological obstacles"—ways of understanding based on some unconscious, culturally acquired schemes of thought and unquestioned beliefs about the nature of mathematics and fundamental categories such as number, space, cause, chance, infinity, . . . inadequate with respect to the present day theory—marked the development of the concept in history, and remained somehow "implicated," to use Bohm's term, in its meaning.

It is then on these obstacles that research concentrated: a "hunt" for epistemological obstacles started at the same time as an effort of precisation, of a better explication of the term was undertaken. The question was posed: on what grounds do we claim that a student's thinking suffers an epistemological obstacle? Is an epistemological obstacle an error, a misunderstanding, or just a certain way of knowing that works in some restricted domain but proves inadequate when the domain is transcended? Or is it an attitude of mind that allows to take opinions for facts, a few cases for evidence of general laws, . . .?[22]

This entire book has, in a sense, been about such obstacles. The words I have used to describe certain situations—words like ambiguity, contradiction, and paradox—and people's reaction to these situations testify to the fact that the situations have been experienced as obstacles. In general, mathematics is full of words—like irrational, imaginary, negative—that indicate that the concepts behind them were obstacles at a given point in time. Thus the notion of an epistemological obstacle applies not only to the learning of mathematics, but to the culture of mathematics in general. There is always resistance to the introduction of new concepts and ideas. For example, the introduction of cardinal numbers had to break down the obstacle of "the whole is greater that the part." The general idea of irrational number had to break down the obstacle that an infinite decimal could be thought of as a single number. The history of mathematics can be approached as the encounter with epistemological obstacles of various kinds. However, this sort of encounter is not restricted to the past. It is going on today and it is the life-blood of mathematics.

Sierpinska asked whether such an obstacle is just an error or misunderstanding. If an obstacle were merely an error, this would imply the existence of a "correct" way to understand the concept, that is, that there is some definitive, objective view of the concept against which a person's understanding can be measured. But does such a definitive view exist? This leads us to a question that is related to the above discussion of mathematical truth: "Are the obstacles that arise in learning as well as in mathematical research objective or subjective?"

To be consistent with what I have said earlier in this chapter, I must and do claim that the obstacles to understanding and doing mathematics have both objective and subjective characteristics. When the word "epistemological obstacle" is used, it is usually the objective side of the obstacle that is being stressed. When students have difficulty with the concept of a real number or with the definition of a continuous function, does this imply that they are lacking in the ability to do mathematics? Or is it true that these concepts are intrinsically difficult as evidenced by the fact that historically even the best mathematical minds struggled with them. The obstacles are objectively there in the substance of the subject. To learn these concepts it is necessary to go through the obstacle; there is no easy way to master a concept except by struggling with it. The student must authentically deal with the difficulties, the same difficulties that arose as the mathematical community tried to unravel these ideas historically.

Therefore history teaches us something about the difficulties inherent in mathematics. This explains the semihistorical approach I have taken in this book. But clearly if we look at the process as completely historical we get lost in the morass of the historical development, it is only in retrospect, in the creative reconstruction of history in the manner of Lakatos, that the great themes and ideas shine forth.

Of course obstacles to the learning of mathematics also have an obvious subjective dimension. Different students bring different talents and abilities to the table. Nevertheless, the idea that the obstacles to the learning of mathematics are purely subjective and only related to something called "mathematical ability" is simplistic. Again, mathematical obstacles have both objective and subjective dimensions. The elements that have hitherto been seen as purely individual and subjective have an element in them of what was called "objective subjectivity." If this were not true, there could be no discipline called "mathematics education."

CONCLUSION

In summary, the notion of mathematical truth and, in particular, the extent to which it is objective—even what is meant by the word objective—lie at the heart of mathematics and any possible philosophy of mathematics. The essence that differentiates the various points of view that have been discussed lies in their differing approaches to mathematical truth. I am claiming here that without a reconsideration of what is meant by objectivity there can be no progress in the philosophy of mathematics. The mythology of "absolute objectivity" is what stands in the way of progress. The challenge is to modify the notion of "absolute objectivity" without eliminating objectivity. Though meaning and understanding may be constructed, this does not imply that there is nothing objective and systematic that is going on.

Finally, it must be emphasized again that these considerations are important for mathematics and for mathematics education but their importance is not restricted to the mathematical community. As I will discuss in the next chapter, society's philosophy of mathematics—our sense of what mathematics is and of mathematical truth—has a profound effect well beyond the boundaries of mathematics, per se. Mathematics is the model for how the natural world works, it is the model for how human beings work, for how we think, for our notion of truth. Our view about the nature of mathematics has a profound effect on our view of ourselves and therefore on the kind of people we shall become in the future.

Conclusion: Is Mathematics
Algorithmic or Creative?

THE OBJECTIVE OF this concluding chapter is to summarize and draw out the implications of the view of mathematics I have been developing in the preceding chapters. I have been describing mathematics as an activity that is dynamic and creative, pulsating with the life of the mind. It is a "way of knowing" that is quite unique.

But mathematics is not off in some obscure corner of human activity; it is central to human experience and human culture. Thus it is not surprising that many of the great questions of the day have a certain reflection within mathematics. The question that I have chosen to explore in this chapter is the relationship between computing and mathematics, between mathematical thought and computer simulations of thought processes. This discussion will allow me to summarize much of what has been discussed in the book so far and demonstrate why its importance extends beyond mathematics itself. It will highlight the difference between the point of view I have been taking and the views one finds among many consumers of mathematics in the scientific and technological community as well as the public at large.

Normally using mathematics to investigate some subject means creating a mathematical model, using mathematical techniques to draw out the mathematical consequences of that model, and, finally, translating the conclusions back to the original situation. This is not the way I shall be using mathematics. The inferences I shall discuss come not from the content of some mathematical model or theory; they will come from the nature of mathematics itself.

CAN COMPUTERS DO MATHEMATICS?

The question that will be a springboard into my discussion will be the naive question of whether computers are capable of doing mathematics—today or conceivably in the future.

What do I mean by the question, "Can a computer do mathematics?" I am not asking whether computers can be helpful, even indispensable assistants to a human mathematician as, for example, in the case of the proof of the four-color theorem (the theorem that states that any map can be colored with four different colors in such a way that no two adjacent regions have the same color) when the computer checked the validity of the conjecture in a large but finite number of cases. After all, it was mathematicians who had shown that the problem could be reduced to the verification of these cases.

Nor am I asking whether a computer can be used to generate data or graphics that can give the human investigator a "feel" for some mathematical situation. Again, I am not asking if the computer can be used to attempt to ascertain the likelihood (formally or informally) that a certain conjecture is or is not the case.

I am asking whether it is conceivable that at some time in the future computers could completely take over the show, whether a machine could be programmed to "do" mathematics from start to finish. This would involve (among other activities) examining mathematical situations or situations that potentially could be mathematized, producing data about these situations, generating conjectures, and demonstrating the validity or invalidity of these conjectures. Put in this way, the answer to the question of whether a computer could ever do mathematics is clearly "No!" (The discussion about whether a computer can do mathematics is usually restricted to the last of these activities, namely, demonstrating the validity of certain mathematical statements.)

William Thurston said, "In practice, mathematicians prove theorems in a social context. It [mathematics] is a socially conditioned body of knowledge and techniques."[1] An increasing number of mathematicians would agree with the statement that mathematics is what (human) mathematicians do. If we take this view, then it is tautological that only mathematicians can do mathematics. Ironically, it is the very success of computers in profoundly infiltrating the day-to-day activities of many working mathematicians that has raised the question of the role of proof in mathematics.[2] If mathematics is not defined merely as a proof-generating activity, then it is difficult to see it being developed by autonomous machine intelligence (if there ever is such a thing.) Thus the more extreme version of the question of computers and mathematics is not very profound.

A more interesting question is the following: What aspects of mathematical activity can computers duplicate and, on the other hand, what aspects of mathematics (if any) cannot be duplicated by machines? Success in isolating some of these characteristics would be important. It would have implications for other related questions like "Is intelligence algorithmic? What is creativity and what is its relationship to deductive thinking? And even what is the nature of Mind?"

There is today a powerful point of view that the mind is a computer and that thinking and problem solving can now, or will in the future, be done by "thinking machines." This point of view is not restricted to the artificial intelligence community but is common in the entire field of cognitive science. More important, the idea that thinking is the kind of thing that computers do or simulate is a pervasive influence throughout our entire culture. Many people have the idea that computers have a kind of infinite potential—if there is some task that they cannot accomplish today, it seems inevitable that they will be able to do it tomorrow or the day after. This is almost an article of faith for many people.

What this book has emphasized is how our current expectations for computers and algorithmic thought arise out of such elements as formalism and the older "dream of reason." This is such a basic element of the cultural heritage handed down to us from the Greeks that it is understandable that as a society we are unwilling to relinquish this dream even in the face of all kinds of evidence to the contrary. Relinquishing this dream means also relinquishing the hope that all of our problems will be eliminated by computer technology. In a way the dream expressed by formalism has morphed into today's dreams of artificial intelligence. These dreams have in common the faith that it is possible to banish ambiguity and, therefore, a certain form of complexity from human life.

Roger Penrose has taken strong issue with the belief in the unlimited potential of algorithmic thought in a series of books that have become very popular.[3] For Penrose there is something other than complicated calculation involved in mathematical activity—there is an aspect of mathematical truth that goes beyond anything that can be accessed by algorithmic thinking. This book has, in a sense, been a search for this noncomputational factor in mathematical thought and therefore, by extension, in all thought.

The idea that "the mind is a computer" is a dangerous one, and it is important that it be refuted. However, any refutation will not come easily precisely because we have all become part of the culture of information technology, and in that culture the notion that just about everything is programmable is a seductive and powerful dream—a modern myth. Why do I say that modeling thought on computer-generated activity is dangerous? It is because such ideas can easily become self-fulfilling prophesies. We may soon come to *define* thought as computer-like activity. Steven Pinker, for example, is a well-known cognitive scientist and author of the immensely popular book, *How the Mind Works*. In this book he gives two criteria for intelligence.[4] The first is "to make decisions rationally by means of some set of rules" and the second is "the ability to attain goals in the face of obstacles based on rational truth-obeying rules." What is interesting is Pinker's repeated use of the terms "rational" and "rules." He is *defining* intelligence as a kind of algorithmic rationality such as may be found in a mathematical proof or a computer algorithm. If we accept this kind of definition then the question of whether a computer is intelligent is a circular one. In this way computer-thought may well have become a metaphor for human thought.

An all-pervasive metaphor becomes its own reality. In the sense that the metaphor "time is money" conditions the way in which we experience time,[5] so "the brain is a computer" is conditioning the way we look at the human mind, even the manner in which we look at what it means to be human. The danger is that our infatuation with the computer as a model for ourselves will result in our forgetting about other ways of using the mind. The danger is that we shall lose touch with the deeper aspects of what it means to be human. This is the larger context of this discussion.

Computers do what they do according to the algorithms that are programmed into them. Questions about the theoretical capabilities of artificial intelligence are really questions about the nature and limits, if any, of algorithmic thought. Now mathematics has a great deal to say about algorithmic thought. In fact, for my purposes the question "Can a computer think?" which is equivalent to the question "Is thought algorithmic?" could be replaced by "Can a computer do mathematics?" or "Is mathematics algorithmic?" Mathematics is the part of our culture that has investigated algorithmic thought in the most profound and systematic way. The results of that investigation should be available to inform the contemporary discussion of the nature of human thought. Unfortunately, many people who address such questions know little of the culture of mathematics. On the other hand, mathematicians, as a rule, are not inclined to stand back from their subject—from the day-to-day problems of their research—and consider the larger metaquestions that are raised, at least implicitly, by the nature of their work. For this reason, there is scarcely any input from mathematics or mathematicians on these questions, even though mathematics provides much of the cultural context within which such questions arise.

Most claims that computers can do mathematics are really claims that computers can generate correct proofs. Thus it may be worthwhile to spend another moment discussing the relationship between proof and mathematics. What is the value of a proof? The irony is that, as was mentioned above, it is precisely the computational and graphical abilities of computers that have thrown into doubt the centrality of proof in the mathematical firmament. Nevertheless we would certainly not want to argue that the computer is incapable of spinning out lines of code that could be interpreted as formal proofs. Indeed, if one is a formalist, that is, if one believes that there is no deeper content to mathematics other than the formal set of axioms, definitions and proofs, then a computer *can* do mathematics. However Chapter 8 contained a discussion of the severe limitations of that way of looking at mathematics.

The difference between a computer and a human mathematician can be demonstrated by asking the question, "What stands behind the formal mathematical proof?" For the computer the formal proof *is* mathematics. For the mathematician there is a whole universe of intuition and understanding that lies behind the formal proof. Proofs, or more precisely the ideas on which the proofs are based, come out of this domain of informal mathematics. But for the computer this domain does not exist.

Mathematics is about understanding! Proofs are important to the extent that they help develop an understanding of some mathematical situation. What a computer can simulate is the proof-generating aspect of mathematics but not the understanding. What computers mimic is a secondary mathematical activity, not the primary activity.

Suppose that a computer succeeded in generating a set of mathematically valid results. What criteria would enable the computer to decide which were important, what directions of inquiry were worth pursuing, and so forth? Mathematics is not a game like bridge or chess. It has no obvious beginning and no well-defined ending. Mathematics is an open-ended exploration. Unlike physics, no one will ever proclaim the "end of mathematics." Mathematics is a human activity[6] intimately connected with the human need to discern patterns in their environment. As such it is related to properties of the human mind and our need to draw conceptual maps that facilitate our understanding of the natural world and ourselves.

Is Mathematical Thought Algorithmic?

Is mathematical thought algorithmic? In keeping with the theme of basing my discussion of mathematics on what mathematicians actually do, I shall approach this question through a discussion of two terms that have great currency in the informal discussions of mathematicians—the terms "trivial" and "deep." The highest compliment a mathematician can give to a piece of mathematics is to say that it is "deep." Although most people would be hard pressed to make precise what is meant by "depth" in mathematics (or in anything else), nevertheless most good mathematicians would have no hesitation in classifying a particular result as "deep" as opposed to superficial or "trivial." In fact, an instinct for which problems and results are deep is often taken as a criterion of how good a mathematician a particular individual is. An excellent mathematician works on "deep" problems and produces "deep" results.

In fact this entire book has given us various hints, not of what to take as a definition of depth, but of conditions that may accompany depth or even produce it. Consider the definition of ambiguity and its essential ingredient, namely, a double perspective. This dualism may be seen to produce depth by analogy with the way visual depth is produced in binocular vision. Two visual perspectives each of which is consistent but which are not identical combine to produce a single, unified perspective that is richer than either of the originals. This is analogous to what happens when you go from a surface appreciation of a particular mathematical phenomenon to a deeper understanding. Understanding seems to carry with it a sense of depth, a multidimensionality or sense of multiple perspectives. Thus the entire discussion of ambiguity is connected to the notion of depth.

Furthermore, the depth of a particular idea seems to sometimes be related to that other aspect of ambiguity—the degree of incompatibility between the original frameworks. Concepts that are difficult in this way—zero, irrational numbers, infinity—turn out to be important, turn out to have a great deal of mathematical content. The resolution of paradoxical situations depends on the development of significant mathematical ideas. This leads to the conclusion that has been mentioned a number of times now: depth resides in ambiguity but only when the situation is resolved by an act of creativity. One might say that the depth of a mathematical situation is a measure of the creativity that accompanies its birth.

"Trivial" results, on the other hand, are results that follow in a mechanical way from the premises; that are superficial, formal, and require no "idea"[7] or act of understanding. I am suggesting that algorithmic thinking is trivial. Though users of mathematics are required to master a certain number of algorithms, the advantage of this mode of thinking is precisely that it can be applied mechanically.

The difference between the "deep" and the "trivial" is crucial to our conception of mathematics. The great French mathematician Henri Poincaré was well aware of what was at stake here. He said:

the very possibility of mathematical science seems an insoluble contradiction. If this science is only deductive in appearance, from whence is derived that perfect rigour which is challenged by none? If, on the contrary, all the propositions which it enunciates may be derived in order by the

rules of formal logic, how is it that mathematics is not reduced to a gigantic tautology? The syllogism can teach us
nothing essentially new, and if everything should be capable of being reduced to the principle of identity, then everything should be capable of being reduced to that principle.
Are we then to admit that the enunciations of all the theorems with which so many volumes are filled, are only indirect ways of saying that A is A?[8]

Poincaré rejects the vision that mathematics is no more than
mere tautology. He goes on to say, "it must be granted that mathematical reasoning has of itself a kind of creative virtue, and is
therefore to be distinguished from the syllogism. *The difference
must be profound*." Poincaré compares the creative to the syllogis-
tic, whereas I use the term algorithmic, but the point is the
same—there is a creative depth to mathematics. This is the mystery of mathematics which cannot be understood through a kind
of logical reductionism.

"Trivial" Mathematics: "I Follow It but I Don't Understand It!"

Mathematics is usually approached in one of two ways. The first
approaches it *instrumentally* as a body of useful results, techniques, formulas, equations, and so on. that are assumed to be
valid and can be applied to solve various kinds of problems.
This is how engineers use mathematics, for example, or psychologists use statistics. One doesn't delve too deeply into the deri-
vation of the techniques or the question of why they work. One
accepts that they work and one moves on.

The second approach to mathematics ostensibly addresses the
question of *why* mathematics works. The question of "why" is
generally answered in theoretical mathematics, the approach
you will find in books written by mathematicians and in courses
taught by them. The question of why mathematics is valid is
answered by embedding that bit of mathematics into a deductive system.

The development of an area within pure mathematics is characterized by a certain approach—the deductive, axiomatic ap-

proach. As in Euclidean geometry, this involves starting with axioms and definitions and building toward results that follow rigorously from these assumptions. One begins with an abstract characterization of the system to be considered. Consider the following example.

Example 1: Groups

A (mathematical) *group* consists of a collection of elements, that is, a set G,[9] and an operation, *, that satisfy the following rules:

(i) If g, h are any elements of G, then $g * h$ is also an element of G.

(ii) If g, h, k are any elements of G, then $g * (h * k) = (g * h) * k$.

(iii) There is a neutral element of G (usually denoted by e) satisfying $g * e = e * g = g$ for any element g in G.

(iv) For every g in G, there is another element, denoted by g^{-1} (the inverse of g) such that $g * g^{-1} = g^{-1} * g = e$.

An example of such a situation is given by taking G to be the integers, $\{0, \pm 1, \pm 2, \pm 3, \ldots\}$ and * to be the usual addition. In this case $e = 0$ and the inverse of the integer n is $-n$.

A second example would consist of all nonzero fractions with * = multiplication. Then $e = 1$ and the inverse of m/n is n/m.

Now a subcollection H of G is called a *subgroup* if it is a group in its own right (using the same operation * as G). For example, the group of integers under addition has subgroups consisting of all multiples of 2, all multiples of 3, and so on. The set of even numbers (multiples of 2) is a subgroup, for example, because (i) the sum or difference of two even numbers is even; (ii) 0 is even; (0 is equal to two times 0); and (iii) the inverse of an even number is also even.

Now if we go back to the completely general situation of *any* group we are in a position to prove a valid and perfectly general mathematical result:

Proposition: *The collection of all the elements common to two subgroups itself forms a subgroup.*

The argument is very simple. The set of elements that is common to the subgroups H and K is usually written as $H \cap K$. We must demonstrate that the set $H \cap K$ satisfies conditions (i)–(iv) above. Let h and k be any two elements of $H \cap K$. Since h and k are elements of H, h^*k is an element of H. Again, since h and k are elements of K, h^*g is an element of K. Thus h^*k belongs to $H \cap K$. Thus condition (i) is true for $H \cap K$. (ii) is true for $H \cap K$ because it is true for all of G. (iii) is true because the element e belongs to H and K and therefore $H \cap K$. Finally if g is any element of $H \cap K$, it is an element of H and K, so its inverse is also. Thus (iv) holds for $H \cap K$. This shows that $H \cap K$ is a subgroup of G. ∎

That is the end of the argument. If this argument frustrates you, if your reaction is that the argument is much ado about nothing or if the argument is too abstract for you, you have a point. In fact I am taking a chance in using this example because I may just succeed in "turning off" my reader just as so many students are turned off in theoretical courses in pure mathematics.

It is difficult to understand what is going on in this proof precisely because this kind of argument is all at the formal, logical level. The point of the example is that one can follow the logic of the definitions and argument without knowing anything about the subject, with no experience with anything connected to the subject, and therefore certainly no intuition or "feel" for the subject. In response to this argument one of my students made the observation, "I follow it but I don't understand it." This is precisely the point that I am making! With a little care and effort one can verify that the argument is logically correct, but, because one has no feeling for the context or ideas that are involved, one feels that one does not understand what is going on. Verification is one thing, understanding quite another. Verification is superficial, restricted to the surface of things. This is what I mean by "trivial."

Whereas an argument that is structured around some mathematical idea (see Chapter 5) can be understood by grasping that idea, a formal argument like the one above cannot really be understood because there is, in a sense, nothing to understand. All that can be done with an argument like this is to verify it, and a computer could do this verification. This is what it means for an argument to be "trivial"!

Almost all of the mathematics that I have discussed in the earlier chapters has been nontrivial to a certain extent at least. However I will now repeat a very simple result just to contrast it with the previous one.

Example 2: Adding integers

Consider the formula for the sum of the first n integers,

$$1 + 2 + \cdots + n = \frac{n(n+1)}{2}.$$

Here is a simple argument for this result: Let

$$S = 1 + 2 + \cdots (n-1) + n;$$

reversing the order of addition gives

$$S = n + (n-1) + \cdots + 2 + 1.$$

Adding we see that $2S$ is given by adding up the n columns each of which sums to $n + 1$. That is,

$$2S = n(n+1).$$

Thus $S = n(n+1)/2$ as required.

This is a very simple argument, but it is not completely trivial because it contains an idea, namely reversing the sum and adding by columns. This idea shows us not only that the formula is valid but also *why* it is true. We understand this result in a way that we do not understand the previous formal argument.

One of the main goals of this work has been to differentiate between the algorithmic and the profound in mathematics. How, that is, can we distinguish between "trivial" thought and "complex" thought? In the above examples a distinction was tentatively drawn between the "trivial" and the "simple"—the first argument was called "trivial," the second is simple. Mathematical thought can be simple and it can also be complex but mostly it is nontrivial. Computer thought, on the other hand, even though it may be very lengthy and complicated, is essentially trivial.

Is the distinction between the "trivial" and the "deep" of any value? Is mathematics trivial or nontrivial? Is science? Is human thought in general nontrivial? Does the algorithmic subsume all other thought, that is, is all thought trivial? In the case of mathematics, the famous attempt of Russell and Whitehead (1925) to show that mathematics could be reduced to logic was therefore an attempt to *prove* that mathematics was trivial. Gödel showed that arithmetic was nontrivial and therefore the attempt of Russell and Whitehead was a failure. This is why Penrose evokes the work of Gödel to support his claim. He knows in his gut that mathematics is nontrivial.[10]

Mathematics is a wonderful vantage point from which to take up the question of "triviality" versus profundity. Return for a moment to the origins of systematic mathematical thought. In Chapter 2 I discussed the "dream of reason" that emerged in the civilization of ancient Greece, the discovery of a new way of thinking that seemed to have the potential of allowing human beings access to a permanent and objective truth. This vision is still a powerful cultural force in our society. In the terminology of Chapter 7, it is a "great idea." In that chapter I discussed formalism as an example of a great idea, but really formalism is just a particular development of a much older idea—the idea of reason itself. This is a core idea for our civilization which has been at work changing human beings and their cultures, their sense of themselves and of the natural world for thousands of years.

Today, with the dawning of the age of information technology we stand at a new and crucial stage in the development of this core idea. What has changed is our technology. The computer is the tool through which this great idea is being implemented. All technology takes a human characteristic, makes it autonomous, and extends it. Behind every technological advance there is an idea—the machine is really this idea given a physical form. The telescope extends the act of seeing; the telephone the act of speaking. The computer extends a certain way of using the mind—it extends rationality, logic, and, in particular, what I have called algorithmic thinking. Now the idea of the computer has all the characteristics of a great idea. It is an extraordinary development and we are all, to a certain extent at least, caught up in the rush of enthusiasm that accompanies the birth of this great idea. However, remember that a great idea is always wrong in that it always overreaches itself when it claims universality. A technology can only objectify a certain aspect of the human potential. It inevitably approaches human beings from a certain perspective and inevitably ignores other perspectives. Human beings are not only logical animals; we are also creative animals. Thus the dream of the human being as computer, of human thought as computer thought, is false even if the computers in question are supercomputers whose computational powers dwarf those of present day machines. Nevertheless, the "dream of reason" in its most modern incarnation is refashioning modern culture in its own image. Yet, in a way, this modern version of the "dream of reason" is the dream that thought is "trivial" or can be made "trivial."

I claim that there is something that is going on in mathematical thought that is nontrivial. What is this other factor? Mathematics is deep; mathematics is basically a creative activity; mathematics is one of the most profound ways of using the mind. Much of this book has been an attempt to get a handle on what it means for a piece of mathematics to be "deep." In a sense the entire book is a discussion of the distinction between the "trivial" and the "nontrivial" within mathematical thought. The question of distinguishing between "depth" and "triviality" is vital! Is profundity a matter of quality or quantity? The proponents of artificial intelligence might argue that it is one of quantity and that the advent of bigger and faster machines will make all of human thought accessible to machine replication. I maintain that the difference is qualitative.

One of the great mysteries of mathematics is contained in this question about the nature of "depth." It is similar to the question of what makes a work of art great. The best practitioners agree that "depth" refers to something that is real, but what is it exactly? This has become an important question ever since the advent of formalism in mathematics and even more so with the advent of computers. Could a computer be programmed to distinguish between the trivial and the deep? Or is depth merely subjective, completely in the mind of the beholder, and thus not really there at all? This is a very important question, a deep question, I am tempted to say. The introduction of the idea of "ambiguity" was an attempt to get a handle on this question.

In his introduction to Davis and Hersh, Gian-Carlo Rota says, "The mystery of mathematics is that . . . conclusions originating in the play of the mind do find striking physical applications." It is indeed this surprising correspondence that pushes us to take a more careful look at what is going on in mathematics. In his famous paper I referred to earlier on the "Unreasonable Effectiveness of Mathematics in the Natural Sciences," Wigner claims that the power of mathematics lies in the ingenious definitions that mathematicians have developed. These definitions, he feels, capture some very deep aspects of reality and are therefore the secret of why mathematics works as well as it does. This is fine as far as it goes, but it cries out for additional clarification. For one thing, what is the nature of these ingenious definitions? What is it, exactly, about mathematical concepts that make them so fruitful?

In my attempt to get a handle on "depth," I found it necessary to reconsider the very things that mathematical reasoning seems to be delivering us from, to reconsider the myth of reason itself. I did this by introducing the notion of "ambiguity" in all of its many guises as another aspect of mathematics that complements the logical structure. Ambiguity and paradox are aspects of mathematical thought that differentiate the "trivial" from the "deep." The "trivial" arises from the elimination of the ambiguous. The "deep" involves a complex multidimensionality such as those evoked by the successful resolution of situations of ambiguity and paradox. Even the word "resolution" is misleading in this context because it usually implies the reduction of the ambiguous to the logical and linear. What really happens is that the ambiguity gives birth to a larger context, a unified framework that contains the various potentialities that were inherent in the original situation.

Mathematics, as I have been describing it, is an art form. The words ambiguity and metaphor are much more acceptable in the arts than they are in the sciences. But ambiguity and metaphor are the mechanisms through which that ultimate ambiguity, the one that divides the objective from the subjective, the natural world from the mind, is bridged. It is the same mechanism that operates at all levels—from a simple mathematical concept like "variable" to the mathematics of "string theory." The same mechanism even applies to a discussion of mathematics as a whole.

Mathematics as Complexity[11]

One of the truly remarkable things about mathematics is the manner in which it has provided human beings with models and metaphors that have been used to make sense of the world. Euclidean geometry is important, not just as geometry but as the model of reason it provided for other areas of mathematics and for human thought in general. It operates in a double way—as scientific theory and as metaphor. Differential equations provided the model of the machine universe in which the past and the future are absolutely determined from knowledge of the present state of affairs. The list goes on and on.

In Chapter 7, I briefly discussed the modern theories of chaos and complexity. These theories have created a great stir in recent years—not because they have succeeded in making original and verifiable predictions—but because they provide a new metaphor, a new framework for thinking about the world, including the world of science. Viewed in this way, "chaos and complexity theory" can be thought of as self-referential—it describes aspects of the natural world and it provides a way of thinking about mathematics of which it forms a part. In a sense this book is a part of this new way of thinking. It is an attempt to develop a new way of thinking about mathematics—an attempt to think out what it would mean to apply the metaphor of complexity to mathematics itself.

What would a description of "mathematics as complexity" be like? The following paragraphs are a brief indication of some of the characteristics that such a description might include.

First of all, one must draw a distinction between the terms "complex" and "complicated." As the terms are used in this book, "complicated" refers to the quantitative, "complex" to the qualitative. As I have noted, it is possible for logical mathematical reasoning to be complicated without being "deep"—without containing very interesting mathematical ideas. "Complexity" refers primarily to the world of mathematical insights. It refers to mathematics both as a body of knowledge and as a practice—the extraordinary mixture of subtle reasoning and profound intuition that characterizes both mathematics and the manner in which the mind is used in mathematics. Mathematics is both complicated and complex, but one's choice of which property to emphasize as essential will reveal one's orientation toward mathematics. Is it toward the algorithmic or the creative?

The theory of chaos arises from the study of *nonlinearity*. Complexity is fundamentally nonlinear. If mathematics is nonlinear, then its essence cannot be captured by algorithmic procedures or by the linear strings of reasoning that characterize both mathematical proofs and deductive systems. Mathematics is a world of dynamic change—an extraordinarily complex world with to its own ecological structure. In this world new concepts are continually coming into being while others are sinking into irrelevance. Knowledge is continually being reorganized and reevaluated in the light of new interests and new ways of thinking. One metaphor for nonlinear mathematics comes from the theory of turbulence, like that describing the flow of water in a river. In a situation of turbulence there are features, such as whirlpools, that are stable but can disappear if the rate of flow of the water changes. A turbulent system is constantly reorganizing itself depending on the rate of flow of the fluid involved. In the same way, mathematics can be thought of, not as permanent and absolute, but as in a continual state of dynamic rearrangement. There are whole areas of mathematics that were once of great interest but that today are not pursued and sometimes not even remembered.

Another aspect of complex systems is that they are *open*. For example, a biological system is in continuous interaction with other systems including the ambient environment. Thus a biological organism, like a human being, for example, can only be considered in isolation as a temporary expedient, not as an absolute reality. Even though it may sound strange, mathematical theories also cannot be considered in isolation from other parts of mathematics, from the sciences and computer science, and from human culture in general. Mathematics is a human endeavor with roots in the natural world and human biology. In particular, mathematics changes and evolves. However it is often described as though it were absolute and timeless.

Complex systems inevitably involve an element of *contingency*, an element of uncertainty. "Randomness" is discussed in Chapter 7 as basic to two of the most far-reaching scientific theories of our time—quantum mechanics and evolution. Human life as we experience it involves a healthy dose of the accidental and the unpredictable. Where is the uncertain in our description of mathematics? Conventionally mathematics is thought of as the antithesis of the uncertain. It is the epitome of the certain and the absolute. In my approach, the uncertain exists as a key element of mathematics yet in a manner that is unique and quite characteristic of the unique way in which mathematics approaches the world. The French philosopher Edgar Morin says, "Logical positivism could not avoid taking the role of an epistemological policeman forbidding us to look precisely where we must look today, toward the uncertain, the ambiguous, the contradictory." In mathematics the role of logical positivism is taken by formalism, which renders invisible the properties of uncertainty and ambiguity that form an indispensable aspect of mathematics.

If mathematics cannot be accurately represented as isolated from its environment, it follows that an adequate description of mathematics will have to include the related processes of creating and understanding mathematics. The normal view is that there is an objective body of mathematical theory on the one side and, on the other, the mathematician who creates the theory or the consumer who uses it. The idea that there could be a point of view that contains both elements, that is, the introduction of this human element, changes everything. Mathematics is no longer a strictly "objective" theory. "Objectivity" is merely a description of one dimension of mathematics; that is, objectivity is merely an approximation to what is going on. It is one point of view—a useful one—but not the definitive one.

If mathematics includes the mathematician, then it is reasonable to see intelligence as an essential ingredient of mathematics. Mathematics is a form of intelligence in action; that is, it is not only the objective result of an act of intelligence but rather a demonstration of the nature of intelligence itself. It is a major way in which intelligence functions. When we study mathematics we are not so much absorbing some predetermined set of facts as we are studying the manner in which the mind works— the manner in which it produces mathematics. (In saying this I am aware that I am using "mathematics" in an ambiguous way, both as content and as process.)

One of the essential properties of intelligence is flexibility. What does it mean to be an expert at something? What makes someone an expert is her ability to respond to a completely novel situation—to solve original problems, for example. Expertise does not only involve having command of a huge amount of factual knowledge—it does not mean being a human data bank. It involves the capacity for flexible response. It is a form of creativity. This description applies equally to the notion of "understanding." Understanding is a kind of expertise. A true measure of intelligence is this capacity for flexible and original response. How does this flexibility arise in mathematics? As was mentioned above, the essence of algorithmic thought is the elimination of flexibility by providing predetermined responses to any given eventuality. Biological systems that are successful in an evolutionary sense have, many believe, the flexibility to adapt to a changing environment. Complex thought in mathematics also contains a kind of flexibility that is part of any reasonable definition of intelligence.

There is a profound relationship between complexity and simplicity. For example, the well-known mathematician and popularizer of "catastrophe theory" E. C. Zeeman said that "Technical skill is the mastery of complexity, while creativity is mastery of simplicity." What then is simplicity? What is it that makes a piece of mathematics simple? It is a commonly observed phenomenon that a piece of mathematics is simple if you understand it and difficult if you do not. This is not a joke or a definition of the word "simple," but it reflects the way the word is often used. This usage of the word implies that the mathematical situation is simplified, if you will, by an act of intelligence, that is, of understanding. Thus a picture of mathematics emerges, a picture that was discussed in the chapters on "mathematical ideas," of mathematics as a hierarchy of the simple. Data look complicated and "hard" until the emergence of a mathematical idea structures the data and even makes them "simple." The ideas that make things simple are not objective in any absolute sense. It makes more sense to call them optimizations of the potential contained in the original situation. These ideas often solidify into new mathematical objects that form the data that may be organized by new, higher order ideas. This is the way in which "complex" situations develop. Both the process of mathematical thought and its product are now unified in a complex system that on one hand is an intricate structure and on the other is flexible and capable of dynamic change. In one dimension there is the order and permanence of the formal, deductive system. The other dimension is open to ambiguity and contradiction, open to insight and creativity, open to change.

ARE HUMAN BEINGS TRIVIAL?

The question for the proponents of "algorithmic intelligence" is the following. Do the creative uses of the mind that I have discussed throughout this book arise from subtle algorithmic processes, or is the reverse the case, namely, that algorithmic processes arise from another and more basic way of using the mind? This book has been an argument for the existence of other intelligent ways of thinking, ways of thinking that do not reduce to Pinker's "rational truth-obeying rules." What happens in a situation of ambiguity? You may have one set of rational rules, one context, that pushes you in one direction and a second that pushes you in a different direction. It is precisely at the level of rational rules that things break down. How does one operate in a situation of conflict and incompatibility? Is intelligence inoperative in such situations? Of course not. A situation of conflict is precisely where we most need the intervention of creative intelligence.

Thus the larger question concerning the centrality of algorithmic thought comes down to whether or not you believe that at the most basic level things are ordered, rational, and algorithmic—what I think of as a kind of Platonic vision of the human mind. Now there is much to be said for this vision of the power and importance of the algorithmic—it is, in fact, a great idea. We are in the inevitable period of inflation that comes with any great idea—the time when it is claimed that this particular insight will explain everything. Thus this book is merely saying, "No, this is not the only way of using the mind. There is another way." In this other way, from this different perspective, disorder and conflict are never definitively eliminated. In fact the very attempt to describe human activity in an algorithmic, rule-based way leads to the problematic situations that I have called ambiguous, contradictory, and paradoxical. Working with these situations, attempting to understand them in the face of their problematic aspects, gives rise to acts of intelligence and creativity that produce the order that we observe.

The implications of this discussion are considerable. What is at stake is nothing less than our conception of what it means to be a human being. To claim that thought is "trivial" is to claim that human beings are "trivial." In my opinion to hold that human beings are trivial is to miss something vital, something that all previous human cultures would have seen as self-evident. *Human beings are not machines! Human beings are not trivial!* What could be more important than this question? It is perhaps ironic that mathematics, the area out of which computer science developed, and the discipline with which it has the greatest affinity, should be the domain in which the distinction between algorithmic and non-algorithmic should be the most accessible.

In other cultures this kind of question about the nature of the human mind would have religious implications. It would be answered with reference to "God" or "spirit" or the "soul." In our secular culture these questions are addressed by computer scientists, cognitive psychologists, and philosophers of science. Nevertheless the stakes remain high and involve everyone for what is at issue is human self-definition—the nature of the beings we tell ourselves that we are.

MATHEMATICS IN THE LIGHT OF AMBIGUITY— A GREAT IDEA

Maybe this is a good place to come back to the beginning and summarize the point of view that is developed in the preceding pages. I have attempted to lay the foundations for a great idea, an idea of human nature as fundamentally creative. This creative process that we call mathematics, for example, has no end, no ultimate objective, and therefore will never be completed. It follows that the problematic situations such as ambiguity and contradiction, are never completely eliminated. In fact the creative process thrives on such situations. Now in order for such a point of view to be considered seriously it has been necessary to put limitations upon a competing great idea—the idea of the human mind as a logical machine. Mathematics in its largest sense is the arena in which this particular battle has been fought, but its implications apply to the worlds of science and technology and to the culture that these have spawned.

In a sense the questions that are raised in this book involve matters of life and death. Not life and death in a literal sense but in the metaphorical sense of living systems versus mechanical ones. In one sense biology is *about* living systems, whereas physics is *about* nonliving ones. However in another sense, physics, biology, and mathematics can be alive or dead. At the research level these subjects are all alive, all growing and changing. The practice of a scientific discipline has many of the characteristics of a living organism. On the other hand, a discipline may stop developing, may become moribund because it is thought of as a completed, theoretical "truth" to which nothing will ever be added or taken away. This attitude signals the end of mathematics and science. If a description of mathematics is to have any value it must describe a discipline that is living and growing, not a subject that is frozen.

One of the aspects of a living system is that it is creative. Life is continually being confronted with problems and the necessity to resolve these problems. Problems, in life, in art, and in science, are inevitable. Not only can they not be avoided but they are the very things that spur development, that spur evolution. The solutions that these problematic situations bring forth are unpredictable, a priori. The solution to such problems often involves an element that is entirely unexpected—the creative element. A creative solution is not mechanical—it does not involve juggling a number of predetermined elements according to predetermined rules. It involves the emergence of a novel way of looking at the original situation. This new way of seeing is often generated by very incompatible tendencies within the original situation that made it problematic in the first place. This is the essence of a living system as it is the essence of mathematics. This living essence is intimately connected to what I have been calling "the light of ambiguity." The essence of mathematics is that it is nontrivial, creative, and alive.

✳ *Notes* ✳

Introduction: Turning on the Light

1. Singh (1997), p. xi.
2. Singh, p. 211.
3. Graham, Knuth, and Patashnik (1989).
4. Davies (1999).
5. Lightman (2005), pp. 10–11. Lightman straddles the humanities and the sciences. He is a novelist and physicist and so is unusually well placed to talk about the creativity of science.
6. Cf. Koestler's definition in Chapter 1.
7. Bernstein (1976), pp. 39–41.
8. Dehaene (1997), for example.
9. Wigner (1960).

Chapter 1: Ambiguity in Mathematics

1. Davis and Hersh (1981), p. 34.
2. Thurston (1994).
3. Albers and Alexanderson (1985), p. 19.
4. Koestler (1964).
5. Uhlig (2002), p. 1.
6. Thurston (1990), p. 846.
7. Bodanis (2000), p. 26.
8. It may be that some cultures and therefore languages are more "thing" oriented (i.e., static) whereas others may be more action oriented. For example, think of the expression "I am going for a run." The noun "run" is a reification of the verb. Which is more basic in general, nouns or verbs? Cf. Boman (1960).
9. Shanks (2001), p. 126.
10. Quoted in ibid.
11. Dunham (1991).
12. My colleague Anna Sierpinska made the observation that even abandoning one of the contexts may still lead to creative mathematics.
13. Grothendieck (1986) quoted in Jackson (2004), p. 1051.
14. This sort of learning problem is sometimes called an "epistemological obstacle." For more on this see Chapter 9.
15. Of course this is not the way things happened historically. It is included in order to emphasize the metaphoric potential of equations

to evoke new mathematical structures that go beyond the structures within which the equation is formulated.

16. Kline (1972), p. 338.

17. For example, in the circle, $x^2 + y^2 = 1$, which variable is dependent?

18. Kline (1972), p. 339.

19. For example, $f(x) = 0$ if x is rational and $f(x) = 1$ if x is irrational.

20. Andrew Wiles (1995) pp. 443–551.

21. Quoted in the engrossing account *Fermat's Enigma* by Simon Singh (1997).

22. David Hilbert delivered a lecture to the Second International Congress of Mathematicians in the year 1900 in which he enumerated a series of key problems that the nineteenth century left to the twentieth century to solve.

23. Foreword to the English translation of Matiyasevich (1993).

24. Shanks, p. 2

25. Tomás Oliveira e Silva, "New Goldbach Conjecture Verification Limit," NMBRTHRY@ listserv.nodak.edu (February 2, 2005)

26. Ramaré (1995)

27. See the discussion of Gödel's incompleteness theorem and Goodstein's theorem in Chapter 6.

28. Campbell (1949).

29. Herbert (1985), p. 33.

30. The parallel between electromagnetism and the Taniyama-Shimura conjecture is brought out clearly in Singh.

31. Greene (2000), p. 3. Subsequent quotes in this section are also taken from Greene.

32. Sfard (1989), p. 151.

33. Gray and Tall (1994), p. 121.

34. Ibid., pp. 120–21.

35. Sfard (1989, 1994); quote (1994), p. 53.

36. Ibid.

37. Dubinsky (1991).

38. Sfard (1994), p. 44.

39. Lakoff and Núñez (2000) (cf. Lakoff and Johnson 1980).

40. Lakoff and Núñez, p. 53.

41. Sfard (1994), p. 46.

42. Sfard, p. 48.

43. Koestler (1964), pp. 35–36.

44. Low (2002).

45. Bernstein (1976).

46. Ibid., p. 195.

47. Ibid., pp. 39–40.

48. Ibid., p. 107.

CHAPTER 2: THE CONTRADICTORY IN MATHEMATICS

1. Bohm (1980), p. 4.
2. Hersh (1997, 1998); Lakatos (1976).
3. Einstein (1930).
4. Dostoevsky (1864), p. 25.
5. Taken from Dunham (1990), p. 33.
6. Dunham, p. 35.
7. Lakatos (1976).
8. Smale (1967).
9. The quotations in this section are taken from Barrow (2000), pp. 37–42.
10. Historical references in this section are taken from Kline (1972), pp. 426–34.
11. In even more precise language, for every $\varepsilon > 0$, there exists a number N such that if $x > N$, then $1/x < \varepsilon$.

CHAPTER 3: PARADOXES AND MATHEMATICS

1. Cf. Becker (1973). This entire book is in a sense about the paradox of living in the face of death. Especially noteworthy is his discussion of Kierkegaard and "normal neurosis," p. 81.
2. Wigner (1960).
3. Davis and Hersh (1981), p. 153.
4. Moore (1990), pp. 1–2.
5. Davis and Hersh (1981), p. 153.
6. Wagon (1985)
7. Heschel (1951), p. 4.
8. Weyl quoted in Davis and Hersh (1981), p. 108.
9. Cf. Aczel (2000).
10. Albers (2004), p. 9.
11. Quoted in Dunham (1990).
12. Russell (1926), p. 183.
13. Zeno's Paradoxes, *Stanford Encyclopedia of Philosophy*.
14. It is interesting that the Greeks came very close to this way of seeing things when they reworked their theory of ratio and proportion in reaction to the crisis precipitated by the proof of the existence of irrational numbers (as discussed in Chapter 1).
15. Recall that *converge* means that the numbers in the sequence give a (mathematical) approximation to some specific number. *Diverge* means "does not converge."
16. Cf. Dunham (1990), Chapter 7.

17. A sequence $x_1, x_2, x_3, \ldots$ is said to be a *Cauchy sequence* if for every number $\varepsilon > 0$ there is some integer N such that $|x_m - x_n| < \varepsilon$ whenever $m, n > N$.

18. Courant (1988).

Chapter 4: More Paradoxes of Infinity

1. We are assuming that a family of parallel lines meets at *one* point at infinity and not two.

2. The usual example is Desargues' theorem; see Maor (1987), p. 112.

3. Of course this duality is a form of mathematical ambiguity. Interestingly, it does not unify points and lines in one "higher" concept but maintains that "points are equivalent to lines" and vice versa while each maintains its identity as a distinct object.

4. Ibid., p. 117.

5. Lakoff and Johnson (1980).

6. Newton also believed in the existence of an absolute time as a one-dimensional continuum.

7. It is interesting that the most recent experimental observations in cosmology imply that the universe is flat, that is, the angles of a triangle add up to 180 degrees (Hu and White 2004).

8. Cf. Kline (1980).

9. Cf. Aczel (2000), p. 133. "Kronecker could not accept the idea that irrational numbers existed."

10. Quoted in Randall (1976), p. 237.

11. William Blake, *Auguries of Innocence*.

12. Self-similarity may be observed in figures 3.1 and 7.2.

13. Thanks to Hershy Kisilevsky for this suggestion.

14. This is discussed in more detail in Chapter 8.

15. Aczel, Chapter 8.

16. The material in this section owes a great deal to the discussion in Davis and Hersh (1981) and two papers by Tall (1980, 2001).

17. See Tall (1980).

18. Joseph Dauben's biography (1995) of Robinson is highly recommended.

19. Compare to Davis and Hersh (1981), p. 243.

Chapter 5: The Idea as an Organizing Principle

1. Ruelle (2004), p. 275

2. Helena Verrill, "Monstrous Moonshine and Mirror Symmetry," http://www.mast.queensu.ca/~helena/seminar/seminar1.html

3. Cited in Briggs (1990).

4. Elliot (1957) p. 97.

5. Mazur (2004).

6. Devlin (1998), p. 3.

7. Davis and Hersh (1981), p. 97.

8. Plotinus, *Stanford Encyclopedia of Philosophy,* http://plato.stanford .edu/contents.html.

9. Low (2002), Chapter 1.

10. This point is stressed by Maturana and Varela (1998).

11. Sacks (1996), "To See and Not See."

12. Pinker (1997).

13. Cf. Von Franz (1974).

14. Dehaene (1997).

15. Pinker, p. 13.

16. Thanks to Joe Auslander for bringing this to my attention.

17. Even the idea embedded in $x\mathbf{R}x$, that is, everything is equivalent to itself, deserves some serious thought. It assumes a stable world of mathematical objects that does not change over time. This is not so much an obvious property of the natural and mathematical world as an *assumption* we make about things. It is a fundamental assumption, one that makes mathematical discourse possible.

18. For example, .5 would be related to 1.5, 2.5, 3.5, etc. so we could write $.5\mathbf{R}2.5$.

19. Two spaces are homeomorphic if one can be continuously deformed into the other, where continuous deformation includes stretching and twisting but not tearing. Thus the symbol $\subset$ is homeomorphic to a line segment or a coffee cup is homeomorphic to a doughnut. In more precise language, two spaces X and Y are homeomorphic if there is a continuous mapping from X to Y which is one-to-one and onto, and whose inverse from Y to X is also continuous.

20. Notice that the proof uses the notion of potential infinity not of actual infinity since what we have shown is that the number of primes is larger than any finite number.

21. Complex numbers have the form $z = x + iy$, where x and y are real numbers. Ordinary real functions can be converted to complex functions be replacing the real x by a complex z. Thus $f(x) = x^2$ becomes $f(z) = z^2$, where $z^2 = (x + iy)^2 = (x^2 - y^2) + i(2xy)$.

22. Pólya (1954), p. 197.

23. Cf. figure 3.1.

24. An example of such a function might be $F(f) = f'$, the function that maps an ordinary function onto its own derivative.

25. Again we have a dual representation for the derivative both as a number and as a (tangent) line.

26. Thurston (1994).

27. The result here was first proved by Sarkovskii (1964). One part of the result was then independently rediscovered by Li and Yorke (1975), who were the first to apply the term "chaos" to the properties of such systems. The idea behind the proof that I discuss was first isolated in the paper of Block, Guckenheimer, Misiurewicz, and Young (1980).

28. Attractors are discussed in Chapter 7.

CHAPTER 6: IDEAS, LOGIC, AND PARADOX

1. Davis and Hersh (1981), pp. 139–40.
2. Albers and Alexanderson (1985), p. 19.
3. Thurston (1994).
4. Obituary Notices of Fellows of the Royal Soc., 4, 1044, pp. 547–53, quoted in Kline (1972), p. 1210.
5. We may have to rethink what we mean by "truth" in mathematics (cf. Chapter 8).
6. Bohm (1980), p. 4.
7. Singh (1997), p. 174.
8. Kline (1974).
9. Arendt (1966), p. 439.
10. Davis and Hersh (1981), pp. 346–47.
11. Cauchy (1813), in Lakatos (1976), p. 7.
12. Lakatos, p. 13.
13. Cf. Kauffman (1995), for example.
14. Gödel (1940).
15. P. J. Cohen (1963, 1964).
16. Chaitin (2002b).
17. Nagel and Newman (1964), p. 61.
18. Ibid., p. 63.
19. Ibid., p. 78.
20. Penrose (1989), p. 106.
21. A typical such proposition might be $1 + 2 + 3 + \cdots + k = k(k+1)/2$.
22. Penrose (1989), pp. 107–8.
23. Penrose (1997), app. 1.
24. Goodstein (1944).
25. Kirby and Paris (1982).
26. Penrose (1997), app. 1, p. 191.
27. Compare the discussion of randomness in Chapter 7.
28. Legarias (1966).

CHAPTER 7: GREAT IDEAS

1. Weil (1986), p. 260.
2. Gardner (2001).

3. Chaitin (2002b).

4. Thurston (1994).

5. See the discussion in Herbert (1985), for example.

6. Aczel (2000), p. 132.

7. Cohen, *Anthem* in *The Future*.

8. Mathematics educator and physicist Gerry Goldin mentioned this example to me in a conversation about the way mathematics develops.

9. The letter z is conventionally used to represent complex numbers.

10. The fundamental theorem of algebra states that a polynomial of order n has n roots; for example, $x^3 + x^2 + x + 1 = 0$ has the roots $x = -1, i, -i$.

11. 4-dimensional numbers of the form $a + ib + jc + kd$, where $i^2 = j^2 = k^2 = -1$ and $ij = k$, $jk = i$, $ki = j$. Multiplication of such numbers is noncommutative: the order of multiplication matters.

12. Stewart (1992), p. 11.

13. Stewart (1992), pp. 11–12.

14. Davis and Hersh (1981), p. 163.

15. Bennett (1998), p. 28.

16. Bennett (1998), p. 78.

17. Bennett (1998), pp. 3–5.

18. Bennett (1998), p. 9.

19. Devlin (1998), p. 3.

20. Peitgen and Richter (1986)

21. Beltrami (1999), pp. 66–81.

22. Writing numbers in base 2 and base 3 is discussed on pages 177 and 178.

23. Beltrami, p. 71.

24. Thus if the solution through the point x is given by $f_t(x)$, we have $f_t(x) \to A$ as $t \to \infty$ for all x in some neighborhood of A.

25. For example, Roger Lewin's book (1992) is called *Complexity: Life at the Edge of Chaos*. Also consider Stuart Kauffman (1995), *At Home in the Universe: The Search for the Laws of Self-Organization and Complexity*.

26. This section is based on Chaitin (1975), reprinted in Gregerson (2003).

27. Chaitin (2002a).

28. Herbert (1885), pp. 66–67.

29. See the discussion in Cohen and Stewart (1994).

CHAPTER 8: MATHEMATICAL TRUTH

1. Russell, *Portraits from Memory*, quoted in Davis and Hersh (1981), p. 333.

2. Quotes are from Koestler (1964), pp. 115–16; italics added.

3. The postmodernist might claim that there is no objective reality—that reality is constructed. The position developed here has some elements in common with this point of view, but, it is hoped, it is precisely "creative certainty" that will save us from the excesses of complete relativism.

4. Banesh Hoffman with Helen Dukas (1972), p. 18.

5. Einstein (1954).

6. Koestler (1964), p. 103.

7. Bohm (1980), p. 4 says much the same thing when he insists that "a theory is primarily a form of insight, i.e., a way of looking at the world."

8. Lakoff and Johnson (1980), chaps. 25–30.

9. Mlodinow (2001), p. 81.

10. Herbert (1985), pp. 161–62.

11. Penrose (1989), p.111.

12. Davis and Hersh (1981), p. 358.

13. Possibly the sense of mystery arises from the dual objective/subjective nature of mathematics. The sense of mystery is not very different from the sense of "depth" that will be discussed in Chapter 9. The sense of depth is latent in even the most objective accounts of mathematics—it is the sense that there is more here than meets the eye. Mystery and a sense of depth are our "subjective" response to the profundity of mathematics.

14. Gödel (1964).

15. Thom (1971).

16. Penrose (1997), pp. 1–2.

17. Davis and Hersh (1981), p. 337.

18. Maturana (1988), p. 30.

19. von Glasersfeld (1991), p. 18.

20. Thurston (1994).

21. Bachelard (1938).

22. Sierpinska (1994), p. xi.

CHAPTER 9: CONCLUSION: IS MATHEMATICS ALGORITHMIC OR CREATIVE?

1. Thurston (1994).

2. Consider the article with the provocative title, "The Death of Proof" by John Horgan (1993).

3. Penrose (1989, 1994, 1997).

4. Pinker (1997), pp. 61–62.

5. Lakoff and Johnson (1980).

6. See the articles by Thurston and Hersh in the Bibliography.

7. Chapters 5 and 6 contain a discussion of mathematical ideas.

8. Poincaré (1952).

9. For our purposes a *set* is any collection whatsoever of objects, usually numbers or geometric points.

10. Penrose (1989, 1994, 1997).

11. Edgar Morin, emeritus research director of the French INRS, has written extensively on the subject of complexity. Morin emphasizes that there is an urgent need for "a reorganization of what we understand under the concept of science."

* Bibliography *

Aczel, Amir D. (2000). *The Mystery of the Aleph: Mathematics, the Kabbalah, and the Search for Infinity.* Pocket Books, Simon & Schuster, New York.

Albers, Don (2004). "In Touch with God: An Interview with Paul Halmos." *College Mathematics Journal* 35, no. 1: 1–14.

Albers, D. J., and G. L. Alexanderson (1985). *Mathematical People— Profiles and Interviews.* Contemporary Books, Chicago.

Arendt, Hannah (1966). *The Origins of Totalitarianism.* Harcourt Brace, New York.

Bachelard, G. (1938). *La formation de l'esprit scientifique.* Presses universitaires de France, Paris.

Barrow, John D. (1992). *Pi in the Sky: Counting, Thinking and Being.* Clarendon Press, Oxford.

—— (2000). *The Book of Nothing.* Vintage, Random House, London.

Becker, Ernest (1973). *The Denial of Death.* Free Press, New York.

Beltrami, Edward (1999). *What Is Random? Chance and Order in Mathematics and Life.* Springer-Verlag, New York.

Bennett, Deborah, J. (1998). *Randomness.* Harvard University Press, Cambridge, MA.

Bernstein, Leonard (1976). *The Unanswered Question: Six Talks at Harvard.* Harvard University Press, Cambridge, MA.

Block, L., J. Guckenheimer, M. Misiurewicz, and L.-S. Young (1980). "Periodic Points and Topological Entropy of One Dimensional Maps." In *Global Theory of Dynamical Systems,* ed. Zbigniew Nitecki and R. Clark Robinson. Lecture Notes in Mathematics 819. Springer, New York, 18–34.

Bodanis, David (2000). $E = mc^2$: A Biography of the World's Most Famous Equation. Anchor Books, Random House, New York.

Bohm, David (1980). *Wholeness and the Implicate Order.* Routledge, London.

Boman, Thorleif (1960). *Hebrew Thought Compared with Greek.* W.W. Norton, New York.

Briggs, John (1990). *Fire in the Crucible.* Jeremy Tarcher, Los Angeles.

Buber, Martin (1958). *I and Thou.* Scribner's, New York.

Byers, Bill (1983). *Beyond Structure: Some Thoughts on the Nature of Mathematics. Proceedings of PME-NA, Montreal,* 2, pp. 31–40.

—— (1984). "Dilemmas in Teaching and Learning Mathematics." *For the Learning of Mathematics* 4, no. 1: 35–39.

Byers, Bill (1999). "The Ambiguity of Mathematics." *Proceedings of the International Group for the Psychology of Mathematics Education, Haifa.*

Campbell, Joseph (1949). *The Hero with a Thousand Faces.* Bollingen Series 17. Princeton University Press, Princeton.

Cauchy, A. L. (1813). "Recherches sur les polyèdres." *Journal de l'École Polytechnique* 9: 68–86.

Chaitin, Gregory J. (1975). "Randomness and Mathematical Proof." *Scientific American* 232, no. 5: 47–52.

—— (2002a). "Paradoxes of Randomness." *Complexity* 7, no. 5: 14–21.

—— (2002b). "Computers, Paradoxes and the Foundations of Mathematics." *American Scientist* 90: 164–71.

Chinn, William G., and N. E. Steenrod (1966). *First Concepts of Topology: The Geometry of Mappings of Segments, Curves, Circles, and Disks.* Mathematical Association of America, Washington, DC.

Cohen, Jack, and Ian Stewart (1994). *The Collapse of Chaos: Discovering Simplicity in a Complex World.* Viking, New York.

Cohen, Paul J. (1963, 1964). "The Independence of the Continuum Hypothesis." *Proceedings of the National Academy of Sciences of the U.S.A.* 50: 1143–48; 51: 105–10.

Conway, John H. (1972). "Unpredictable Iterations." *Proceedings of the 1972 Number Theory Conference, University of Colorado, Boulder,* 49–52.

Courant, Richard (1988). *Differential and Integral Calculus.* Wiley Classics. John Wiley, New York.

Dauben, J. W. (1995). *Abraham Robinson: The Creation of Nonstandard Analysis, a Personal and Mathematical Odyssey.* Princeton University Press, Princeton.

Davies, Paul (1999). *The Fifth Miracle: The Search for the Origin and Meaning of Life.* Touchstone, Simon & Schuster, New York.

Davis, Philip J., and Reuben Hersh (1981). *The Mathematical Experience.* Birkhäuser, Boston.

Dehaene, Stanislas (1997). *The Number Sense: How the Mind Creates Mathematics.* Oxford University Press, London.

Devlin, Keith (1998). *The Language of Mathematics: Making the Invisible Visible.* W.H. Freeman, New York.

Dostoevsky, Fedor (1864). *Notes from the Underground.* Penguin Books, London.

Dubinsky, Ed (1991). "Reflective Abstraction in Advanced Mathematical Thinking." In *Advanced Mathematical Thinking,* ed. David Tall. Kluwer, Dordrecht, pp. 95–123.

Dunham, William (1990). *Journey Through Genius: The Great Theorems of Mathematics.* John Wiley, New York.

Ekeland, Ivar (1988). *Mathematics and the Unexpected.* University of Chicago Press, Chicago.

Einstein, Albert (1930). "Religion and Science." *New York Times Magazine*, November 9, 1–4.

Elliot, T. S. (1957). *On Poetry and the Poets*. Faber & Faber, London.

Gardner, Martin (2001). *The Colossal Book of Mathematics*. W.W. Norton, New York.

Gödel, Kurt (1940). *The Consistency of the Continuum Hypothesis*. Princeton University Press, Princeton.

——— (1964). "What Is Cantor's Continuum Problem?" In *Philosophy of Mathematics, Selected Readings*, ed. Paul Benacerraf and Hilary Putnam. Prentice-Hall, Englewood Cliffs, NJ, 258–73.

Goodstein, R. L. (1944). "On the Restricted Ordinal Theorem." *Journal of Symbolic Logic* 9: 33–41.

Graham, Ronald, Donald Knuth, and Owen Patashnik (1989). *Concrete Mathematics*. Addison Wesley, Reading, MA.

Gray, Eddie, and David Tall (1994). "Duality, Ambiguity and Flexibility: A 'Proceptual' View of Simple Arithmetic." *Journal for Research in Mathematics Education* 25, no. 2: 116–40.

Greene, Brian (2000). *The Elegant Universe: Superstrings, Hidden Dimensions, and the Quest for the Ultimate Theory*. Vintage Books, Random House, New York.

Grothendieck, Alexandre (1986), *Récoltes et semailles: Réflexions et témoignages sur un passé de mathématicien*. Université des Sciences et Techniques du Languedoc, Montpellier et Centre National de la Recherche Scientifique.

Herbert, Nick (1985). *Quantum Reality: Beyond the New Physics*. Anchor Books, Doubleday, New York.

Hersh, Reuben (1997). *What Is Mathematics Really?* Oxford University Press, New York.

——— (1998). "Some Proposals for Reviving the Philosophy of Mathematics." In *New Directions in the Philosophy of Mathematics: An Anthology*, ed. Thomas Tymoczko. Princeton University Press, Princeton.

Heschel, Abraham Joshua (1951). *Man Is Not Alone*. Farrar, Straus, and Giroux, New York.

Hilbert, David (1900). *Mathematical Problems*. http://aleph0.clarku.edu/~djoyce/Hilbert/ problems.html.

Hoffmann, Banesh, with Helen Dukas (1972). *Albert Einstein, Creator and Rebel*. Viking, New York.

Hofstadter, Douglas R. (1979). *Gödel, Escher, Bach: An Eternal Golden Braid*. Basic Books, New York.

Horgan, John (1993). "The Death of Proof." *Scientific American* 269, no. 4: 93–103.

Hu, Wayne, and Martin White (2004). "The Cosmic Symphony." *Scientific American*, 290, no. 2: 44–53.

Jackson, Allyn (2004). "Comme appelé du Néant—As if Summoned from the Void: The Life of Alexandre Grothenkieck." *Notices of the American Mathematical Society* 51, nos. 9, 10.

Kaplan, Robert (1999). *The Nothing That Is: A Natural History of Zero*. Oxford University Press, London.

Kauffman, Stuart (1995). *At Home in the Universe: The Search for the Laws of Self-Organization and Complexity*. Oxford University Press, London.

Kirby, L. A. S., and J. B. Paris (1982). "Accessible Independence Results for Peano Arithmetic." *Bulletin of the London Mathematical Society* 14: 285–93.

Kline, Morris (1972). *Mathematical Thought from Ancient to Modern Times*. Oxford University Press.

——— (1973). *Why Johnny Can't Add: The Failure of the New Math*. St. Martin's, New York.

——— (1980). *Mathematics, the Loss of Certainty*. Oxford University Press, New York.

Koestler, Arthur (1964). *The Act of Creation*. Picador Pan Books, London.

Kieren, C. (1981). "Concepts Associated with the Equality Symbol." *Educational Studies in Mathematics* 12: 317–26.

Lagarias, Jeffrey C. (1966). *The $3x + 1$ Problem and Its Generalizations*. AT&T Bell Laboratories, Murray Hill, NJ.

Lakatos, Imre (1976). *Proofs and Refutations: The Logic of Mathematical Discovery*. Cambridge University Press, New York.

Lakoff, George, and Mark Johnson (1980). *The Metaphors We Live By*. University of Chicago Press, Chicago.

Lakoff, George, and Rafael E. Núñez (2000). *Where Mathematics Comes From: How the Embodied Mind Brings Mathematics into Being*. Basic Books, New York.

Lewin, Roger (1992). *Complexity: Life at the Edge of Chaos*. Macmillan, New York.

Li, T. Y., and J. A. Yorke (1975). "Period Three Implies Chaos." *American Mathematical Monthly* 82: 985–92.

Lightman, Alan (2005). *A Sense of the Mysterious: Science and the Human Spirit*. Pantheon Books, New York.

Low, Albert (2002). *Creating Consciousness: A Study of Consciousness, Creativity, Evolution, and Violence*. White Cloud Press, Ashland, OR.

Maor, Eli (1987). *To Infinity and Beyond: A Cultural History of the Infinite*. Birkhäuser, Boston.

Matiyasevich, Yuri (1993). *Hilbert's Tenth Problem*. MIT Press, Cambridge, MA.

Maturana, Humberto R. (1988). "Reality: The Search for Objectivity or the Quest for a Compelling Argument." *Irish Journal of Psychology* 9, no. 11: 25–82.

402

Maturana, Humberto R., and Francisco J. Varela (1998). *The Tree of Knowledge: The Biological Roots of Human Understanding*. Shambhalas, Boston.

Mazur, Barry (2004). "Perturbations, Deformations, and Variations (and 'Near-Misses') in Geometry, Physics, and Number Theory." *Bulletin of the American Mathematical Society* 41, no. 3: 308.

Mlodinow, Leonard (2001). *Euclid's Window: The Story of Geometry from Parallel Lines to Hyperspace*. Simon & Schuster, New York.

Moore, A. W. (1990). *The Infinite*. Routledge, London.

Morin, Edgar, *On Complexity* (unpublished manuscript).

Nagel, Ernest, and James R. Newman (1964). *Gödel's Proof*. New York University Press, New York.

Peitgen, H.-O., and P. H. Richter (1986). *The Beauty of Fractals: Images of Complex Dynamical Systems*. Springer-Verlag, Berlin.

Penrose, Roger (1989). *The Emperor's New Mind: Concerning Computers, Minds, and the Laws of Physics*. Oxford University Press, London.

——— (1994). *Shadows of the Mind: A Search for the Missing Science of Consciousness*, Oxford University Press, London.

——— (1997). *The Large, the Small and the Human Mind*. Cambridge University Press, Cambridge.

Pinker, Steven (1997). *How the Mind Works*. W.W. Norton, New York.

Poincaré, Henri. (1952). *Science and Hypothesis*. Dover, New York.

Pólya, George (1954). *Mathematics and Plausible Reasoning*. 2 vols. Princeton University Press, Princeton.

Ruelle, David (2004). "Application of Hyperbolic Dynamics to Physics: Some Problems and Conjectures." *Bulletin of the American Mathematical Society* 41, no. 3: 275.

Robinson, Abraham (1966). *Nonstandard Analysis*. North Holland, Amsterdam.

Russell, Bertrand (1926). *Our Knowledge of the External World*. Allen and Unwin, London.

Russell, Bertrand, and Alfred North Whitehead (1925). *Principia Mathematica*. Cambridge University Press, Cambridge.

Sacks, Oliver (1996). *An Anthropologist on Mars*. Vintage Books, New York.

——— (2001). *Uncle Tungsten: Memories of a Chemical Boyhood*. Alfred Knopf, New York.

Saramago, José (2003). *The Cave*. Harcourt, New York.

Searle, John R. (1992). *The Rediscovery of Mind*. MIT Press, Cambridge, MA.

Seife, Charles (2000). *Zero: The Biography of a Dangerous Idea*. Penguin, New York.

Sfard, Anna (1989). "Transition from Operational to Structural Conception: The Notion of Function Revisited." In *Proceedings of the Thirteenth Conference for the Psychology of Mathematics Education*, ed. G. Vergnaud, J. Rogalski, and M. Artigue. Centre National de la Recherche Scientifique, Paris, pp. 151–58.

——— (1994). "Reification as the Birth of Metaphor." *For the Learning of Mathematics* 14, no. 1.

Shanks, Daniel (2001). *Solved and Unsolved Problems in Number Theory*. AMS Chelsea, Providence.

Sharkovskii, A. N. (1964). "Co-existence of the Cycles of a Continuous Mapping of the Line into Itself." *Ukrain. Math. Z.* 16: 61–71.

Sierpinska, Anna (1994). *Understanding in Mathematics*. Faber Press, London.

Singh, Simon (1997). *Fermat's Enigma: The Epic Quest to Solve the World's Greatest Mathematical Problem*. Penguin Books, New York.

Smale, Steven (1988). "The Newtonian Contribution to Our Understanding of the Computer." In *Newton's Dream*, ed. Marcia Sweet Stayer. McGill-Queen's Press, Montreal.

Stewart, Ian (1992). *The Problems of Mathematics*. 2nd ed. Oxford University Press, London.

Tall, David (1980). "The notion of infinite measuring number and its relevance in the intuition of infinity." *Educational Studies in Mathematics* 11: 271–84.

——— (2001). "Conceptual and Formal Infinities." *Educational Studies in Mathematics* 48: 199–238.

Thom, René (1971). "Modern Mathematics: An Educational and Philosophical Error?" *American Scientist* 59: 695–99.

Thurston, W. P. (1994). "On Proof and Progress in Mathematics." *Bulletin of the American Mathematical Society* 30, no. 2: 161–77.

——— (1990). "Mathematical Education." *Notices of the American Mathematical Society* 37: 844–50.

Uhlig, Frank (2002). *Transform Linear Algebra*. Prentice-Hall, Englewood Cliffs, NJ.

Verrill, Helena, *Monstrous Moonshine and Mirror Symmetry*. http://www.mast.queensu.ca/~helena/seminar/seminar1.html

Von Franz, Marie-Louise (1974). *Number and Time: Reflections Leading toward a Unification of Depth Psychology and Physics*. Northwestern University Press, Evanston, IL.

Von Glasersfeld, Ernst (1991). "Knowing without Metaphysics: Aspects of the Radical Constructivist Position." In *Research and Reflexivity*, ed. Frederick Steier. Inquiries into Social Construction. London: Sage. Reprinted in Karl Jaspers Forum, http://douglashospital.qc.ca/kjf/17-TAGLA.htm

Wagon, Stan (1985). "Is Pi Normal?" *Mathematics Intelligencer* 7, no. 3: 65–67.

Weil, Simone (1986). *Simone Weil: An Anthology.* London: Virago.

Weyl, Hermann (1932). *God and the Universe: The Open World.* New Haven: Yale University Press.

——— (1949). *Philosophy of Mathematics and Natural* Science. Trans. Olaf Helmer. Princeton University Press, Princeton.

Wigner Eugene (1960). "The Unreasonable Effectiveness of Mathematics in the Natural Sciences." *Communications in Pure and Applied Mathematics* 13, no. 1.

Wiles, Andrew. *Nova* Interview. http://www.pbs.org/wgbh/nova/proof/wiles.html.

——— (1995). "Modular Elliptic Curves and Fermat's Last Theorem." *Annals of Mathematics* 141, no. 3: 443–551.